The Role of Microalgae in Wastewater Treatment

Lala Behari Sukla • Enketeswara Subudhi
Debabrata Pradhan
Editors

The Role of Microalgae in Wastewater Treatment

Springer

Editors
Lala Behari Sukla
Biofuels and Bioprocessing Research Center, Institute of Technical Education and Research
Siksha 'O' Anusandhan (Deemed to be University)
Bhubaneswar, Odisha, India

Enketeswara Subudhi
Biomics & Biodiversity Laboratory, Centre for Biotechnology
Siksha 'O' Anusandhan (Deemed to be University)
Bhubaneswar, Odisha, India

Debabrata Pradhan
Biofuels and Bioprocessing Research Center, Institute of Technical Education and Research
Siksha 'O' Anusandhan (Deemed to be University)
Bhubaneswar, Odisha, India

ISBN 978-981-13-1585-5 ISBN 978-981-13-1586-2 (eBook)
https://doi.org/10.1007/978-981-13-1586-2

Library of Congress Control Number: 2018956307

This Springer imprint is published by the registered company Springer Nature Singapore Pte Ltd.
The registered company address is: 152 Beach Road, #21-01/04 Gateway East, Singapore 189721, Singapore

Contents

About the Editors

Professor Lala Behari Sukla is at present research professor and director in Biofuels and Bioprocessing Research Center, Institute of Technical Education and Research, Siksha 'O' Anusandhan (Deemed to be University), Bhubaneswar, India. He has worked as chief scientist and was heading the Bioresources Engineering Department in CSIR-IMMT, Bhubaneswar, India. He has more than 40 years of R&D experience in the area of algal biotechnology, bio-mineral processing and hydrometallurgy, and has contributed over 194 research articles in international and national journals in the area of biological science and other related fields. Further, he has published 4 books and has 11 patents. Under his guidance, seven students have been awarded Ph.D. degree. Prof. Sukla is the recipient of several prestigious awards including Prof. S.R. Vyas Memorial Award for the year 2010 by the Association of Microbiologists of India (AMI) toward his significant contribution to the development of microbiology in India, IIME Mineral Beneficiation Award: Academic/R&D for the year 2009 for his outstanding professional contribution to mineral engineering, and Sita Ram Rungta Memorial Award – 2007 of the Society of Geoscientists and Allied Technologists (SGAT) for the year 2007 for outstanding work in the field of bio-mineral processing. He is the editorial board member of *Scientific Reports*, *World Environment journal* (Scientific & Academic Publishing), and *International Journal of Nonferrous Metallurgy*. He is senior editor of *INGLOMAYOR* magazine, Santiago (Chile). He is the national expert in Envis Centre on Environmental Biotechnology. Prof. Sukla has implemented different projects from Alcoa, USA; Ministry of Mines, DST, DBT, and MOEF of Government of India; and OMC, Government of Odisha. He has also served as emeritus scientist, CSIR, and emeritus professor, AcSIR, New Delhi.

Dr. Enketeswara Subudhi an alumnus of IIT, Kharagpur, is currently working as associate professor in the Centre for Biotechnology, Siksha 'O' Anusandhan (Deemed to be University), Bhubaneswar, India. He was awarded visiting fellowship by DST-SERB, New Delhi. He was designated as biopesticide expert to the Government of Odisha by DFID. He has been handling various externally funded projects supported by DST and DBT, New Delhi. Dr. Subudhi has worked with the rural community for livelihood generation through biotechnology transfer. He was invited by Chinese Academy of Sciences to present his research findings at Dalian,

China. He has been working in microbial genomics and metagenomics of environmental samples like hot springs, functional genomics-based establishment of the mechanism behind the multidrug resistance in clinical isolates, and isolation and characterization of bioactive principles in medicinal plants. He has published more than 100 articles in journals, book chapters, and seminars of international repute and submitted more than 100 new gene and genome sequences to the public databases. To his credit he has published one patent.

Dr. Debabrata Pradhan is presently working as Assistant Professor in the Biofuels and Bioprocessing Research Center of Siksha 'O' Anusandhan (Deemed to be University), Bhubaneswar. He received his PhD from Chungnam National University, Daejeon, South Korea, in 2011. After completing his PhD, he worked as Postdoctoral Research Fellow in different institutions like Korea Institute of Geoscience and Mineral Resources (KIGAM) and Chungnam National University, South Korea. He is an expert in algal technology, wastewater treatment, ceramic materials synthesis, fuel cell, solid waste management, leaching (chemical and bio), and solvent extraction. He has more than 40 SCI journal publications and 1 granted patent to his credit. He is a life member of various professional societies.

Cyanobacteria in Reducing Pollution Load from Wastewater and Laboratory Bioassay of Heavy Metals on Ecotoxicity Study: A Review

1

Aditya Kishore Dash, Mira Das, and Abanti Pradhan

Abstract

Cyanobacteria, also named as blue-green algae, are the only known prokaryotes capable of oxygenic photosynthesis. Treatments of both industrial and domestic wastewater through physico-chemical methods are invariably cost-intensive to be employed in industries especially in developing and underdeveloped countries. Therefore, in recent years, the importance of low-cost biological wastewater treatment by using the cyanobacteria compared to the conventional wastewater treatment plants has attracted the attention of the researchers. It has been reviewed that there is a reduction of about 70% calcium, 46% chloride, 100% nitrate, 88% nitrite, 100% ammonia, 92% total phosphorus, 12.5% magnesium, 85% BOD and 85% COD from different wastewater by application of different species of cyanobacteria. Further, the metals like Cu, Al, Cd, Zn, Hg, Cr, Ni, Pb, etc. play an important role in the growth and development of cyanobacteria under laboratory culture conditions. Toxicity on growth, effect on photosynthesis, damage of cell, algaecide effect, toxicity at sublethal concentration, ultrastructural changes, cell division and movement, changes in cellular components, etc. are some of the observations in cyanobacteria under laboratory bioassay for metal toxicity study. Besides these, cyanobacteria also show growth effect when grown in wastewater containing different types of pesticides, herbicides and other toxic chemicals. In the present review, an attempt has been made to review the role of different

A. K. Dash (✉) · A. Pradhan
Biofuels and Bioprocessing Research Center, Institute of Technical Education and Research, Siksha 'O' Anusandhan (Deemed to be University), Bhubaneswar, Odisha, India
e-mail: adityadash@soa.ac.in; abantipradhan@soa.ac.in

M. Das
Department of Chemistry, Institute of Technical Education and Research, Siksha 'O' Anusandhan (Deemed to be University), Bhubaneswar, Odisha, India
e-mail: miradas@soa.ac.in

L. B. Sukla et al. (eds.), *The Role of Microalgae in Wastewater Treatment*,
https://doi.org/10.1007/978-981-13-1586-2_1

species of cyanobacteria in reducing the pollution load from different wastewater and also the laboratory bioassay of heavy metals on ecotoxicity of aquatic cyanobacteria.

1.1 Introduction

Rapid industrialization and urbanization coupled with an increased awareness about the need for a clean and green environment have forced the environmentalists, industrialists and governments to find out for efficient, lasting and cost-effective solutions to wastewater treatment and recycling. Treatments of both industrial and domestic wastewater through physico-chemical methods are invariably cost-intensive to be employed in industries particularly in developing and underdeveloped countries [1]. Therefore, in recent years, the importance of low-cost biological treatment of wastewater by using the cyanobacteria, compared to the conventional wastewater treatment plants have attracted the attention of the researchers all over the globe. Microalgae have attracted the attention considerably due to their unique advantages of fast growth, high oil content, synergy with CO_2 biofixation and bioremediation of wastewater [2].

The algal system is useful in treating the wastewater [1, 3–7] and also produces a number of useful byproducts from their biomass [8]. Due to the primary producer and widely occuring nature in almost all auqatic habitats, algae can be served as an indicator of habitat condition [9, 10]. Use of cyanobacteria in wastewater treatment could prove beneficial in many ways since they are useful in bringing about oxygenation and mineralization in addition to being a food source for aquatic species [11]. Algal-bacterial symbiosis has been proved to be an inexpensive process for reclamation of wastewater [8, 12–15]. Toxicity on growth [16], effect on photosynthesis [17], damage of cell [18], algaecide effect [19], toxicity at sublethal concentration [20], ultrastructural changes [21], cell division and movement and changes in cellular components [22] are some of the observations in cyanobacteria under laboratory bioassay for metal toxicity study. Cyanobacteria also play an important role in reducing pollution load from different wastewater [23–26]. Besides these, cyanobacteria also show growth effect when grown in wastewater containing different types of pesticides [27–29], herbicides [30, 31] and other toxic chemicals. In the present review, an attempt has been made to review the role of different species of cyanobacteria in reducing the pollution load from different wastewater and also the laboratory bioassay of heavy metals on ecotoxicity of aquatic cyanobacteria.

1.2 Cyanobacteria

Cyanobacteria, also named as blue-green algae, are considered as the only known prokaryotes which are helpful for oxygenic photosynthesis. A unique and common characteristic of all species of cyanobacteria is their dual nature of a prokaryotic cell structure and an O_2 evolving photosynthesis which is typical for all green plants. They are Gram-negative bacteria which are oxygenic photosynthetic autotrophic

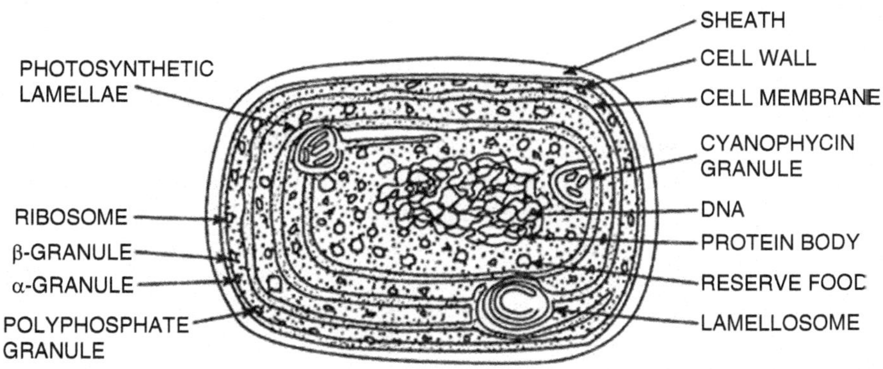

Fig. 1.1 Typical ultra-structure of a cyanobacterial cell

organisms, and they are considered as among the most successful and oldest forms of life in nature. Cyanobacteria are important primary producers globally which play a crucial role in the bio-geochemical cycles of oxygen, carbon and nitrogen. Cyanobacteria are also recognized for their high potential in a variety of biotechnological applications all over the world.

About 3 billion years ago, oxygenic photosynthesis started, when ancient cyanobacteria evolved apparatus which are capable of capturing and utilizing visible solar radiation (300–700 nm). By using electrons that are extracted from H_2O, the reductions of CO_2 to energy-rich carbohydrates, with concomitant release of O_2, had become possible [32–34]. Different species of cyanobacteria originates around 20–30% of world primary productivity, which corresponds to the annual CO_2 fixation of about 20–30Gt into biomass. This releases about 50–80 Gt of O_2 by these oxygenic prokaryotes into the atmosphere [35]. Different cyanobacterial species are capable of fixing atmospheric N_2 into a biologically accessible form. Thus they play a key role in the nitrogen cycle of biosphere [36]. Figure 1.1 shows the typical ultra-structure of a cyanobacterial cell.

1.3 Role of Cyanobacteria in Reducing Pollution Load from the Wastewater

The ability of cyanobacteria to reduce pollution load from different types of wastewater has been studied by different scientists [2, 6, 26, 37–39]. They play an important role in reducing either one or in combination of different nutrients from algal-based treatment systems [40–46].

Cyanobacteria can uptake and bioaccumulate a variety of environmental contaminants present in the water. The typical capacity of cyanobacteria to bring about changes to suit their requirement was well studied by different authors. Rath and Adhikary [47] found that the maximum growth and chlorophyll a content of *A. fertilissima* were observed in pH range between 8–9 and that of *A. variabilis* at pH 7–8, while at acidic pH (5–6) and at above 9, growth was affected. Misra et al. [48] found that blue-green algae found from waste contaminated with mercury accumulate mercury in their cell. Uma and Subramanian [26] have studied the effective use of cyanobacteria (*Halobacterium* sp. US 101, *Oscillatoria* sp. BDU 10142 and *Aphanocapsa* sp. BDU 16) in ossein industry wastewater which has a very high concentration of total dissolved solids, calcium and chloride content. With a serial incubation of these organisms, given retention time of 4 days for each treatment under laboratory batch cultures, there was more than 50% reduction in calcium and nearly 50% reduction in chloride contents. In the field condition, the reduction of calcium was only 40% and chloride was 25%. Stevenson [49] studied the excretion of organic acids by cyanobacteria and their capacity to solubilize magnesium in the wastewater. Cyanobacteria as an agent to remove large amount of phosphorous from industrial wastewater has been carried out by several workers [50, 51]. Reports are also available on reduction of BOD, COD and high levels of nitrogenous compounds by algal cultures in different wastewater [51–53].

Aquatic cyanobacteria have been implicated in the degradation of organic contaminants. Sengar et al. [54] found that on the 30th day of growth, complete removal of NO_3-N occurred by mixed algal culture. Cyanobacteria are known to grow well in sewage [8, 53]; however their precise roles in treating sewage as well as the impact of sewage on these organisms are not known [25]. Domestic wastewater are mainly treated to limit pollution problem and to minimize other hazards which possibly results from the disposal of inadequate treated sewage, through the reduction of number of pathogenic microorganisms as well as the oxidation of organic materials [55, 56]. Role of cyanobacteria to reduce large amount of phosphorus from effluent was demonstrated by Chan et al. [50]. Tam and Wong [51] also reports about the effective removal of high level of nitrogenous compounds by the algae from the wastewater. Manoharan and Subramanian [25] studied the role of cyanobacteria *Oscillatoria pseudogeminata* var. *Unigranulat* in reducing the pollution load from sewage water and found that the maximum reduction of BOD and COD was around 80% and the initial DO increased considerably. The correlation between the initial DO increase and removal of BOD as well as COD observed in the study agree with the observations by Kankal et al. [57]. Manoharan and Subramanian [25] recorded a 100% removal of nitrate and ammonia and 50–100% removal of nitrate from sewage by *Oscillatoria* alone and in combination with natural population of microbes and a total or near total removal of all types of phosphates. Removal of 40 and 20% of calcium in unsterilized and sterilized sewage, respectively, and 76% magnesium in unsterilized sewage by *Oscillatoria* was also reported by Manoharan and Subramanian [24]. In another study, Dash and Mishra [1] have found out that there is a reduction of sodium (68%), potassium (50.04%), calcium (71.23%), chloride (23.27%), sulphate (74.16%), phosphate (90.03%) and COD (78.33%) from paper mill wastewater with basal nutrient medium by using the cyanobacterium, *W. prolifica.* Xiaochen et al. [2] have studied that, by using the microalgae *C. vulgaris* UTEX 2417, there is a reduction of COD (86–88%), phosphorus (63–69%) and nitrogen (43–46%), and the pH has increased from 6.3 to 8.5. Some of the other research works on interaction of different species of algae with wastewater have been reported [58–70]. Table 1.1 shows the percentage reduction of nutrients/

Table 1.1 Percentage reduction of nutrients/pollutants from the wastewater treated with cyanobacteria

Parameter	Type of wastewater	Organisms	% Reduction	References
Sodium	Paper mill effluent	*W. prolifica*	68	[1]
Potassium	Paper mill effluent	*W. prolifica*	50.04	[1]
Calcium	Ossein effluent	*Halobacterium Oscillatoria Aphanocapsa*	22–53	[26]
	Paper mill effluent	*Oscillatoria*	70.5	[24]
	Paper mill effluent	*W. prolifica*	71.23	[1]
Chloride	Ossein effluent	*Halobacterium Oscillatoria Aphanocapsa*	30–46	[26]
	Paper mill effluent	*Oscillatoria*	25	[24]
	Paper mill effluent	*W. prolifica*	23.27	[1]
Nitrate	Sewage	*Oscillatoria*	50	[25]
	Paper mill effluent	*Oscillatoria*	88	[24]
Ammonia	Sewage	*Oscillatoria*	100	[25]
	Paper mill effluent	*Oscillatoria*	80	[24]
pH	Ossein	*Oscillatoria*	Up to 8.09	[23]
	Paper mill effluent	*W. prolifica*	Up to 8.07	[1]
	Centrate	*C. vulgaris*	6.3–8.5	[2]
Nitrogen	Centrate	*C. vulgaris*	43–46	[2]
Total phosphorous	Ossein effluent	*Oscillatoria*	92	[23]
	Centrate	*C. vulgaris*	63–69	[2]
Magnesium	Sewage	*Oscillatoria*	12.5	[23]
Sulphate	Paper mill effluent	*W. prolifica*	74.16	[1]
Phosphate	Paper mill effluent	*W. prolifica*		[1]
BOD	Sewage	*Oscillatoria*	80–85	[23]
	Ossein effluent	*Oscillatoria*	80–85	[23]
COD	Sewage	*Oscillatoria*	80–85	[23]
	Ossein effluent	*Oscillatoria*	80–85	[23]
	Paper mill effluent	*W. prolifica*	78.33	[1]
	Centrate	*C. vulgaris*	86–88	[2]

pollutants from the wastewater treated with different species of cyanobacteria. Photoplate 1.1(a–l) shows the picture of some of the cyanobacterial species whose pollution reduction potential has been reviewed (Fig. 1.2).

Sallal and Babaa [56] has reported that use of sewage effluent containing the heavy metals such as Fe, Zn, Pb, Cd, Ni, Co, Cr, etc. favoured the growth of algae and cyanobacteria. Reports are available about the toxicity and uptake of heavy

Plate (a)-*A. inequalis*

Plate (b)-*Aphanocapsa*

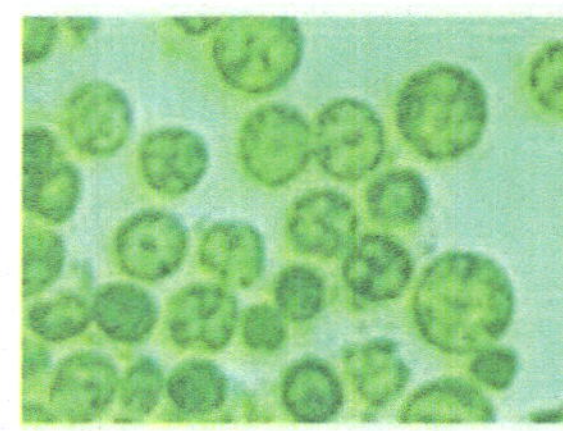

Plate (c)-*C. pyrenoidasa*

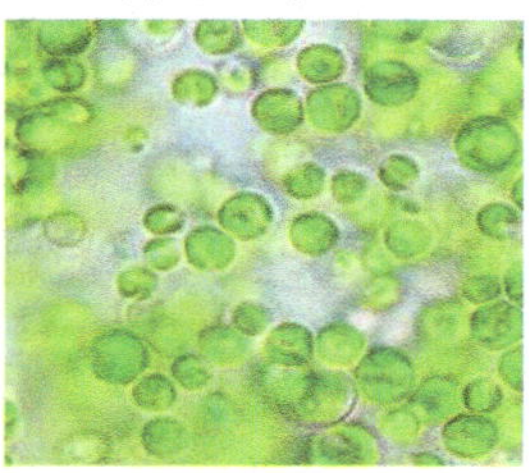

Plate (d)-*C. vulgaris*

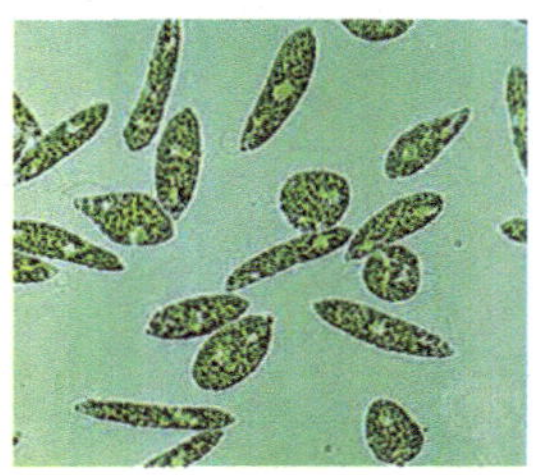

Plate (e)-*E. gracilis*

Plate (f)-*Halobacterium*

Plate (g)-*N. calcicola*

Plate (h)-*Oscillatoria*

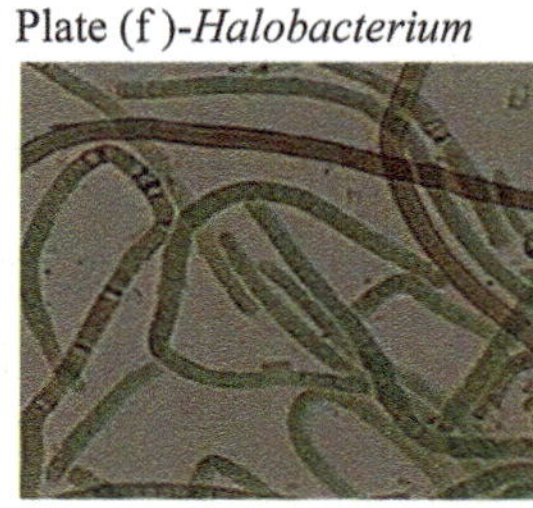

Plate (i)-*P. boryanum*

Plate (j)-*S. platensis*

Plate (k)-*S. quadracauda*

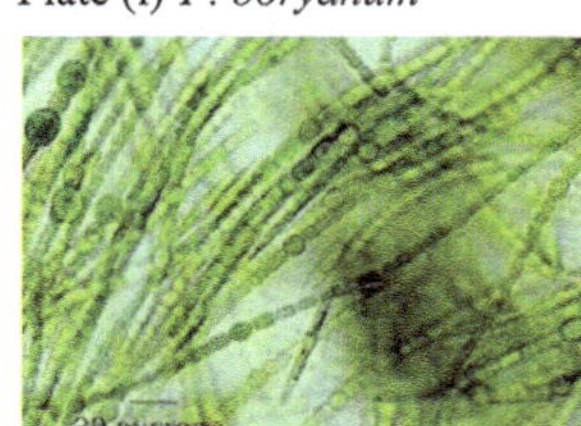

Plate (l)-*W. prolifica*

Photoplate 1.1 (**a–l**) Picture of some cyanobacteria species used for pollution reduction study

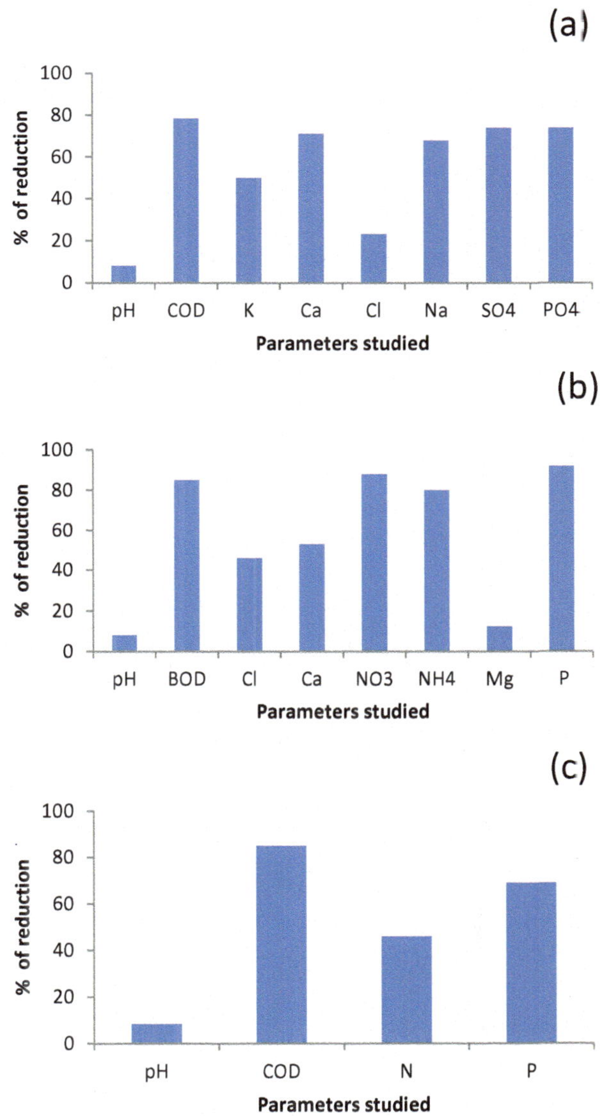

Fig. 1.2 Percentage of reduction of various pollutants from different wastewater by using the cyanobacterial species (**a**) *W. prolifica*, (**b**) *Oscillatoria* and (**c**) *C. vulgaris*

metals like Fe,Zn and Cu by different species of cyanobacteria under laboratory study [71]. The capacity of algae to reduce the heavy metal pollution load in river water has been stated by Sengar et al. [54] where the authors have found that on the 30th day of growth, complete removal of Fe, Zn and Cu was achieved by mixed algal culture. They found that *Chlamydomonas confera* removes Fe and Cu completely, while *Phormidium conium* removes Zn and Cu completely on the 30th day. Some of the other research work on metal toxicity using different species of algae

Table 1.2 Laboratory bioassay for metal toxicity

Metal	Organisms	Observations	References
Cu	*Chlorella vulgaris*	Toxicity of growth	[16]
Cu	*Chlorella pyrenoidosa*	Photosynthesis	[17]
Cu	*Chlorella* sp.	Damage of cell	[18]
Cu	Blue-green algae	Algicide effect	[19]
A	*Chlorella pyrenoidosa*	Toxicity assay	[81]
Cd	*Scenedesmus quadricauda*	Growth	[82]
Zn	*Chlorella vulgaris* and *Plectonema boryanum*	Toxicity assay	[83]
Mn	Blue-green algae	Toxicity assay	[84]
Cu	*Oscillatoria theribauti*	Toxic effect	[85]
Cd	*Euglena gracilis*	Growth	[86]
Hg	*Anabaena inaequalis*	Photosynthesis	[87]
Cd	*Euglena gracilis*	Cytotoxicity	[88]
Cd & Cu	*Anabaena* sp.	Ultrastructural changes	[21]
Cu, Zn	*Euglena exigua*	Resistance	[89]
Heavy metals	*Plectonema boryanum*	Changes in cellular components	[22]
Cd	*Nostoc calcicola*	Resistance	[90]
Mn, Zn, Hg, Pb, Cu, Cd, Co, Ni	*Plectonema boryanum* and *Anabaena flosaquae*	Morphological analysis and nitrogen-fixing ability	[91]
Hg, Zn	*Spirulina platensis*	Toxicity assay	[92]
Ni & Ag	*Nostoc muscorum*	Growth and other biochemical properties	[93, 94]
Heavy metals	Blue-green algae	Toxicity	[95]
Heavy metals	*Anabaena flos-aquae*	Toxicity monitoring	[96]
Heavy metals	*Anabaena doliolum*	Toxicity assay	[97]
Cu, Ni, Fe	*Anabaena doliolum*	Physical and biochemical characteristic of a copper tolerant and wild strain	[98]
Pb	*Nostoc muscorum*	Metal induced inhibition of photosynthetic electron transport chain	[99]
Cd	*Nostoc linckia*	Toxicity to photosynthetic and associated electron transport system	[100]

have been reported [72–80]. Table 1.2 shows the laboratory bioassay of different cyanobacteria species for different metal toxicity.

Algae can also be used for various other applications like biohydrogen production [101], biofuel using a biorefinery approach [102, 103], lipid yield and lipid accumulation [104, 105], etc.

1.4 Conclusion

Conventional chemomechanical method of wastewater treatment is so far one of the best methods but is cost prohibitive requiring a regular maintenance and adequate technical manpower. Therefore, there has been a growth in the development of low-cost biological treatment methods of wastewater which requires less capital investment and minimal operational attention. One of the trends is to search for new photosynthetic organisms in different environment with high biomass yield and growth rate and high utilization potential which could be mass cultured in wastewater and play a dual role of purifying the wastewater and serving as a source of feed and fertilizer. They are ideally suitable to perform these functions by virtue of their high flexibility to adapt to varied environments and of their known nutritional and fertilizer values. Therefore, different species of cyanobacteria can be mass cultured in different wastewater lagoon in combination with the microorganisms to help in the degradation of organic matter, reducing pollution load and also to meet the requirement of nitrogenous fertilizer and fish food with a less capital investment as compared to the conventional wastewater treatment plants.

References

1. Dash AK, Mishra PC (1998) Role of cyanobacteria in water pollution abatement. Indian J Environ Ecoplan 1:1 & 2): 1–1 & 2):11
2. Xiaochen M, Wenguang Z, Zongqiang F, Yanling C, Min M, Yuhuan L, Yunkai Z, Paul C, Roger R (2014) Effect of wastewater borne bacteria on algal growth and nutrients removal in wastewater – based algae cultivation system. Bioresour Technol 167:8–13
3. Dash AK, Mishra PC (1996a) Changes in pigment and protein content of *Westiellopsis prolifica*, a blue-green alga, grown in paper mill wastewater. Microbios 85:257–266
4. Dash AK, Mishra PC (1996b) Changes in biomass, pigment and protein content of *Westiellopsis prolifica*, a blue-green alga, grown in nutrient manipulated paper mill wastewater. Cytobios 88:11–16
5. Dash AK, Mishra PC (1997) Blue –green alga in sewage – amended paper mills waste water. Int J Environ Stud 53:9–10
6. Dash AK, Mishra PC (1999b) Growth response of the blue-green alga, *Westiellopsis prolifica* in sewage enriched paper mill waste water. Rev Int Cont Amb 15(2):79–83
7. Dash AK, Pradhan A (2013) Growth and biochemical changes of the blue-green alga, *Anabaena doliolum* in domestic waste water. Int J Sci Engg Res 4(6):2753–2758
8. Subramanian G, Shanmugasundaram S (1986) Sewage utilization and waste recycling by cyanobacteria. Ind J Environ Health 28:250–253
9. Palmer CM (1980) Algae and water pollution, vol 96. Castle House Publications Ltd, York, pp 31–38
10. Rana BC, Gopal T, Kumar HD (1971) Studies on the biological effect of industrial waste on the growth of algae. Ind J Environ Health 13:138–143
11. Elnabarawy MT, Welter AN (1984) Utilization of algae cultures and assays by industry. In: Shubert LE (ed) Algae as ecological indicators. Academic, pp 317–328
12. Dor I, Svi B (1980) Effect of heterotrophic bacteria on green algae growing in waste water. In: Sheief G, Soeder CJ (eds) Algae biomass. Elsevier/North Holland Biomedical Press, pp 421–429

13. Ganapathi SV, Amin PM (1972) Studies on algal bacterial symbiosis in low cost waste treatment systems. In: Desikachary TV (ed) Taxonomy and biology of BGA, pp 483–493
14. Gotaas HB, Oswald WH (1955) Transactions of the use of solar energy. Arizona, USA 4:95–11
15. Oswald WJ, Gotaas HB (1957) Photosynthesis in sewage treatment. Trans Am Soc Civi Engrs 122:73–105
16. Mc Brien DCH, Hassal KA (1967) The effect of toxic doses of copper upon respiration photosynthesis and growth of *Chlorella vulgaris*. Physiol Plant 20:114
17. Steemann NL, Nielsen K, Wium AS (1969) The effect of deleterious concentration of copper in the photosynthesis of *Chlorella pyrenoidosa*. Physiol Plant 24:480
18. Gross RE, Punso P, Dugger WM (1970) Observation on the mechanism of copper damage in *Chlorella*. Plant Physiol 46:183
19. Gibson CE (1972) The algicidal effect of copper on a green and blue green alga and some ecological implications. J Appl Ecol 9:513
20. De Filippis LF, Pallaghy CK (1976) The effect of sublithal concentration of mercury and Znic on *Chlorella* III. Development and possible mechanism of resistance to metals. Z Pflanzenphysiol 79:323
21. Massalski A, Laube VM, Kushner DJ (1981) Effect of cadmium and copper on the ultrastructure of *Ankistrodesmum braunii* and *Anabaena*, 7120. Microb Ecol 7:183
22. Jensen TE, Baxter M, Rachlin JW, Jain V (1982) Uptake of heavy metals by *Plectonema boryanum* (Chlorophyceae) into cellular components, especially polyphosphate bodies, an X-ray energy dispersive study. Environ Poll 27:119–127
23. Manoharan C, Subramanian G (1993) Feasibility studies on using cynobacteria in Ossein effluent treatment. Ind J Environ Health 5(2):88–96
24. Manoharan C, Subramanian G (1992a) Interaction between paper mill effluent and the cyanobacterium *Oscillatoria pseudogeminata* var, *Urigranulata*. Poll Res 11(2):73–84
25. Manoharan C, Subramanian G (1992b) Sewage cyno-bacterial interaction, a case study. Indian J Environ Prot 12(4):251–258
26. Uma L, Subramanian G (1990) Effective use of cynobacterium in effluent treatment, proc, Natl. Symp. Cynobacterium in Nitrogen fixation, IARI, New Delhi, Jan 29-31:437–443
27. Panigrahi KC (1984) Physiological and genetical effect of pesticides on blue green algae (effect of carbamate pesticides). Ph.D. dissertation, Berhampur University
28. Schauberger CW, Wildman RB (1977) Accumulation of aldrin and dieldrin by blue-green algae and related effects on the photosynthetic pigments. Bull Envion Contain Toxicol 17:534
29. Snyder CE, Sheridan RP (1974) Toxicity of the pesticide zectran on photosynthesis, respiration and growth in four algae. V Phycol 10:137
30. Cullimore DR, Mc Cann AE (1977) Influence of four herbicides on the algal flora of a prairie soil. Plant Soil 40:455
31. Hawxby K, Tubea B, Ownby I, Baseler E (1977) Effect of various herbicides on four species of algae. Pest Biochem Physiol 7:203
32. Marshall K (1994) Microbial adhesion in biotechnological process. Curr Opin Biotechnol 5:296
33. Nakagawa T, Fukui M (2002) Phylogenetic characterization of microbial mats and streamers from a Japanese alkaline hot spring with a thermal gradient. J Gen Appl Microbiol 48:211
34. Whitton BA (1992) Diversity, ecology and taxonomy of the cyanobacteria. In: Mann NH, Carr NG (eds) Photosynthetic prokaryotes. Plenum Press, New York, pp 1–51
35. Ward DM, Weller R, Bateson MM (1990) 16S rRNA sequences reveal numerous uncultured microorganisms in a natural community. Nature 344:63
36. Fuhrman JA, McCallum K, Davis AA (1993) Phylogenetic diversity of subsurface marine microbial communities from the Atlantic and Pacific Oceans. Appl Environ Microbiol 59:1294
37. Adhikary SP, Bastia AK, Tripathy PK (1992) Growth response of the nitrogen fixation cyanobacterioum *Westiellopsis prolitica* Janet, to fertilizer factory effluents. Bull Environ Contam Toxicol Springer – Verlag, New York, Inc 49:137–144

38. Dash AK, Mishra PC (1999a) Role of the blue – green alga, *Westiellopsis prolifica* in reducing pollution load from paper mill wastewater. Indian J Environ Protec 19(1):1–15
39. Sengar RMS, Sharma KD (1986) Role of algae in the assessment of pollution in river Yamuna. J Ind Bot Soc 66:352–334
40. Kumar KS, Dahms HU, Won FJ, Lee JS, Shin KH (2015) Microalgae—a promising tool for heavy metal remediation. Ecotoxicol Environ Saf 113:329–352
41. Mahapatra D, Chanakya HN, Ramachandra TV (2013) Treatment efficacy of algae-based sewage treatment plants. Environ Moniron Asses 185:7145–7164
42. Priya M, Gurung N, Mukherjee K, Bose S (2014) Microalgae in removal of heavy metal and organic pollutants from soil. Microb Biodegrad Bioremedn 23:519–537
43. Raj S, Patel PK, Vaishya RC (2017) Microalgae based treatment system efficient for wastewater: a review. 49th annual convention of IWWA on "Smart Water Management" January, 2017, pp 19–21
44. Rasoul-amini S, Montazeri-najafabady N, Shaker S, Safari A, Kazemi A, Mousavi P (2014) Removal of nitrogen and phosphorus from wastewater using microalgae free cells in bath culture system. Biocat Agrl Biotech 3:126–131
45. Weiming H, Yin J, Deng B, Zhiqiang H (2015) Application of nano TiO2 modified hollow fiber membranes in algal membrane bioreactors for high-density algae cultivation and wastewater polishing. Bioresour Technol 193:135–141
46. Yang J, Cao J, Xing G, Yuan H (2015) Lipid production combined with biosorption and bioaccumulation of cadmium, copper, manganese and zinc by oleaginous microalgae *Chlorella minutissima* UTEX2341. Bioresour Technol 175:537–544
47. Rath B, Adhikary SP (1996) Effect of pH, irradiance and population size on the toxicity of Furadon to two species at Anabaena. Biol Plant 38(4):563–570
48. Misra BB, Nanda DR, Mishra BN (1985) Reclamation with blue green algae, mercury uptake by algae cultured in solid waste of a chlor – alkali factory and its effect on growth and pigmentation. J Environ Biol 6(4):223–231
49. Stevenson FJ (1967) Organic acids in soil. In: Mc Laren AD, Paterson GR (eds) Soil biochemistry. Marcel
50. Chan KY, Wong KH, Wond PK (1979) Nitrogen and phosphorus removal from sewage effluent with high salinity by *Chlorella salina*. Environ Poll 18:139–146
51. Tam NFY, Wong YS (1989) Wastewater nutrient removal by *Chlorella pyrenoidosa* and *secenedesmus* sp. Environ Poll 58:19–34
52. Govindan VS (1984) Studies on algae in relation to treatment of diary wastewater. Ind J Environ Health 26:261–263
53. Saxena PN, Tewari A, Khan MA (1974) Effect of *Anacystis nidulans* on the physico-chemical and biological characteristics of raw sewage. Proc Ind Acad Sci 79:139–146
54. Sengar RMS, Sharma KD, Mittal S (1990) In vitro studies for the publication of river water by algal treatment. Geobios 17:77–81
55. Feachem RE, McGarry M, Mara D (1977) Water, waste and health in hot climates. Wiley, New York
56. Sallal AKJ, Babaa MM (1982) Enumeration and identification of pathogenic bacteria from sewage in Kuwait. Zentbl Microbial 137:603–610
57. Kankal NC, Neema P, Gokhe BH, Mehata CG (1987) Dissolved oxygen and detention time as process parameters in sewage treatment by aerated lagoon. I.A, C.P.W. Tech Annual 14:61–66
58. Abdel-Raouf N, Al-Homaidan AA, Ibraheem IBM (2012) Microalgae and wastewater treatment. Saudi J Biol Sci 19:257–275
59. Chen Z, Ma W, Han M (2008) Biosorption of nickel and copper onto treated alga (*Undaria pinnatifida*), application of isotherm and kinetic models. J Hazard Matter 155:327–333
60. Deng L, Su Y, Su H, Wang X, Zhu X (2006) Sorption and desorption of lead (g) from wastewater by green algae *Cladophora fascicularis*. J Hazard Matter, Dio 10:1016
61. Lim S, Chu W, Phang S (2010) Use of *Chlorella vulgaris* for bioremediation of textile wastewater. J Bioresour Technol 101:7314–7322

62. Liu J, Sun Z, Gerken H (2014) Potential applications of microalgae in wastewater treatments. In: Recent advances in microalgal biotechnology. OMICS Publ, Gull Ave, pp 1–8
63. Lourie E, Gjengedal E (2011) Metal sorption by peat and algae treated peat: kinetics and factors affecting the process. Chemosphere 85:759–764
64. Pandi M, Shashirekha V, Swamy M (2009) Biosorption of chromium from retan chrome liquor by cyanobacteria. Microbiol Res 164:420–428
65. Rajhia SAL, Raut N, Qasmi FAL, Qasmi M, Saadi AA (2012) Treatment of industrial wastewater by using microalgae. Int Conference on Environ Biomedical and Biotech IPCBEE 41:217–221
66. Samhan AF (2008) Assessment of the ability of microalgae in removal of some industrial wastewater pollutants. M.Sc. thesis, Botany Department, Faculty of Science, Beni-Suef University, Egypt, pp 1–16
67. Shi J, Podola B, Melkonian M (2007) Removal of nitrogen and phosphorus from wastewater using microalgae immobilized on twin layers: an experimental study. J Appl Phycol 19(5):417–423
68. Tastan BE, Duygu E, Donmez G (2012) Boron bioremoval by a newly isolated *Chlorella* sp. and its stimulation by growth stimulators. Water Res 46:167–175
69. Tuzen M, Sari A (2010) Biosorption of selenium from aqeous solution by green algae (*Cladophora hutchinsiae*) biomass: equilibrium, thermodynamic and kinetic studies. Chem Eng J 158:200–206
70. Zhu G, Peng Y, Li B, Guo J, Yang Q, Wang S (2008) Biological removal of nitrogen from waste water. Rev Environ Contam Toxicol 192:159–195
71. Mittal S, Sengar RMS, kavshik BD (1992) Uptake and toxicity of heavy metals to algae. Ind J Microb Environ 32(1):51–55
72. Afkar A, Ababna H, Fathi AA (2010) Toxicological response of the green alga *Chlorella vulgaris* to some heavy metals. Am J Environ Sci 6(3):230–237
73. Chakraborty N, Banerjee A, Pal R (2011) Accumulation of lead by free and immobilized cyanobacteria with special reference to accumulation factor and recovery. Bioresour Technol 102:4191–4195
74. Gupta VK, Rastogi A (2008) Biosorption of lead (II) from aqueous solution by non-living algal biomass Oedogonium sp. and Nostoc sp.- a comparative study. Colloids Surf B Bioint 64:170–178
75. Gupta VK, Rastogi A, Nayak A (2010) Biosorption of nickel onto treated alga (*Oedogonium hatei*): application of isotherm and kinetic models. J Colloids Surf Sci 342:533–539
76. Kumar D, Rai J, Gaur JP (2012) Removal of metal ions by *Phormidium bigranulatum* (cyanobacteria)-dominated mat in batch and continuous flow systems. Bioresour Technol 104:202–207
77. Liu Y, Cao Q, Luo F, Chen J (2009) Biosorption of Cd^{2+}, Cu^{2+}, Ni^{2+} and Zn^{2+} ions from aqueous solutions by pretreated biomass of brown algae. J Hazard Matter 163:931–938
78. Pahlavanzadeh H, Keshtkar AR, Safdari J, Abadi Z (2010) Biosorption of nickel (II) from aqueous solution by brown algae: equilibrium, dynamic and thermodynamic studies. J Hazard Matter 175:304–310
79. Sari A, Tuzen M (2008) Biosorption of Pb (II) and Cd (II) from aqueous solution using green alga (*Ulva lactuca*) biomass. J Hazard Mater 152:302–208
80. Singh L, Pavankumar AR, Lakshmanan R (2012) Effective removal of Cu^{2+} ions from aqueous medium using alginate as biosorbent. Ecol Eng 38:119–124
81. Foy CD, Gerloff GC (1972) Response of *Chlorella pyrenoidosa* to aluminum and low pH. J Phycol 8:268
82. Klass E, Rowe DW, Massaro EJ (1974) The effect of cadmium on population growth of the green alga, *Scenedesmus quadracanda*. Bull Environ Cont Toxicol 12:442
83. Rana BC, Kumar HD (1974) The toxicity of zinc to *Chlorella vulgaris* and *Plectonema boryanum* and its protection by phosphate. Phykos 13:60
84. Singh SP, Kashyap AK (1978) Manganese toxicity and mutagenesis in two blue green algae. Environ Expt Bot 18:47

85. Ructer JG, Mc Carthy JJ, Carpenter EJ (1978) The toxic effect of copper on *Oscillatoria theirbautii*. Limnology and Ocenography 24:558
86. Nakano Y, Kamota JKO, Tod S, Fuwa F (1978) Toxic effect of cadmium on *Euglena gracilis* grown in zinc deficient and zinc sufficient media. Agric Biol Chemis 42:901
87. Stratton GW, Hubber AL, Corker CT (1979) Effect of mercury ion on *Anabaena inaqualis*. Appl Environ Microbiol 38:537
88. Bonaly J, Bariand A, Durel S, Mester JC (1980) Cadmium cytotoxicity and variation in nuclear content of DNA in *Euglena gracilis*. Physiol Plant 49:286
89. Whitton BA, Daiz BM (1981.) Verh. Int Verein) Int verein Theo Angew Limnol 21:459
90. Singh SP, Pandey AK (1982) Cadmium mediated resistance to metals and antibiotics in a cynobacterium. Mol Gen Genet 187:247
91. Rachlin JW, Jensen TEM, Jain V (1982) Utilization of morphometric analysis in evaluating response of *Plectonema boryanum* (Chlorophyceae) to expose to light heavy metals. Arch Environ Conta Toxicol 11:233
92. Kotangate LR, Sarkar R, Krishnamurthy KP (1984) Toxicity of mercury and zinc to *Spirulina platensis*. Indian J Environ Health 26(1):41
93. Rai LC, Raizada M (1985) Effect of nickel and silver ions on survival, growth, carbon fixation and nitrogenase activity of *Noctoc muscorum* regulation of toxicity by EDTA and calcium. J Gen Appl Microbiol 31:329
94. Rai LC, Raizada M (1986) Nickel induced stimulation of growth heterocyst differentiation, CO_2 uptake and nitrogenase activity in *Nostoc muscorum*. New Phytol 104:111
95. Rai LC (1989) Silver toxicity in a nitrogen fixing cynobacterium, interaction with chromium, nickel and lead. Biol metals 2:122
96. Rai LC, Jenson TE, Rechlin JW (1990) A morphometric and X ray energy dispersive approach to monitoring pH altered cadmium toxicity in *Anabaena flosaquae*. Arch Environ Contam Toxicol 119:479
97. Dubey SK, Rai LC (1990) Heavy metal toxicity in a nitrogen cynobacterium *Anabaena dolioum,* regulation toxicity by certain environmental factor. Biomed Environ Sc 3(2):240
98. Rai LC, Mallick N, Singh JB, Kumar HD (1991) Physiological and biochemical characteristics of a copper tolerant and wild type train of *Anabaena dolioum* under copper stress. J Plant Physiol 138:68
99. Prasad SM, Singh FB, Rai L, Pelczar MJ, Reid RD, Chan ECS (1977) Microbiology. Tata Mc Graw – Hill Publication Company, New Delhi
100. Husaini Y, Singh AK, Rai LC (1991) Cadmium toxicity to photosynthesis and associated electron transport system of *Nostoc linckia*. Bull Environ Cont Toxicity 46:146
101. Batista AP, Ambrosano L, Graca S, Sousa C, Marques PA, Ribeiro B (2015) Combining urban wastewater treatment with biohydrogen production—an integrated microalgae-based approach. Bioresour Technol 184:230–235
102. Singh B, Guldhe A, Singh P, Singh A, Rawat I, Bux F (2015) In: Kaushik G (ed). *Applied environmental biotechnology: Present scenario and future trends*Sustainable production of biofuels from microalgae using a biorefinary approach, pp 115–128
103. Vanthoor-Koopmans M, WIjffels RH, Barbosa MJ, Eppink MH (2013) Bio-refinery of microalgae for food and fuel. Bioresour Technol 135:142–149
104. Han L, Pei H, Hu W, Han F, Song M, Zhang S (2014) Nutrient removal and lipid accumulation properties of newly isolated microalgal strains. Bioresour Technol 165:38–41
105. Ramanna L, Guldhe A, Rawat I, Bux F (2014) The optimization of biomass and lipid yields of *Chlorella sorokiniana* when using wastewater supplemented with different nitrogen sources. Bioresour Technol 168:127–135

Bioremediation of Municipal Wastewater and Biodiesel Production by Cultivation of *Parachlorella kessleri*-I

2

Amit Kumar Singh, Humaira Farooqi, Malik Zainul Abdin, and Shashi Kumar

Abstract

The continuous growth of human population, industrialization and urbanization has led to the increased release of pollutants into the environment. The water pollutants are chemically toxic compounds causing harmful effect on human and animal health. Algae are effective in reducing the nutrients and toxic compound from contaminated water reservoirs, and the process is known as bioremediation. The microalgae can play a crucial role in producing the bioenergy if integrated with the remediation of wastewater while growing biomass for biofuel feedstock. This article lays emphasis on the dual role of a robust oleaginous marine microalga *Parachlorella kessleri*-I, which is substantial in bioremediating the wastewater and producing biofuel feedstock.

2.1 Introduction

There are two kinds of water present on earth depending upon the presence of salts, fresh and saline water. The freshwater that sustains human life is about 3% of total water present on earth, and only 0.5% is available for human consumption [1]. Hence, the stock of freshwater is limited on earth, and recycling is the only process to fulfil the demand. Despite the fact, most of the water resources have been polluted due to unplanned urbanization and growing industrialization. The domestic, industrial and agricultural are major sectors that release untreated water into water bodies. The untreated water contains pollutants to varying degrees of organic and

A. K. Singh · S. Kumar (✉)
International Centre for Genetic Engineering and Biotechnology, New Delhi, India
e-mail: skrhode@icgeb.res.in

H. Farooqi · M. Z. Abdin
Department of Biotechnology, Jamia Hamdard, New Delhi, India

L. B. Sukla et al. (eds.), *The Role of Microalgae in Wastewater Treatment*,
https://doi.org/10.1007/978-981-13-1586-2_2

inorganic compounds. There are several water treatment plants that have been established; however, the number is not much and the process is based on usage of hazardous chemicals. In the developing countries, the wastewater treatment plants treat about 28% wastewater, the remaining released in water bodies [2]. The untreated water causes diseases and creates water scarcity issues to humans. Under the pressure of population growth, developing countries are facing water scarcity issues more than developed countries. About 50% of the Chinese and Indian population facing at least 1 month a year of water scarcity live [3]. Further, the discharge of untreated wastewater into seas and oceans is also responsible for deoxygenated dead zones, which are growing rapidly and affect marine ecosystem such as fisheries, livelihoods and food chains [4]. The release of nutrients (e.g. nitrogen and phosphorus) and agrochemicals from intensive agriculture and animal waste can further accelerate the eutrophication of freshwater and also pollute the groundwater. In general, nitrogen and phosphorous are the major nutrient sources responsible for eutrophication that causes serious environmental issues such as formation of toxic algal blooms and declines in the biodiversity [5–8]. The nutrient enrichment through raw sewage into the lake or other stagnant water system may cause various negative effects in the water body such as luxuriant growth of aquatic weeds that imbalance the fauna and flora of aquatic biome. Persistent pesticides, chemical solvents and other substances slowly invade into the environment, bioaccumulating in animals and human food chain [9].

The bioremediation process to treat wastewater is considered to be economical and environment friendly [10]. There are many biological means for bioremediation of wastewater such as bacteria [11], fungi [12, 13], microalgae [14] and higher plants [15]. Microalgae are unique due to their ability of photosynthesis like plant and utilizing nutrient (nitrogen, phosphorous, organic and inorganic carbon substrate) from wastewater while sequestering the CO_2 for photosynthesis and generating biomass for biofuel feedstock. The present article emphasizes on the usage of marine microalga *Parachlorella kessleri*-I to remediate the wastewater and its application in producing the biomass feedstock for biofuel using an integrated approach.

2.2 Coupling of Bioremediation of Wastewater and Biofuel Feedstock Production

The economy of technology is an essential requirement for its sustainability. An integrated system has been proposed that is capable of removing nutrients from wastewater while producing algal biomass as a biofuel feedstock. The terminology of remediation varies with the driver such as phytoremediation where 'phyto' stands for plant-based remediation. However, algae-based remediation is coined as 'phycoremediation' where 'phyco' stands for algae-based remediation [16]. Notably, microalgae are the primary producers in aquatic food chains and are also useful as key indicator in determining the quality of water and extent of pollutant toxicity to the aquatic ecosystem [17]. For decades, microalgae have been utilized as a value-added active biocompounds in pharmaceutical, nutraceuticals and animal feed.

Other benefit of using microalgae in wastewater remediation is disinfection capability by increasing the pH of wastewater as a result of photosynthesis [18].

Oswald and co-worker have investigated the bioremediation of wastewater using microalgae, and the first algal ponding system was established for the municipal sewage treatment [19, 20]. This investigation was instrumental for the development of economical algae biomass production process for the biofuel applications. In the wastewater treatment process, the burden of cost can be imputed to secondary and tertiary treatment. The rationale behind this is the energy costs of oxygen and chemical supply in secondary treatment (biological) and tertiary treatment, respectively. In the wastewater, microalgae and bacteria live in symbiotic relationship. In this association, bacterial population feast on organic wastes to decompose into simple nutrients (nitrogen and phosphorous including CO_2) by using algae-generated oxygen; however, microalgae used these nutrients and other growth-promoting factors (e.g. vitamins) for their growth (Fig. 2.1). Interestingly, Croft et al. [21] have shown that algae use vitamin B12 released from bacteria in a symbiotic relationship.

Recently, the microalgae have received a considerable attention due to their ability to remediate wastewater and simultaneous biofuel production (Fig. 2.2). The integration of biofuel production while treating the wastewater with supplementation of CO_2 was suggested by Oswald group [22].

Using an integrated technology, only a few microalgae strains are explored for the wastewater treatment and subsequent biofuel production. The most common strains were *Chlorella vulgaris* [23], *Chlorella pyrenoidosa* [24], *Chlamydomonas polypyrenoideum* [25], *Scenedesmus obliquus* [26] and *Botryococcus braunii* [27]. Marine microalgal species is also tested for the treatment of wastewater despite their salinity requirements [28]. In our study, we have noticed that a marine green alga *Parachlorella kessleri*-I showed a high growth rate using the wastewater as compared to mixed combination of wastewater and freshwater [29] (Fig. 2.3). This is in agreement to Osundeko et al.'s [30] report, where *P. kessleri* has shown high tolerance to the wastewater environment.

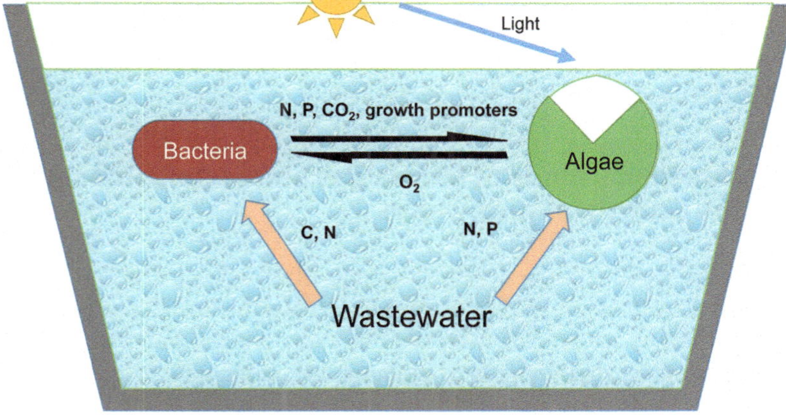

Fig. 2.1 Algal-bacterial species symbiosis during the wastewater treatment

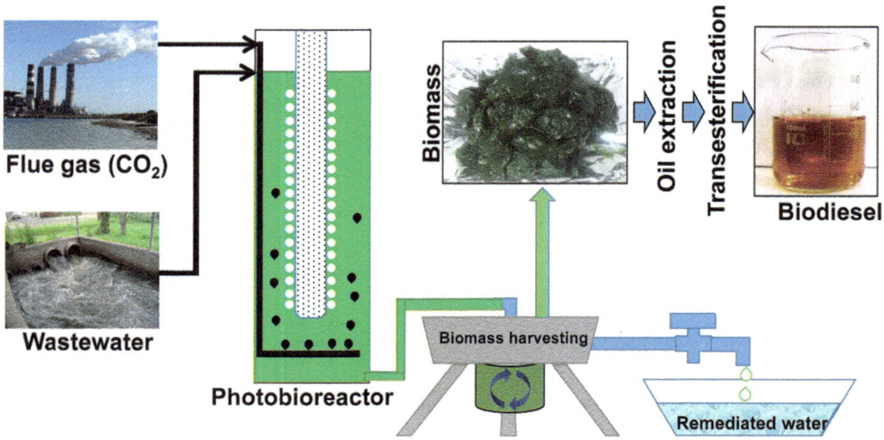

Fig. 2.2 Schematic representation of an integrated microalgal culture system for the bioremediation of wastewater and production of biomass feedstock for biofuel applications

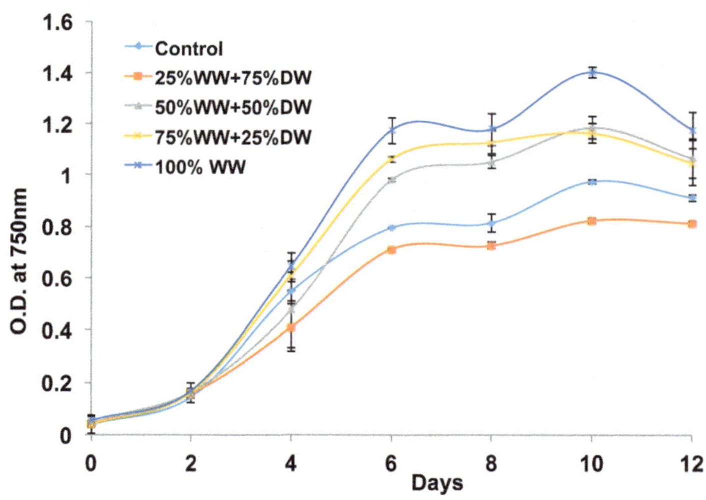

Fig. 2.3 Growth study of microalga *Parachlorella kessleri*-I in wastewater. (*Source*: [29])

2.3 Nutritional Mode of Microalgae and Factors Affecting the Growth in Wastewaters

Microalgae have ability to grow in a wide range of wastewaters such as municipal, industrial and agricultural types [31]. Depending upon the species and environment, microalgae can survive on three different nutritional modes that depend upon carbon assimilation for the synthesis of biomass, viz. autotrophic, heterotrophic and mixotrophic [32], as shown in Fig. 2.4.

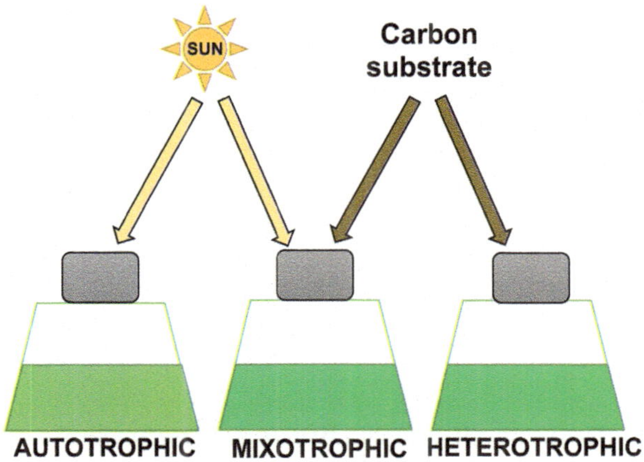

Fig. 2.4 Nutritional mode of microalgae

The heterotrophic and mixotrophic mode of nutrition showed slightly higher growth than autotrophic. The modes of nutrition vary species to species. Autotrophic mode of nutrition requires no carbon source as a substrate. Such algal species depend upon the photosynthetically fixed sugars, whereas heterotrophic mode of nutrition depends on the exogenously present carbon substrate, reported in case of *C. protothecoides*, *Crypthecodinium cohnii* and *Schizochytrium limacinum*. The mixotrophic mode of nutrition is a combination of both auto (photosynthesis) and heterotrophic (carbon substrate) nutrition, shown by *Chlamydomonas reinharatii* and *Chlorella vulgaris*. Microalgae contain multiple metabolic pathways related to this mode of nutrition. Some microalgae have capability to change their mode of nutrition depending on the condition present at that time; however, other microalgae have capability to use both the pathways simultaneously [33].

The average stoichiometric formula of algae biomass is $C_{106}H_{181}O_{45}N_{16}P$ in which carbon (C) has more than 50% contribution [34, 35]. During photosynthesis, algae assimilate inorganic carbon (CO_2) from air and convert it to starch and oils. For experimental or industrial scale algal culture, aeration can be used to provide atmospheric CO_2 [36]. Besides this, algae can also consume organic carbon compounds (e.g. glucose, glycerol, acetate, etc.) during heterotrophic growth ([37, 38]: [39]). Nitrogen (N) is the second important macronutrient essential for algae growth. Nitrogen is present in ionic forms such as NH_4^+, NO_3^-, NO_2^-, etc. Among these forms, algae assimilate ammonium (NH_4^+) as a most preferable source of nitrogen as compared to others, i.e. NO_3^- and NO_2^- [40]. Phosphorus is another important macronutrient to algae and prefers uptake in form of inorganic orthophosphate (PO_4^{3-}) for its growth.

Apart from nutritional requirement, the factors like light and temperature are also an essential requirement for the algal cultivation. Yan et al. [41] have demonstrated that the performance of and growth of *Chlorella vulgaris* in synthetic wastewater varied considerably under the effects of various LED light wavelengths and intensities. Further, they have showed that the light intensity plays important role in the growth of microalgae only when it provided with optimum light intensity [41]. The temperature range (20–40 °C) was used to determine the maximum growth of microalga *Chlorella sorokiniana* and found that the optimum temperature was about 38 °C for maximum growth. However, a drastic decrease in growth was reported at 40 °C [42]. Thus, nutrient removal efficiency of the algae is not only affected by the availability of nutrients but also affected by physico-chemical factors such as pH, light intensity, photoperiod, temperature and biological factors [43]. Also, the biological factor may count particularly on open pond culture condition such as microbial communities and predation [44].

2.4 Percentage of Nutrients Removal from Wastewater by Microalgae

Previous microalgae research mostly focused on removing nutrients from wastewater as a tertiary treatment. However, microalgae are also efficient in removing water contaminants as a secondary treatment process [45, 46]. Microalgal species can be explored further for removing nutrients in wastewater remediation due to their ability to use inorganic nitrogen and phosphorous for their growth [34, 35, 47, 48]. Further, biological processes are ecofriendly compared to the chemical and physical processes, which are costly and used chemicals leading to secondary pollution [49].

The treatment municipal wastewater by microalgae is carried efficiently which contain carbon, nitrogen, phosphorous, etc. However, only few algal species have been found useful for remediating the wastewater. Due to industrialization and urbanization, the wastewater characteristics are also changing. Therefore, discovery of new strains will be important for efficient treatment of wide range of wastewaters. Moreover, due to extension of the process of cultivation of microalgae in wastewater towards mass cultivation, many new microalgae species were reported recently depending upon the wastewater type as shown in Table 2.1. In our study *P. kessleri*-I was observed efficient for removing ammoniacal nitrogen, total nitrogen and phosphorus by 89%, 81% and 98%, respectively, from municipal wastewater (Table 2.1). Notably, ammonia is the main constituent of domestic wastewater and exists in soluble form with equilibrium between ammonia (NH_3) and ionized ammonia (NH_4^+) in water [55].

2.5 Mechanism of Nutrient Removal by Algae

Microalgae have many biochemical pathway involved in nutrient remediation process of water (Fig. 2.5). Microalgae need nutrients for construction of cell structure and formation of basic molecules like protein, carbohydrate, lipid, nucleic acids, etc. As discussed earlier, carbon, nitrogen, phosphorous and sulphur are major constituents of nutrients that are responsible for growth of algae [56].

Table 2.1 Nitrogen and phosphorus removal efficiencies of different microalgal species using wastewater

Microalgae	Wastewater	Removal (%)			References
		NH_4^+	TN	TP	
Scenedesmus dimorphus	Agro-industrial	95%	–	–	[50]
Botryococcus braunii LEM 14	Domestic	–	79.6%	100%	[51]
Chlorella pyrenoidosa	Soybean processing	89.1%	88.8	70.3%	[52]
Neochloris oleoabundans	Synthetic	–	99%	100%	[53]
Euglena sp.	Domestic	98%	93%	66%	[54]
Parachlorella kessleri-I	Municipal	89%	81%	98%	[29]

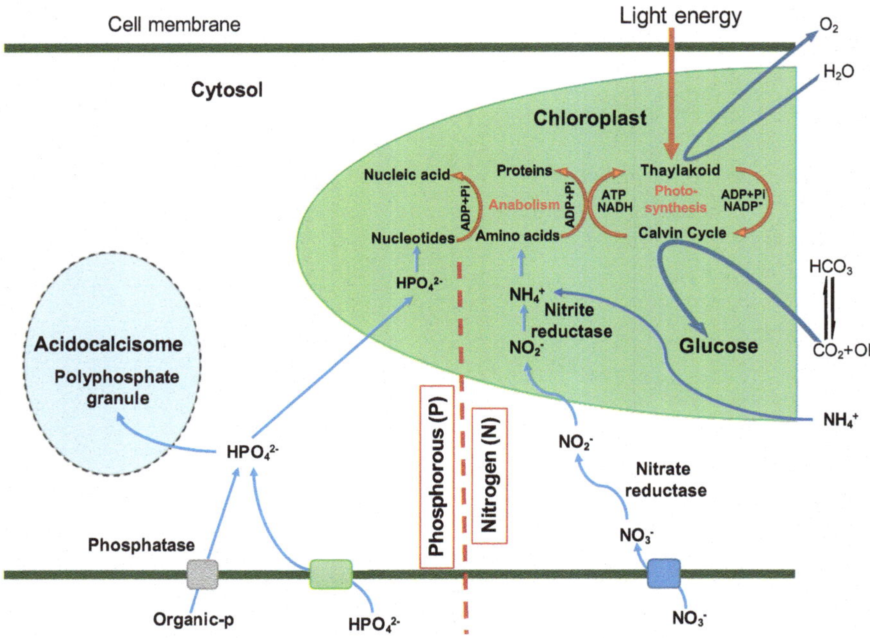

Fig. 2.5 Schematic representation of microalgal remediation mechanism of nitrogen and phosphorus from aquatic media

The nitrogen is the key macronutrient involved in the biosynthesis of all proteins, chlorophylls, ADP (adenosine diphosphate) and ATP (adenosine-5′-triphosphate) [57]. The basic mechanism involved in uptake of nitrogen is based on the availability of inorganic and organic nitrogen. The inorganic nitrogens are available in the form of nitrite (NO_2^-), nitrate (NO_3^-) and ammonia salt as ammonium (NH_4^+) in the wastewater. In uptake process, these ionic molecules cross the cell membrane in the preference of $NH_4^+ > NO_3^- >$ organic nitrogen [20, 58]. Notably, microalgae *C. vulgaris* and *S. obliquus* showed preferences for ammonium uptake as compared to other form of nitrogen present in wastewater [59]. The uptake of ammonium (NH_4^+) takes place through the ammonium transporter which is found to be closely related to a group of protein transporter generally present in bacteria, yeasts and higher plants [60].

The translocation of inorganic nitrogen molecules (NO_3^- and NO_2^-) takes place across the plasma membrane by transporters. Inside cell, these molecules undergo reduction process followed by ammonium (NH_4^+) formation as end product which finally incorporated into protein via protein anabolism (Fig. 2.5). In the nitrogen uptake process, firstly, nitrate (NO_3^-) entered into cytosol, and with the help of nitrate reductase (a NADH-dependent) enzyme, it converted into nitrite (NO_2^-). Thereafter, nitrite converted to ammonium by nitrite reductase (a NADPH-linked) enzyme [61, 62]. The ammonium is precursor for amino acid production inside the cell. However, the direct uptake of ammonium is also taking place due to the reduced energy requirement necessary for reduction and assimilation [56]. Finally, all inorganic forms of nitrogen are incorporated into biomolecule (organic form).

Phosphorus also plays a crucial role in algal cell growth and metabolisms. It is an essential structural and functional component in lipids, proteins and nucleic acids [63]. Phosphorus mainly exists in the forms of phosphate such as dihydric phosphate ($H_2PO_4^-$) and hydrophosphate (HPO_4^{2-}) [64]. In the phosphorylation process, phosphorous is incorporated into organic compounds. Phosphorylation process needs energy that comes from the respiration and photosynthesis processes [56]. The excess phosphorous is stored in form of polyphosphate bodies inside an organelle, the acidocalcisome, and can be used for growth in absence of phosphorous in medium [65].

2.6 Application of Microalgae Biomass

The history of production of microalgal biomass has been described in literature in the late 1800s and early 1900s. The first application of algae has been reported in China that appeared about 2500 years ago, where edible Nostoc (a cyanobacterial species) was used as food by the natives to survive [66, 67]. Another cyanobacterial species *Spirulina* and *Chlorella* species has been grown on large scale up as protein sources for human consumption. The focus now is shifting towards the use of microalgae cultivation for biofuel production. Due to limited stock of fossil fuel and unfavourable environmental consequences caused by burning of fossil fuels, algae-based biodiesel has gained the attention as an ecofriendly and sustainable biofuel. Comparing to petroleum diesel, biodiesel is renewable, and its combustion emits reduced amount of carbon monoxide (CO), particulate matter and unburnt hydrocarbons [68]. It is estimated theoretically that about 100 tons of photosynthetically active microalgal biomass may fix 183 tons of CO_2 from the environment, and, hence, only algae have accounted for about 50% of the world's atmospheric CO_2 fixation [22, 69].

The process of biodiesel production from microalgae is not an economical process due to high cost of algae cultivation and harvesting and producing biodiesel [70, 71]. Use of waste materials such as wastewater (water and nutrients) and CO_2 as flue gas from industry, for growth of microalgae, is considered to be feasible solution to the cost issue in microalgal biomass production. Table 2.2 showed few microalgae biomass yield grown in wastewater under batch culture.

In our study, we have grown *P. kessleri*-I in two different media for comparative analysis and found that increase in biomass and lipid yield by 50% and 115%, respectively, in wastewater as compared to control medium [29]. Further, the biodiesel obtained from *P. kessleri*-I grown in wastewater has the appropriate combination of saturated and monounsaturated fatty acids. Hence, the compatibility of biodiesel to international standard was also analysed and found that the properties such as density, viscosity, high heating value and saponification values were within the limit in wastewater-grown algal biodiesel (Table 2.3).

Table 2.2 Microalgae biomass production using the wastewater

Microalgae	Biomass production (g/L)	References
Scenedesmus sp. LX!	0.11	[72]
Chlorella ellipsoidea YJ1	0.425	[73]
Chlorella sp. 227	0.41–0.67	[74]
Chlorella vulgaris	0.76–0.82	[75]
Chlorella sorokiniana	0.25–0.35	[76]
Parachlorella kessleri-I	0.308	[29]

Table 2.3 Biodiesel properties obtained from wastewater-grown *P. kessleri*-I

Physical properties	Wastewater	EN 14214:2008	ASTM D6751
Density (g/cm^3)	0.877	0.860–0.900	0.875–0.900
Saponification value (mg KOH/g oil)	196	–	–
Iodine value (g I/100 g)	104	<120	–
Cetane number	51	≥51	≥47
Higher heating value (MJ/kg)	40	–	>35
Viscosity (mm^2/s)	4.1	3.5–5.0	1.9–6.0

Source: Singh et al. [29]

2.7 Conclusions

The anthropogenic activities generate large quantities of aqueous waste containing nutrient and toxic heavy metals. Many initiatives have been taken for lowering nutrient concentrations derived from natural resources and anthropogenic activities in the water, and considerable efforts are under progress for developing an efficient and cost-effective technology treating the wastewater. The coupling of microalgae to remove nutrient from wastewater and generation of biofuel could be an important techno-economic strategy for reducing the cost of biofuel production. Although some progress has been made for the neutralization of pollutants from waters via algae, many challenges are remaining there to be addressed. This is a new emerging area of integrated technology, where the main focus is on the development of environmentally friendly technologies with economic feasibility.

Acknowledgements SK acknowledges the Department of Biotechnology (DBT), Government of India and Reliance Industries Limited (RIL) for grants support.

References

1. WBCSD (Editor). 2009. Facts and trends. Water (version 2). Geneva: world business Council for Sustainable Development
2. Sato T, Qadir M, Yamamoto S, Endo T, Zahoor A (2013) Global, regional, and country level need for data on wastewater generation, treatment, and use. Agricul Water Manage 31(130):1–3
3. WWAP (United Nations World Water Assessment Programme) The United Nations World Water Development Report (2017) Wastewater: the untapped resource. UNESCO, Paris

4. Corcoran E, Nellemann C, Baker E, Bos R, Osborn D and Savelli H (eds) (2010) Sick water? the central role of wastewater management in sustainable development. United Nations Environment Programme/United Nations Human Settlements Programme/GRID-Arendal (UNEP/UN- Habitat). www.unep.org/pdf/SickWater_screen.pdf
5. Conley DJ, Paerl HW, Howarth RW, Boesch DF, Seitzinger SP, Karl E, Karl E, Lancelot C, Gene E, Gene E (2009) Controlling eutrophication: nitrogen and phosphorus. Science 123:1014–1015
6. Correll DL (1998) The role of phosphorus in the eutrophication of receiving waters: a review. J Environ Qual 27(2):261–266
7. Lewis WM Jr, Wurtsbaugh WA, Paerl HW (2011) Rationale for control of anthropogenic nitrogen and phosphorus to reduce eutrophication of inland waters. Environ Sci Technol 45(24):10300–10305
8. Shelef G, Soeder CJ (eds) (1980) Algal biomass: production and use. Elsevier/North Holland Biomedical Press, Amsterdam, p 852
9. Stefanidou M, Maravelias C, Spiliopoulou C (2009) Human exposure to endocrine disruptors and breast milk. Endocr Metab Immune Disord Drug Targets 9(3):269–276
10. Valderrama LT, Del Campo CM, Rodriguez CM, de-Bashan LE, Bashan Y (2002) Treatment of recalcitrant wastewater from ethanol and citric acid production using the microalga Chlorella vulgaris and the macrophyte Lemna minuscula. Water Res 36(17):4185–4192
11. Wasi S, Tabrez S, Ahmad M (2013) Use of pseudomonas spp. for the bioremediation of environmental pollutants: a review. Environ Monit Assess 185(10):8147–8155
12. Kaushik P, Malik A (2015) Mycoremediation of synthetic dyes: an insight into the mechanism, process optimization and reactor design. In Microbial Degradation of Synthetic Dyes in Wastewaters, springer, pp. 1–25
13. Mishra A, Malik A (2013) Recent advances in microbial metal bioaccumulation. Crit Rev Environ Sci Technol 43(11):1162–1222
14. Prajapati SK, Choudhary P, Malik A, Vijay VK (2014) Algae mediated treatment and bioenergy generation process for handling liquid and solid waste from dairy cattle farm. Bioresour Technol 167:260–268
15. Salt DE, Smith RD, Raskin I (1998) PHYTOREMEDIATION. Annu Rev Plant Physiol Plant Mol Biol 49:643–668
16. Olguin EJ, Sanchez-Galvan G (2012) Heavy metal removal in phytofiltration and phycoremediation: the need to differentiate between bioadsorption and bioaccumulation. New Biotechnol 30:3–8
17. Stevenson J, Graham L (2014) Ecological assessments with algae: a review and synthesis. J Phycol 50(3):437–461
18. Abdel-Raouf N, Al-Homaidan AA, Ibraheem IBM (2012) Microalgae and wastewater treatment. Saudi J Biol Sci 19:257–275
19. Benemann JR, Oswald WJ (1996) Systems and economic analysis of microalgae ponds for conversion of CO2 to biomass, final report. In: Pittsburgh energy technology Center
20. Oswald WJ, Gotaas HB (1957) Photosynthesis in sewage treatmentTrans. Am Soc Civil Eng 122(1957):73–105
21. Croft MT, Lawrence AD, Raux-Deery E, Warren MJ, Smith AG (2005) Algae acquire vitamin B12 through a symbiotic relationship with bacteria. Nature 438:90–93
22. Chisti Y (2008) Biodiesel from microalgae beats bioethanol. Trends Biotechnol 26(3):126–131
23. Feng Y, Li C, Zhang D (2011) Lipid production of Chlorella vulgaris cultured in artificial wastewater medium. Bioresour Technol 102:101–105
24. Kothari R, Pathak VV, Kumar V, Singh DP (2012) Experimental study for growth potential of unicellular alga Chlorella pyrenoidosa on dairy waste water: an integrated approach for treatment and biofuel production. Bioresour Technol 116:466–470
25. Kothari R, Prasad R, Kumar V, Singh DP (2013) Production of biodiesel from microalgae Chlamydomonas polypyrenoideum grown on dairy industry wastewater. Bioresour Technol 144:499–503

26. Mata TM, Melo AC, Simões M, Caetano NS (2012) Parametric study of a brewery effluent treatment by microalgae Scenedesmus obliquus. Bioresour Technol 107:151–158
27. An JY, Sim SJ, Lee JS, Kim BW (2003) Hydrocarbon production from secondarily treated piggery wastewater by the green alga Botryococcus braunii. J Appl Phycol 15:185–191
28. Ledda C, Villegas GR, Adani F, Fernández FA, Grima EM (2015) Utilization of centrate from wastewater treatment for the outdoor production of Nannochloropsis gaditana biomass at pilot-scale. Algal Res 12:17–25
29. Singh AK, Sharma N, Farooqi H, Abdin MZ, Mock T, Kumar S (2017) Phycoremediation of municipal wastewater by microalgae to produce biofuel. Internat J Phytoremed https://doi.org/10.1080/15226514.2017.1284758)
30. Osundeko O, Dean AP, Davies H, Pittman JK (2014) Acclimation of microalgae to wastewater environments involves increased oxidative stress tolerance activity. Plant Cell Physiol 55(10):1848–1857
31. Bajhaiya AK, Mandotra SK, Suseela MR, Toppo K, Ranade S (2010) Algal biodiesel: the next generation biofuel for India. Asian J Exp Biol Sci 1(4):728–739
32. Mata TM, Martins AA, Caetano NS (2010) Microalgae for biodiesel production and other applications: a review. Renew Sust Energ Rev 14(1):217–232
33. Perez-Garcia O, Escalante FM, de-Bashan LE, Bashan Y (2011) Heterotrophic cultures of microalgae: metabolism and potential products. Water Res 45(1):11–36
34. Oswald WJ (1988a) Micro-algae and waste-water treatment. In: Borowitzka MA, Borowitzka LJ (eds) Micro-algal biotechnology. Cambridge University Press, Cambridge, pp 305–328
35. Oswald WJ (1988b) Large-scale algal culture system (engineering aspects). In: Borowitzka MA, Borowitzka LJ (eds) Micro-algal biotechnology. Cambridge University Press, Cambridge, pp 357–394
36. Fontes AG, Vargas MA, Moreno J, Guerrero MG, Losada M (1987) Factors affecting the production of biomass by a nitrogen-fixing blue-green alga in outdoor culture. Biomass 13(1):33–43
37. Fogg GE (1975) Algal cultures and phytoplankton ecology, 2nd edn. The University of Wisconsin Press, Wisconsin
38. Kawaguchi K (1980) Microalgae production systems in Asia. In: Shelef and Soeder (ed) algae biomass: production and use, pp 25–33
39. Borowitzka MA (1998) Limits to growth. In Wastewater treatment with algae. Springer Berlin Heidelberg, pp. 203–226
40. Bhaya D, Schwarz R, Grossman AR (2000) Molecular responses to environmental stress. In: The ecology of cyanobacteria. Springer, Dordrecht, pp 397–442
41. Yan C, Zhang L, Luo X, Zheng Z (2013) Effects of various LED light wavelengths and intensities on the performance of purifying synthetic domestic sewage by microalgae at different influent C/N ratios. Ecol engine 28(51):24–32
42. Ugwu CU, Aoyagi H, Uchiyama H (2007) Influence of irradiance, dissolved oxygen concentration, and temperature on the growth of Chlorella sorokiniana. Photosynthetica 45(2):309–311
43. Delgadillo-Mirquez L, Lopes F, Taidi B, Pareau D (2016) Nitrogen and phosphate removal from wastewater with a mixed microalgae and bacteria culture. Biotechnol Reports 11:18–26
44. Fouilland E (2012) Biodiversity as a tool for waste phycoremediation and biomass production. Rev Environ Sci Biol 11:1–4
45. Tam NFY, Wong YS (1989) Wastewater nutrient removal by Chlorella pyrenoidosa and Scenedesmus sp. Environ Pollut 58:19–34
46. Wang L, Min M, Li Y, Chen P, Chen Y, Liu Y, Wang Y, Ruan R (2010) Cultivation of green algae chlorella sp. in different wastewaters from municipal wastewater treatment plant. Appl Biochem Biotechnol 162(4):1174–1186
47. Evonne PY, Tang (1997) Polar cyanobacteria versus green algae for tertiary wastewater treatment in cool climates. J Appl Phycol 9:371–381
48. Richmond A (ed) (1986) CRC handbook of microalgal mass culture. CRC Press, Boca Raton, pp 1–528

49. Noüe J, Laliberté G, Proulx D (1992) Algae and waste water. J Appl Phycol 4(3):247–254
50. Gonzalez LE, Canizares RO, Baena S (1997) Efficiency of ammonia and phosphorus removal from a Colombian agro industrial wastewater by the microalgae Chlorella vulgaris and Scenedesmus dimorphus. Bioresour Technol 60(3):259–262
51. Sydney EB, da Silva TE, Tokarski A, Novak AC, de Carvalho JC, Woiciecohwski AL, Larroche C, Soccol CR (2011) Screening of microalgae with potential for biodiesel production and nutrient removal from treated domestic sewage. Appl Energ 88(10):3291–3294
52. Su HY, Zhang YL, Zhang CM, Zhou XF, Li JP (2011) Cultivation of Chlorella pyrenoidosa in soybean processing wastewater. Bioresour Technol 102(21):9884–9990
53. Wang B, Lan CQ (2011) Biomass production and nitrogen and phosphorus removal by the green alga Neochloris oleoabundans in simulated wastewater and secondary municipal wastewater effluent. Bioresour Technol 102(10):5639–5644
54. Mahapatra DM, Chanakya HN, Ramachandra TV (2013) Euglena sp. as a suitable source of lipids for potential use as biofuel and sustainable wastewater treatment. J Appl Phycol 25(3):855–865
55. Clément B, Merlin G (1995) The contribution of ammonia and alkalinity to landfill leachate toxicity to duckweed. Sci Total Environ 170(1–2):71–79
56. Cai S, Park YS, Li Y (2013) Nutrient recovery from wastewater streams by microalgae: status and prospects. Renew Sust Energ Rev 19:360–369
57. Barsanti L, Gualtieri P (2006) Algae: anatomy, biochemistry, and biotechnology. CRC Press, Boca Raton
58. Sandau E, Sandau P, Pulz O, Zimmermann M (1996) Heavy metal sorption by marine algae and algal by-products. Acta Biotechnol 16:103–119
59. Ruiz-Marin A, Mendoza-Espinosa LG, Stephenson T (2010) Growth and nutrient removal in free and immobilized green algae in batch and semi-continuous cultures treating real wastewater. Bioresour Technol 101(1):58–64
60. Wilhelm C, Büchel C, Fisahn J, Goss R, Jakob T, LaRoche J, Lavaud J, Lohr M, Riebesell U, Stehfest K, Valentin K (2006) The regulation of carbon and nutrient assimilation in diatoms is significantly different from green algae. Protist 157(2):91–124
61. Ahmad I, Hellebust JA (1990) Regulation of chloroplast development by nitrogen source and growth conditions in a *Chlorella protothecoides* strain. Plant Physiol 94:944–949
62. Fernandez E, Galvan A (2007) Inorganic nitrogen assimilation in *Chlamydomonas*. J Exp Bot 58:2279–2287
63. Richmond A (2004) Handbook of microalga culture: biotechnology and applied phycology. Blackwel Science Ltd, Oxford
64. Martinez ME, Jimenez JM, El Yousfi F (1999) Influence of phosphorus concentration and temperature on growth and phosphorus uptake by the microalga Scenedesmus obliquus. Bioresour Technol 67(3):233–240
65. Seufferheld MJ, Curzi MJ (2010) Recent discoveries on the roles of polyphosphates in plants. Plant Mol Biol Report 28(4):549–559
66. Andersen RA (2005) Algal culturing techniques. Elsevier Inc, San Diego
67. Spolaore P, Joannis-Cassan C, Duran E, Isambert A (2006) Commercial applications of microalgae. J Biosci Bioeng 101:87–96
68. Canakci M, Sanli H (2008) Biodiesel production from various feedstocks and their effects on the fuel properties. J ind microbial biotechnol 35(5):431–441
69. Field CB, Behrenfeld MJ, Randerson JT, Falkowski P (1998) Primary production of the biosphere: integrating terrestrial and oceanic components. Science 281:237–240
70. Benemann JR (1997) Energy Convers Manag 38:S475–S479
71. Van Beilen JB (2010) Why microalgal fuels won't save the internal combustion machine. Biofuels Bioprod Biorefin 4:41–52
72. Li X, Hu HY, Yang J (2010) Lipid accumulation and nutrient removal properties of a newly isolated freshwater microalga, Scenedesmus sp. LX1, growing in secondary effluent. New Biotechnol 27:59–63

73. Yang J, Li X, Hu HY, Zhang X, Yu Y, Chen Y (2011) Growth and lipid accumulation properties of a freshwater microalga, Chlorella ellipsoidea YJ1, in domestic secondary effluents. Appl Energy 88:3295–3299
74. Cho S, Luong TT, Lee D, Oh YK, Lee T (2011) Reuse of effluent water from a municipal wastewater treatment plant in microalgae cultivation for biofuel production. Bioresour Technol 102:8639–8645
75. Cabanelas ITD, Ruiz J, Arbib Z, Chinalia FA, Garrido-Pérez C, Rogalla F, Nascimento IA, Perale JA (2013) Comparing the use of different domestic wastewaters for coupling microalgal production and nutrient removal. Bioresour Technol 131:429–436
76. Bohutskyi P, Liu K, Nasr LK, Byers N, Rosenberg JN, Oyler GA, Betenbaugh MJ, Bouwer EJ (2015) Bioprospecting of microalgae for integrated biomass production and phytoremediation of unsterilized wastewater and anaerobic digestion centrate. Appl Microbiol Biotechnol 99:6139–6154

Arsenite S-Adenosylmethionine-Producing *Spirulina platensis*: A New Trump Card on the Face of Global Arsenic Poisoning

3

Anshuman Sahu, Ankita Pattanayak, Rajesh Kumar Sahoo, Mahendra Gaur, Kalpana Sahoo, and Enketeswara Subudhi

Abstract

Arsenic is a gray-appearing metalloid which occurs naturally and is the 20th most prolific element in the earth's crust. It is an integral part of more than 200 minerals. These are mostly ores containing sulfides, along with copper, nickel, lead, and other metals. In the environment, arsenic and its compounds are very mobile. Although in its organic form arsenic is nontoxic, it is highly toxic in its inorganic form (arsenite, a free form of arsenic) with arsine gas being the most fatal. The World Health Organization recommends a concentration below 20 mg/l for an individual to be considered free of arsenic poisoning. Accumulation of arsenic in the body beyond this level could adversely affect human health. An individual suffering from chronic arsenic poisoning via contaminated water could suffer from severe skin-related ailments like melanosis (pigmentation of the skin), keratosis (associated with the formation of rough, dry, and popular skin lesions), and leucomelanosis (also known as spotted melanosis) ultimately leading to arsenicosis in the long term. Other than that arsenic poisoning also may lead to other manifestations like neurological disorders, diabetes mellitus, high blood pressure, obstetric problems, disorders of the respiratory system, and cancer in the lung, skin, and bladder. The Indian subcontinent is very rich in arsenic, and countries like India and Bangladesh are a disaster waiting to happen. West Bengal, India, is a state severely affected by arsenic-contaminated water, and a case study showed an astounding 16 sites from one single village with very high concentrations of arsenic. As such, it is the need of the hour for governments to

A. Sahu · A. Pattanayak · R. K. Sahoo · M. Gaur · K. Sahoo · E. Subudhi (✉)
Centre for Biotechnology, Siksha 'O' Anusandhan (Deemed to be University), Bhubaneswar, Odisha, India
e-mail: enketeswarasubudhi@soa.ac.in

L. B. Sukla et al. (eds.), *The Role of Microalgae in Wastewater Treatment*, https://doi.org/10.1007/978-981-13-1586-2_3

be ready with an immediate action plan to tackle such large-scale disasters. Existing solutions to this problem include phytoremediation via hyperaccumulation with plants like *Pteris vittata* and grasses like *A. delicatula* and use of phosphate-based fertilizers. However, a long-term use of phosphate-based fertilizers may ultimately lead to an algal bloom in water bodies, and phytoremediation is a time-consuming process. Planktons, however, have the potential to be a game changer in tackling arsenic-contaminated water bodies by virtue of accumulation and bioremediation. *Spirulina platensis*, a typical plankton, produces an enzyme called arsenite S-adenosylmethionine methyltransferase which has the ability to methylate arsenic making it nontoxic. This enzyme confers *Spirulina platensis* the unique ability to convert the toxic trivalent arsenic to its nontoxic pentavalent form. *Spirulina platensis* produces this enzyme by the virtue of arsenite S-adenosylmethionine methyltransferase (SpArsM) gene.

Isolation and overexpression study of this gene in a heterologous host like *E. coli* followed by pilot-scale study ultimately leading to the industrial mass production of this enzyme is an unexplored and untapped area which has a huge potential to tackle the menace of arsenic contamination in water bodies.

3.1 Introduction

Arsenic (As) is an infamously poisonous metalloid which is extensively scattered in marine, freshwater bodies, and the soil. It has different physiochemical properties and has four primary oxidation states of −3, 0, +3, and + 5 [1]. Arsenic is an active constituent of 245 different types of minerals and is regularly associated with other metals like copper, lead, gold, and ore sulfides [2]. The congregation of arsenic in marine water bodies is said to be about 1.5 μg/l [3]. Arsenic has a very negligible role to play in biological activities [4]. Depending upon various geological features and mobilization of arsenic under the combined effect of natural processes (like mineralization) and anthropogenic emissions [5], variations may occur in the concentration of arsenic. Although arsenic is found in the form of a minute element in earth's crust with an average concentration of 5 μg/g, under the influence of natural processes, arsenic has the potential to become highly concentrated in some parts of the world [6]. Anthropogenic activities like drilling for oil, mining, fossil fuel combustion, and smelting may also release arsenic into the environment [6]. According to a 1999 factsheet of the WHO, the contamination as a result of heavy metal like arsenic was mentioned to be the source of a potential disaster whose avoidance required immediate correction on an emergency basis [7]. More than 20 different nations like Chile, India, Bangladesh, China, and Argentina have suffered from high incidences of arsenicosis over the years [8–10]. Today arsenic has become such a menace because of its presence in the environment and its potentially severe effects upon human exposure that the United States Agency for Toxic Substances and Disease Registry has been forced to give arsenic the topmost priority as a toxic heavy metal ahead of polychlorinated biphenyls, mercury, and lead (http://www.atsdr.cdc.gov/SPL/index.html). Microalgae comprise of both prokaryotic (cyanobacteria) and eukaryotic (Chlorophyta, diatoms, etc.) organisms. Because of their unchanged morphology over the last 3.5 billion years, microalgae have developed the ability to survive under highly stressed and toxic environments [11]. Microalgae are present at the foot of the food chain and are distributed in various types of terrestrial and aquatic environments. They carry out a crucial task in the cycling of arsenic in the environment [12–15]. Microalgae have as a result developed various methods to resist and metabolize toxic arsenicals, because of which they have gained a considerable attention in bioremediation of arsenic-contaminated water bodies. In order to understand the feasibility of the extent to which accumulation and detoxification of arsenic by microalgae can take place, the understanding of speciation and metabolism of arsenic in these organisms gain primary importance [16]. In this chapter the distribution of arsenic globally, its potential sources, and its ill effects have been discussed in brief. The mechanism of arsenic toxicity and a few case studies of arsenic poisoning over the years have also been discussed. The mechanism of detoxification and biomethylation of arsenic by microorganisms has also been discussed. A simple hypothesis for treatment of patients suffering from an acute form of arsenicosis with the help of SpArsM protein and its detailed mechanism has also been given. A few notable works carried out on bioremediation of arsenic and the challenges ahead have also been discussed at the end of this chapter.

3.2 Global Distribution of Arsenic

Today a large number of surface water bodies have been established where the concentration of arsenic is above 50 μg/L. The regions which have notably high concentrations of arsenic lie in parts of Bangladesh, West Bengal (India), China, Taiwan, Vietnam, Nepal, Myanmar, Cambodia, Chile, Argentina, Mexico, and Hungary, and large swaths of the southwestern USA [17]. Smedley et al. [17] gave a beautiful account of the distribution of arsenic in various parts of the world depending on the type of environment prevalent.

3.2.1 Reducing Environment

A reducing environment is one in which there is the removal of oxygen and other oxidizing gases or vapors which prevents oxidation. These types of environments are usually filled with gases which are actively reducing in nature. Among the regions included under reducing environments, Bangladesh and West Bengal (India) represent the regions with extremely high occurring concentrations of arsenic. Arsenic in this region lies within a range of 0.5 to 3200 μg/l. Arsenic concentration more than 50 μg/l has also been mentioned in about 27% of the shallow wells of Bangladesh [18]. Some districts of southeast Bangladesh are among the worst hit areas where more than 90% of the wells are affected with high concentrations of arsenic [17]. More than 35 million people in Bangladesh [18] and up to six million people in West Bengal, India [19], have come in direct contact with arsenic in drinking water. Around 5000 individuals were identified who had arsenic-related health issues in West Bengal although the estimated number of patients suffering from arsenicosis were over 200,000 [20]. Arsenic-related problems in groundwater were first brought to the fore in Taiwan in the 1960s [21]. Tseng et al. [21] reported very high concentrations of arsenic in groundwater in southwestern part of Taiwan, whereas similar reports in the northeastern region of the island were made by Hsu et al. [22]. Arsenic concentration ranging between 10 and 1800 μg/l from southwestern Taiwan was reported by [23]. Kuo [23] also found that more than half of the analyzed samples had an arsenic concentration ranging between 400 and 700 μg/l. Arsenic found in northeast and southwest Taiwan was mostly trivalent (III) which supported the hypothesis that groundwater in these regions is reducing in nature [24]. Arsenic had been found from aquifers from Inner Mongolia as well as Xinjiang and Shanxi province, China, which were above the Chinese permissible limit of 50 μg/l. The first instances of arsenic poisoning were reported in the Xinjiang province in the early 1980s. Similarly arsenic concentration in groundwater bodies in Ba Men region and the Tumet Plain (including Huhhot Basin) in Inner Mongolia was reported to be above the 50 μg/l permissible threshold for arsenic in groundwater bodies allowed in China [25–27]. Luo et al. [25] reported arsenic-related diseases in this region with major health effects being keratosis, pigmentation in the skin and lung, and skin and bladder cancer. Concentrations of arsenic ranging from 1 to 3050 μg/l in Hanoi, the capital city of Vietnam, were reported by Berg et al. [28]. Similarly, an arsenic concentration above 50 μg/l has also been reported in groundwater from alluvial soil sediments found in southern part of the Great Hungarian Plain of Hungary and certain regions of neighboring Romania. The concentration of arsenic measuring 176 μg/l in aquifers of Romania was reported by Gurzau and Gurzau [29].

3.2.2 Arid Oxidizing Environments

An arid oxidizing environment is one which is associated with acute lack of availability of water, and as a result, these types of environments are usually associated with lack of vegetation (http://www.atsdr.cdc.gov/SPL/index.html). Del Razo et al. [30] reported arsenic concentration in the range of 8–624 μg/l (average 100 μg/l, $n = 128$) in groundwaters of the Lagunera region located in North Central Mexico which was associated with high pH and high oxidizing environment. It is believed that the number of people in the Lagunera region exposed to an arsenic concentration greater than 50 μg/l was about 400,000 [30]. In Antofagasta, Chile, the arsenic concentration in groundwater was reported to be around 500 μg/l [31]. Very high arsenic concentration ranging between 6 and 11,500 μg/l was reported in Cordoba, Argentina, by Nicolli et al. [32]. They also reported the high concentration of arsenic in groundwater of Tucuman Province (12–1600 μg/l) with a median concentration of 46 μg/l. Desorption of arsenic occurs under high pH in the presence of metal oxides [17] although the presence of volcanic glass in water bodies is considered to be the root cause behind this high concentration of arsenic [32].

3.2.3 Mixed Oxidizing and Reducing Environments

Arsenic in the range 1 μg/l to 2600 μg/l was found in groundwaters from Tulare Basin of the San Joaquin Valley, California, USA [33]. Robertson [34] also reported the very high concentration of arsenic in alluvial aquifers in the Basin and Range Province of Arizona with the groundwater in that region being highly oxidizing in nature. High arsenic concentration is a feature of that basin which was supported by the continuous presence of dissolved oxygen in the aquifers at significant depths of 600 meters despite the old age of groundwater (up to 10 Ka old).

3.3 Sources of Arsenic Contamination

The ubiquitous distribution of arsenic in the environment makes it a genuine threat. It is an interesting fact that although the distribution of arsenic varies greatly, it is considered to be the 20th most lavish element in the crust of the earth [35]. Anthropogenic arsenic (around 18,000 tonnes per year) was estimated to comprise about 70% of the arsenic released globally which was mostly in the form of flux [36]. The concentration of arsenic in seawater varies greatly between 0.006 and 0.03 parts per million [37]. Arsenic may be present in groundwater in excess where it is associated with ore sulfides predominantly in the form of pyrite and arsenopyrite [38]. Smelting industries and burning of fossil fuels serve as the primary source of anthropogenic release of arsenic directly into the environment where arsenic primarily exists in the form of particles of dust [3]. Andreae [39] reported arsenic concentration around 0.5 μg/l in rainfall under the potential effect of smelting and coal burning, although higher concentration (average 16 μg/l) has been reported in Seattle, USA, located 35 km downwind of a copper smelting unit [40]. Geothermal activities may also lead to the direct release of arsenic in water bodies. Arsenic concentration up to 370 μg/l in Madison River was reported by Nimick et al. [41] under the influence of geothermal release of arsenic from the Yellowstone geothermal system. High concentrations of arsenic under geothermal influences have also been found in new hot springs from parts of Chile, France, New Zealand, Argentina, Japan, the USA, and Dominica [5] (Fig. 3.1). Similarly, high arsenic concentration ranging between 85 and 153 μg/l under geothermal influences in Hot Creek, a tributary of Owens River, California, was reported by Wilkie and Hering [42]. A noticeable increase in the arsenic concentration of river waters may occur under the influence of industrial or sewage effluents. Arsenic concentration up to 30 μg/l was reported in waters from Zenne River, Belgium, as a result of arsenic release from industries and urban areas with sewage being the most prominent source [43]. Arsenic concentration as high as 200–300 μg/l was reported in surface waters of Moira River, Ontario, primarily under the effect of Sn and Au mining activities

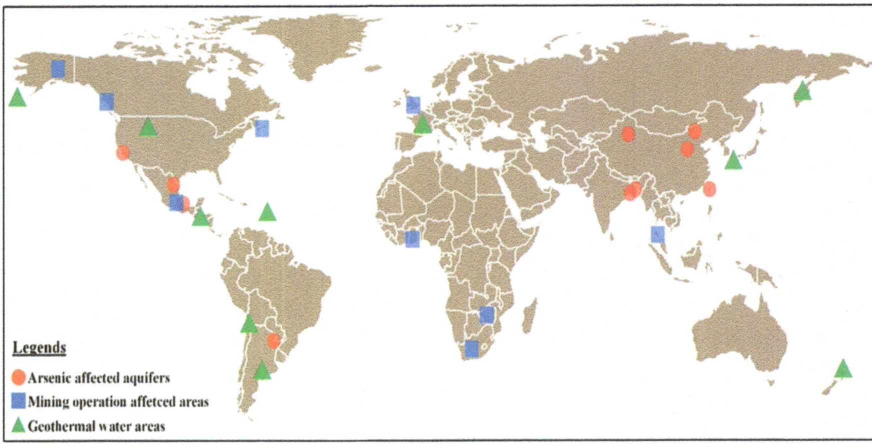

Fig. 3.1 Figure depicting the distribution of arsenic globally under natural and anthropogenic influences [3]

[44, 45]. High concentrations of arsenic are found in a wide range of environments, which include oxidizing (high pH) and reducing aquifers and areas affected by mining and geothermal activity. High concentration of a wide range of solutes like arsenic has been reported under extremely acidic conditions which tend to build up as a result of acid mine drainage [46]. An acidic seepage in Richmond Mine at Iron Mountain, California, resulted in the highest ever reported concentration of arsenic at 850,000 μg/l [47]. Arsenic may also be released into the environment from oil fields and brines. Although available data of such occurrences are very low, a few reports have also been made which suggest high concentrations of arsenic under such influences. A dissolved arsenic concentration of 230 μg/l was reported in a $NaHCO_3$-rich groundwater at a depth of 1000 meters in an oil field pool from Ellis Pool, Alberta, Canada [48]. They also reported a NaCl-dominated brine from Tisakurt, Hungary, with arsenic concentration as high as 5800 μg/l. The presence of arsenic was also reported in tobacco (mean concentration of 0.15 μg/g) and cigarettes (mean concentration of 0.11 μg/g) with reported arsenic concentration per pack of tobacco being 6 μg [49].

3.4 Ill Effects Pertaining to Arsenic Poisoning

Accumulation of arsenite and MA(III) in the body as a result of ingestion of arsenic through contaminated water may lead to their accumulation in vital organs of the body which may go on to cause atherosclerosis, hypertension, hepatotoxicity, ischemic heart diseases, diabetes, nephrotoxicity, and skin, bladder, and lung cancer on the long run. In this section, the mechanisms which are involved in the pathogenesis of arsenic-induced toxicity which ultimately leads to organ damage have been discussed.

3.4.1 Arsenic-Mediated Cardiovascular Dysfunction

Exposure to inorganic arsenic, in the long run, may lead to cardiovascular disorders like atherosclerosis, hypertension, ischemic heart diseases, and ventricular arrhythmias [50–52]. Stimulation of NADPH oxidase in the presence of arsenite causes an increase in production of reactive oxygen species (ROS) like hydrogen peroxide and superoxides in vascular endothelial cells and vascular smooth muscle cells [53, 54]. The coupling of nitric oxide with ROS results in the formation of peroxynitrite, a strong oxidant which is responsible for the upregulation of inflammatory mediators like cyclooxygenase-2 [55]. The ROS generated during arsenite exposure results in increased expression of genes related to atherosclerosis such as interleukin-6 (IL-6), heme oxygenase (HO-1), and monocyte chemoattractant protein (MCP-1) which promote the penetration, attachment, and migration of monocytes in VSMC [56]. Arsenite regulates the vasoconstriction of blood vessels by phosphorylation of myosin light chain kinase (MLCK) and increasing sensitization to calcium which ultimately results in hypertension [57]. Long-term exposure to arsenic may induce oxidative stress and alter the release of vasoactive mediators in the blood vessels which may ultimately result in elevation of blood pressure [58] (Fig. 3.2).

3.4.2 Arsenic-Induced Diabetes Mellitus

Long-term arsenic exposure decreases the expression of PPAR-γ. This reduces the sensitivity of insulin response for type II diabetes induction by arsenic [59]. Arsenic replaces a phosphate group from ATP resulting in the formation of ADP-arsenate, which decreases the rate of glucose metabolism, interrupts with energy production,

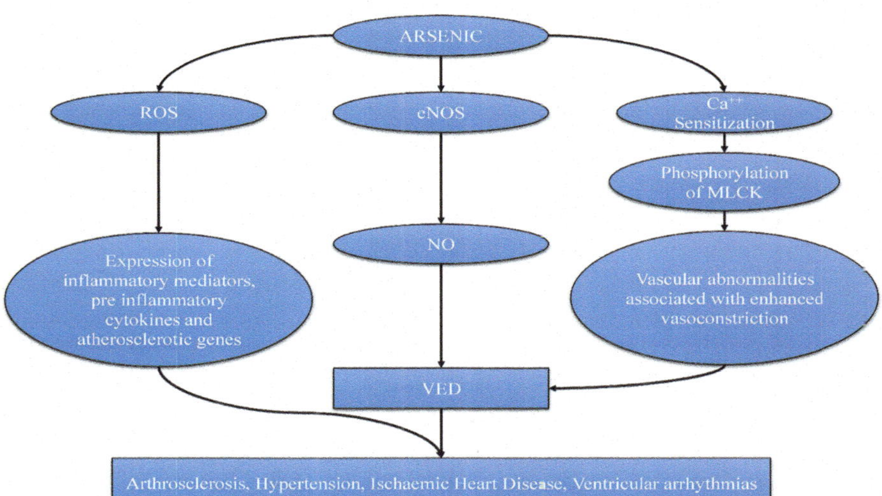

Fig. 3.2 Arsenic-mediated cardiovascular dysfunction

and interrupts with the secretion of ATP-dependent insulin [60] (Fig. 3.3). The high affinity of arsenite toward sulfhydryl groups results in the formation of covalent bonds with disulfide groups of the glucose transporter, insulin receptors, insulin, and enzymes involved in the metabolism of glucose [60]. Chronic exposure to arsenic may also result in hypoglycemia due to a notable decrease in glucose-6-phosphate activity in both the liver and kidneys [61]. Besides sodium arsenite has also been reported to suppress the expression of mRNA specific to insulin expression [62].

3.4.3 Arsenic-Induced Neurotoxicity

The ability of arsenic to easily trespass the blood-brain barrier makes the brain a very soft target for arsenic toxicity. Long-term exposure to arsenic poses various neurological implications like poor concentration, impaired memory, Parkinson's disease, speech disorders, encephalopathy, and peripheral neuropathy [63–67] (Fig. 3.4). The mechanism of arsenic-induced neurotoxicity mainly involves oxidative stress associated with an increase in the number of reactive oxygen species and lipid peroxidases along with a decrease in levels of superoxide dismutase and glutathione [68]. In recent studies, it has been revealed that chronic exposure to arsenic results in a significant drop in levels of adrenaline, noradrenaline, and serotonin in the corpus striatum, hippocampus areas, and frontal cortex of the brain [69]. Protein kinase (p38MAPK) and JNK3 pathways activated by p38 mitogen result in the induction of apoptosis in the cerebral neurons as a result of arsenic-mediated neurotoxicity [70].

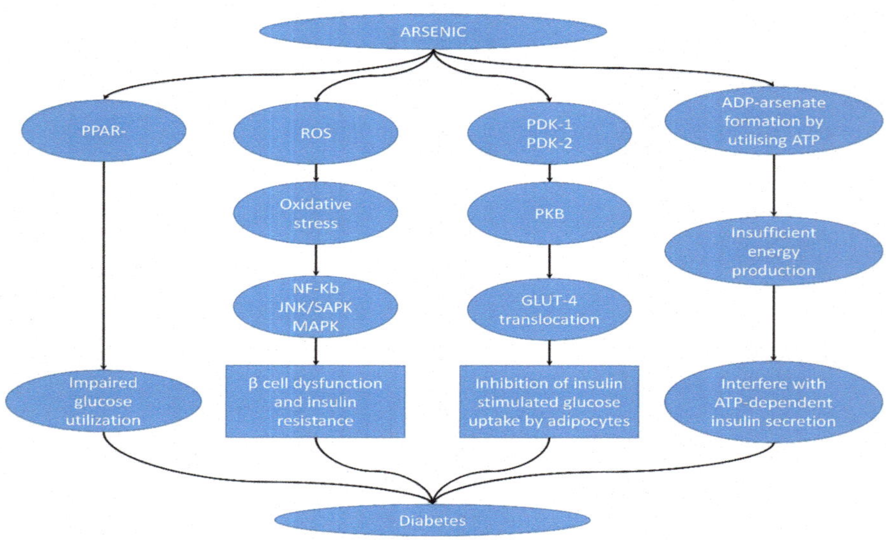

Fig. 3.3 Flowchart of arsenic-induced diabetes mellitus

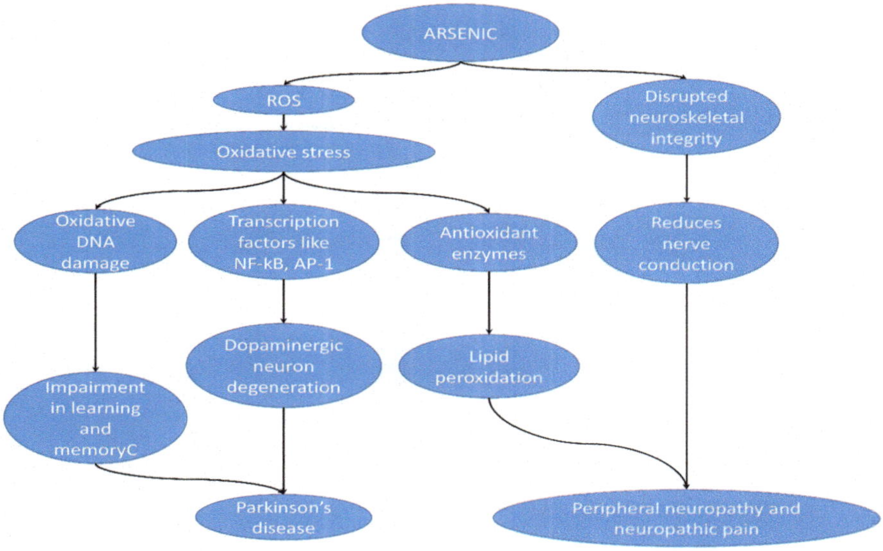

Fig. 3.4 Flowchart of arsenic-induced neurotoxicity

3.4.4 Arsenic-Induced Hepatotoxicity

The expression of HO-1 and MAPK increases due to oxidative stress induced by arsenic. HO-1 and MAPK, in turn, regulate transcription factors such as Elk-1, activating transcription factor-2 (ATF-2) and activator protein-1 (AP-1) which ultimately result in renal toxicity [71, 72] (Fig. 3.5). A study also revealed that the liver and kidneys serve as the primary targets for arsenic-induced toxicity with the levels of arsenic found in the liver being much higher than in the kidneys [73]. Activation of JNK and p38 MAPK under the influence of chronic arsenic-mediated oxidative stress results in the induction of apoptosis in the hepatocytes [74–76]. Further, upregulation of pro-apoptotic proteins under the influence of arsenic-induced oxidative stress results in hepatic apoptosis [68, 77]. A study also showed that the levels of alanine aminotransferases, total bilirubin, malondialdehyde, and aspartate in the body increased under the influence of arsenic exposure which strongly supported arsenic-induced hepatotoxicity [78].

3.4.5 Arsenic-Induced Pyruvate Dehydrogenase Complex Inhibition

Pyruvate dehydrogenase complex is a multienzyme complex comprising of three enzymes, namely, pyruvate dehydrogenase (E1) (decarboxylation of pyruvate), dihydrolipoyl transacetylase (E2) (formation of acetyl-CoA by transfer of acetyl group to CoA), and dihydrolipoyl dehydrogenase (E3) (re-oxidation of dihydrolipoamide for continuous conversion of pyruvate to acetyl-CoA). The formation of

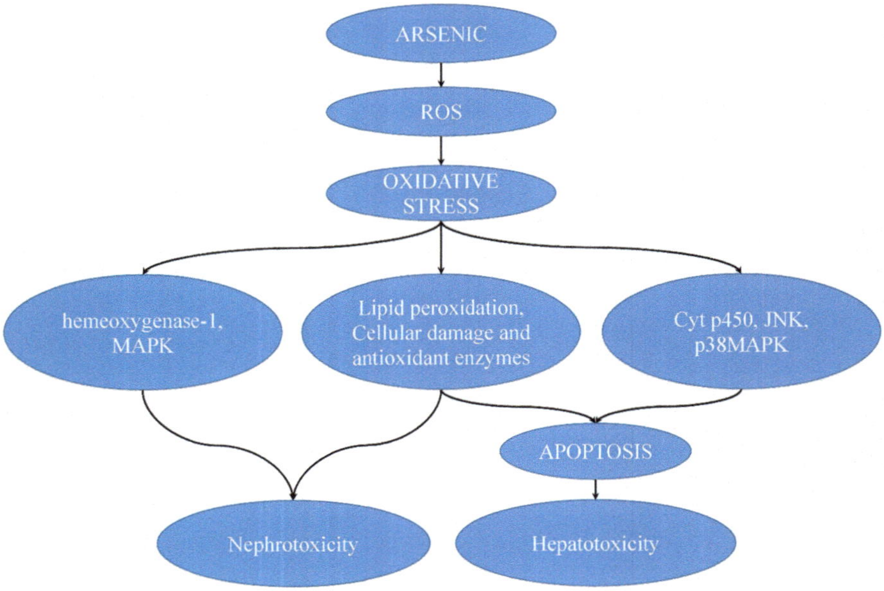

Fig. 3.5 Flowchart of arsenic-induced nephrotoxicity and hepatotoxicity

acetyl-CoA ceases as regeneration of dihydrolipoamide is blocked under the influence of arsenic which forms a stable complex with dihydrolipoamide [79].

3.4.6 Arsenic-Induced Inhibition of α-Ketoglutarate Dehydrogenase Complex of TCA Cycle

α-Ketoglutarate dehydrogenase complex is another multienzyme complex comprising of α-ketoglutarate dehydrogenase (E1), dihydrolipoyl succinyltransferase (E2), and dihydrolipoyl dehydrogenase (E3). The function of E3 is quite identical to that of pyruvate dehydrogenase complex which oxidizes dihydrolipoamide [80]. Arsenite and its organoarsenicals block the activity of the enzyme in a manner similar to that of pyruvate dehydrogenase complex, as a result of which succinyl-CoA is stopped.

3.4.7 Arsenic-Induced JAK-STAT Pathway Inhibition

The phosphorylation of key tyrosine and serine residues of STAT3 were analyzed to study the effect of arsenite by immunoblotting. The results so obtained were consistent with EMSA results where exposure to arsenite caused inhibition of tyrosine (Y705) residue, whereas addition of IL-6 to the system alone caused phosphorylation of Y705. This showed that arsenic blocked tyrosine phosphorylation of the Y705 residue which in turn led to the blockade of IL-6-induced activation of STAT-3 [81].

3.4.8 Arsenic-Induced ATP Suppression in Glycolysis

Arsenic renders the process of glycolysis useless. This is because although the production of pyruvate takes place, there is no net production of ATP. Glyceraldehyde phosphate dehydrogenase (GAPDH) is an enzyme which converts glyceraldehyde-3-phosphate to 1, 3-biphosphoglycerate. In the presence of arsenic, GAPDH converts glyceraldehyde-3-phosphate to 1-arseno-3-phosphoglycerate which is an analog of 1, 3-biphosphoglycerate. Each 1-arseno-3-phosphoglycerate molecule breaks down into arsenate and 3-phosphoglycerate molecule without any formation of ATP [82].

3.5 Mechanism of Arsenic Poisoning

The indisputable authority of arsenic as a toxic substance stamps from its ability to inhibit about 200 enzymes involved in the production of energy in various cellular pathways, synthesis, and repair of DNA [83]. Toxicity developed due to both trivalent and pentavalent arsenical species has been discussed here.

3.5.1 Mechanism of Pentavalent Arsenic Toxicity

Dixon [84] reported the ability of phosphates in a variety of biochemical reactions because of its structural similarity and similarity in properties. Lagunas [85] and Gresser [86] also further validated Dixon in their work by showing the ability of arsenate to react with glucose and gluconate to form glucose-6-arsenate and 6-arsenogluconate, respectively. Glucose-6-arsenate and 6-arsenogluconate show structural similarity with glucose-6-phosphate and 6-phosphogluconate. Glucose-6-arsenate serves as a substrate for glucose-6-phosphate dehydrogenase and can inhibit hexokinase [85]. In the human RBC, phosphate in the sodium pump and the anion exchange transport system can also be replaced by arsenate [87]. During the process of glycolysis, ATP is generated in the presence of phosphate (substrate-level phosphorylation), but the reverse is true in the presence of arsenate [82, 88]. Delnomdedieu et al. [89] and Winski and Carter [90] showed reduced levels of ATP in human erythrocytes and rabbits, respectively, following in vitro exposure to arsenate (human 0.1–10 mM; rabbit 0.8 mM).

3.5.2 Mechanism of Trivalent Arsenic Toxicity

Certain receptors, enzymes, and coenzymes contain thiols or vicinal sulfhydryls as specific functional groups. These thiols and vicinal sulfhydryls play a very crucial role in the proper functioning of these molecules. Trivalent arsenicals have a very high affinity toward such functional groups and react readily in vitro with molecules such as GSH and cysteine which contain thiols as their functional groups [91, 92]. Moreover, the affinity of arsenite for dithiols is far greater than monothiols. This was shown successfully by Delnomdedieu et al. [93] where the transfer of arsenite from $(GSH)_3$-arsenic complex to the dithiol 2,3-dimercaptosuccinic acid, a highly favorable reaction, takes place [91]. A study by Styblo et al. [94] has shown that trivalent organic arsenicals like MMA(III) and DMA(III) undergo in vitro binding with proteins with a far greater affinity than pentavalent arsenicals. Methylated organic arsenicals like MMA(III) interact with thiol groups of GSH reductase and thioredoxin reductase ultimately resulting in the suppression of these molecules [95, 96].

3.6 Arsenic Contamination (A Few Case Studies)

Over the years incidences associated with arsenic-related contamination have been reported globally. A few such cases reported over the last two decades have been discussed here. Das et al. [97] carried out a survey of groundwater bodies in six districts of West Bengal, India, for a period of 5 years. The six districts spanned an area of 34,000 km^2 which encompassed a population of 30 million people. In their studies, they found that arsenic concentration in the groundwater bodies of these districts had exceeded the maximum threshold of 0.05 mg/l as recommended by the WHO. Their studies revealed about 800,000 people from 312 villages in 37 blocks drank water that was contaminated with arsenic, out of which about 175,000 people showed arsenic-related lesions in the skin which were a representative of later stages of arsenic toxicity. The affected individuals also showed most of the three stages of clinical symptoms closely related to arsenic poisoning, out of which melanosis, depigmentation, keratosis, and hyperkeratosis were the most common. Upon analysis of water samples from these regions, they detected a combination of arsenite and arsenate but could not detect methyl arsenic or dimethyl arsenic acid.

The distribution of arsenic species in different organs from the dead body of a 28-year-old man who died after fatal intoxication with arsenic trioxide was studied. The cause of death was reported to be massive ingestion of arsenic trioxide (~8 g) orally. Autopsy studies revealed As(III) to be the most dominant species found in various organs with the liver and kidneys showing the highest arsenic concentration. Detoxification of arsenic by methylation and its subsequent elimination was reported to be the primary reason behind such high concentration of arsenic in these organs. The affinity of As(III) toward vicinal dithiols in hepatic cytosolic proteins was proposed to be another reason behind the high concentration of arsenic in the liver (5–50-folds higher than most other organs). The distribution of arsenic was however almost uniform throughout most other organs [98].

Rahman et al. [99] reported arsenic levels higher than the WHO recommended limit of 50 μg/l in groundwater bodies in 50 districts in Bangladesh and nine districts in West Bengal, India, which covered areas of 118,849 km^2 with populations of 104.9 million and 42.7 million, respectively. Upon clinical examination of about 18,000 individuals from Bangladesh and 86,000 persons from West Bengal involving these arsenic-affected districts, 3695 (20.6% including 6.11% children) in Bangladesh and 8500 people (9.8% including 1.7% children) in West Bengal showed arsenic-related skin lesions.

Liu et al. [100] have reported some 3000 cases of arsenic poisoning in the Southwest Prefecture of Guizhou, with the burning of high arsenic-rich coal for fuel being the root cause of this problem. Mineralization of coal in this region has led to release of high concentrations of arsenic, and burning of coal in open pits has led to an increase in arsenic concentration in indoor air. Due to the cumulative effect of all these factors, over 200,000 people of this region are at a serious risk of arsenic overexposure.

The levels of exposure to arsenic in 10-year-old school children selected randomly from a high-exposure area and a low-exposure area of Ron Phibun subdistrict, Ron Phibun District, and Nakhon Si Thammarat Province, Thailand, were studied over a period of 1 year. The students belonging to high-exposure area and low-exposure area were then compared to their counterparts belonging to a control area. The concentration of inorganic arsenic and its metabolites in the urine samples of the students belonging to the high-risk area and low-risk area were considerably higher than those who belonged to the control area with drinking water and surface soil being the primary sources of exposure. A risk analysis study of these children revealed a higher chance of cancerous development in children belonging to the high-exposure area [101].

The people of Chile get exposed to arsenic via drinking water and air pollution primarily due to mining activities. The early effects of arsenic poisoning such as vascular diseases, bronchiectasis, and lesions in the skin were observed in children and adults. Long terminal effects of arsenic poisoning such as lung and bladder cancers were reported 20 years after highest levels of arsenic exposure [102].

The groundwater in Terai regions of Nepal was analyzed to determine the extent of arsenic contamination. Mineralization was reported to be the primary cause of arsenic contamination. Levels of arsenic concentration higher than 50 μg/l were detected in about 0.5 million people of that region, whereas arsenic concentration between 10 and 50 μg/l was reported in about 3.5 million people which comprised of about 31% population of that region. These individuals repeatedly suffered from chronic arsenic poisoning as a result of consumption of arsenic-affected water from groundwater bodies [103].

Arsenic concentration in drinking water (119–310) μg/l, alfalfa hay (1.9–6.9) μg/l, cultivated soil (46.7–819.9) μg/l, wool (1.56–10.79) μg/l, and blood samples (86.3–656) μg/l was evaluated in sheep from Ebrahimabad and Baba Nazar villages in Kurdistan Province of Iran. These reports suggested the presence of a very high concentration of arsenic above recommended levels in these areas which also goes on to suggest that arsenic has penetrated into the biogeochemical cycle by direct or indirect pathways. The conclusion drawn from the study was that the sheep from the contaminated areas suffered from anemia which was supported by a decrease in packed cell volume and hemoglobin in sheep that reared in these arsenic-contaminated zones [104].

3.7 *Spirulina platensis* in Large-Scale Bioremediation of Arsenic-Contaminated Water Bodies

Cyanobacteria are ubiquitously distributed in different types of aquatic environments and are considered to be ancient phototrophic organisms [105]. They are generally considered to be blue-green, and these colors are conferred upon them by their photosynthetic pigments. Some cyanobacteria have a nitrogen-fixing ability and so have the potential to play an important role in rice paddy water [106]. They may exist in the form of colonies [107] and individual cells [108] and may form coccoid [109] or filamentous structures [110]. The morphology of cyanobacteria has remained unaltered over the last 3.5 billion years because they are highly resistant to contamination and, as such, they have evolved different mechanisms to survive in the harsh arsenic-contaminated environments [11]. However, characterization of microbes based upon morphological features is highly subjective and speculative in nature. Today, however, more weight is being given on genome-based characterization, and a very good review of the different types of genetic tools available for characterization has been beautifully described by Koksharova and Wolk [111]. Cyanobacteria produce compounds which are protective in nature which in turn shield them from the harshness of the environment [112]. Some species of cyanobacteria even possess the ability to produce molecules which are anticancerous and antimicrobial in nature [113, 114]. Spirulina originated about 3.5 billion years ago, and during that course of time, it has successfully developed the ability to derive nutrition for the purpose of reproduction from dissolved carbon dioxide in seawater. It is a photosynthesizing blue-green alga and shows high growth under high temperature, high alkaline conditions, and strong sunlight. Among the countless number of cyanobacteria, *Arthrospira* (spirulina) *platensis* is a blue-green alga which has the ability to successfully thrive under raised alkaline pH [115]. *Spirulina platensis* can be easily recognized because of its peculiar shape of cylindrical trichomes. These cylindrical trichomes are arranged in a left-handed helix throughout the filament [116]. Three different species of spirulina were cultured, viz., *Spirulina platensis*, *Spirulina laxissima*, and *Spirulina lonar*, to study their biochemistry and evaluate their growth in different types of media containing organic and inorganic nutrients. Out of the three species, the highest growth rate, biomass production, pigmentation, and low accumulation of intracellular phenolics (results in toxicity) were shown by *Spirulina platensis* [117]. This result also cemented the importance of *Spirulina platensis* as a favorable strain for large-scale cultivation because of its short doubling time and high growth rate [117]. Addition of ammonium nitrate to the culture provides an easily assimilable source as well as a reserve of nitrogen that not increases the biomass of *Spirulina platensis* but also reduces the cost of production by reducing the cost of cultivation media [118]. All these favorable factors associated with the ability of *Spirulina platensis* to successfully and efficiently detoxify arsenic from water bodies by biomethylation [119] makes it a top-notch candidate for the large-scale treatment of wastewater bodies contaminated with arsenic.

3.8 Mechanism of Biomethylation of Arsenic

The process of methylation as a means of detoxification of inorganic arsenic by living organisms is a complex process, and there exists a lack of information about how this process works. Moreover, the methods of methylation are not universal and as such show small to large variations from organism to organism. Arsenic which is primarily present in the water bodies in inorganic form can be biotransformed into methylated arsenicals, arsenosugars, and arsenolipids. This biotransformation can be carried out by oxidation, reduction, and methylation. Oxidation of arsenite (As (III)) to arsenate (As (V)) is considered to be a detoxification pathway as As (V) is less toxic as compared to As (III) [120, 121]. Similarly, arsenic methylation is also considered to be a detoxification pathway as intermediate products formed during this process like methyl arsenite (III) and dimethyl arsenite (III) are considered to be more toxic [16, 122]. Alternate reduction and methylation are also very good methods to detoxify arsenic (III) which ultimately leads to production of less toxic methylated arsenicals like methyl arsenate (V), dimethyl arsenate (V), and trimethyl arsine oxide (TMAO (V)) [123, 124].

However, new findings have shed more light on the fact that alternative pathways for detoxification of arsenic also exist in which DMA (III) and MMA (III) persist in their trivalent state during the entire course of the catalytic cycle [125, 126].

3.8.1 The Work of Challenger and Associates (Leeds School)

The formation of trimethyl arsine (a less toxic form of arsenic) occurred in accordance with a mechanism proposed by [124] and associates of the Leeds School. This mechanism has managed to hold its own for a very long time and is considered to be one of the very few universally accepted mechanisms of arsenic detoxification in living organisms. According to this group, a positive methyl group must be generated which will then be transferred to the metalloid. S-Adenosylmethionine or SAM was considered to be this methyl donor. For the formation of trimethyl arsine from arsenate, four stepwise $2e^-$ reductions are necessary. Each reduction would then result in the formation of one lone pair of electrons on the arsenic atom with the last step being the only exception. At the end of each reduction step, arsenic is methylated by SAM.

3.9 Tackling Arsenicosis (A Hypothesis)

Guo et al. [119] isolated a SpArsM gene from *Spirulina platensis* which could successfully detoxify As(III) by methylation followed by conversion of MMA(V) and DMA(V). CmarsM7 and CmarsM8 genes from *Cyanidioschyzon*, a eukaryotic microalga which was isolated from Yellowstone, had the ability to methylate arsenic and conferred high arsenic resistance to the transformed strain of *E. coli*. There are many more reports which support the existence of such genes throughout the world. These reports encourage the use of proteins encoded by these genes and its applications in the field of drug development to successfully tackle acute arsenicosis in patients. In this section, a hypothesis has been developed to treat such patients suffering from arsenicosis with arsenic methylating proteins.

Step 1: Isolation of the SpArsM Protein

- Collection of *Spirulina platensis* from various sources (brackish water bodies, freshwater, and marine water bodies).
- Growth in Zarrouk's medium.
- Screening for *Spirulina platensis* in as(III)-supplemented media.
- Collecting information about the flanking sequence of SpArsM gene and design of primers using bioinformatics tools.
- Cloning of the PCR product into a pUC vector by using E. coli as a heterologous host.
- Plasmid isolation with plasmid isolation kit.
- Plasmid digestion with specific restriction enzymes.
- Insertion of the fragment into the pET vector.
- The pET vector with the SpArsM gene is used to transform E. coli that lack the ability to methylate as(III) where screening is subsequently carried out in an LB media supplemented with as(III) and IPTG to induce the expression of his-tagged SpArsM protein.
- This his-tagged SpArsM protein that is produced may then be easily separated by affinity chromatography.
- Determination of protein concentration at 280 nm followed by characterization and in-silico analysis of the protein.
- Development and formulation of the protein into a suitable drug dosage form.

Step 2: The Process of Treatment

- Under normal conditions (case I), lipoic acid bound to the E2 subunit of pyruvate dehydrogenase complex undergoes FAD-dependent oxidation of lipolate. This lipolate then undergoes a reduction in the presence of thiamine triphosphate (TPP) by binding to the CH_3COO group attached to TPP. This CH_3COO group bound to lipoic then binds to the CoA group of pantothenic acid resulting in the formation of acetyl-CoA which serves as the starting material for Krebs cycle (Fig. 3.6).
- In the case of a person suffering acute arsenicosis (case II), arsenite (III), the trivalent form of arsenic, directly binds itself very strongly to the thiol groups of lipoic acid. This prevents FAD-dependent oxidation of lipolate which renders the TPP inept in reducing this structure. Ultimately there is no formation of acetyl-CoA as a result of which the Krebs cycle loses its efficiency because of the loss of its starting material. This ultimately may result in the death of that individual because of acute shortage of ATP (Fig. 3.6).
- However, when the SpArsM protein (now developed into a drug) is administered into the body of the patient suffering from arsenicosis, the SpArsM protein methylates the trivalent form of arsenic. This prevents arsenite from binding to the thiol groups of lipoic acid. The lipoic acid is now free to carry out its normal function, and this enables the pyruvate dehydrogenase complex to utilize the

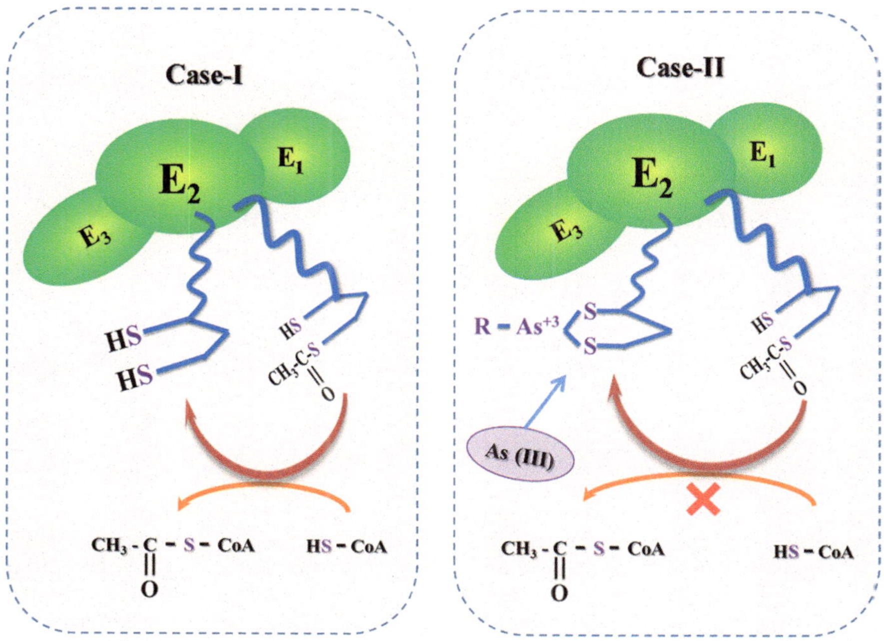

Fig. 3.6 Figure depicting the effect of arsenic on the functioning of pyruvate dehydrogenase complex. Case I represents a healthy individual with properly functioning pyruvate dehydrogenase complex which ultimately produces acetyl-Co A. Case II represents the effect of arsenic on the functioning of pyruvate dehydrogenase complex. In Case II arsenic binds to the two thiol groups of lipoic acid, thus blocking the proper functioning of this enzyme. This prevents the formation of acetyl-CoA which ultimately hinders the Krebs cycle. (E1, pyruvate dehydrogenase; E2, dihydrolipoyl transacetylase; E3, dihydrolipoyl dehydrogenase)

pyruvate generated at the end of glycolysis and produce acetyl-CoA as the end product. Now, because of the availability of acetyl-CoA (the starting material of Krebs cycle), the Krebs cycle retains its normal functioning which results in the normal production of ATP as a result of which the patient gradually recovers and survives (Fig. 3.7).

3.10 A Few Notable Works on Remediation of Arsenic

Katsoyiannis et al. [127] devised a method that could successfully remove arsenic from groundwater bodies with an efficiency of about 80% by using fixed bed upflow bioreactors without the use of any additional chemicals during simultaneous biological oxidation of iron and manganese from groundwater by using *Gallionella* and *Leptothrix* species. *Gallionella ferruginea* and *Leptothrix ochracea* oxidized soluble iron(II) and manganese(II) which then served as effective adsorbents for removal of As(V) and As(III).

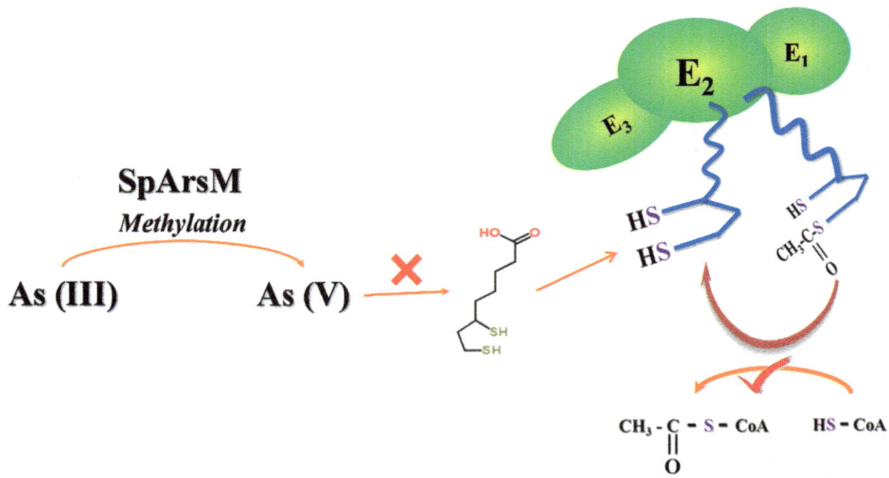

Fig. 3.7 Figure depicting the effect of SpArsM (when delivered to an arsenicosis patient in a drug dosage form) on the functioning of pyruvate dehydrogenase complex. In this case, SpArsM protein methylates As(III) to As(V) as a result of which arsenic is no longer able to bind to the thiol groups of lipoic acid. This enables pyruvate dehydrogenase complex to retain its original function. (E1, pyruvate dehydrogenase; E2, dihydrolipoyl transacetylase; E3, dihydrolipoyl dehydrogenase)

Dhankher et al. [128] genetically transformed *Arabidopsis thaliana* with arsenic reductase (arc C) gene and γ-glutamylcysteine synthetase (γ-ECS) gene from *E. coli*. The arc C gene was expressed by light-inducible soybean rubisco promoter (SRS1p) which highly expresses arc C only in the leaves. Similarly, γ-ECS was expressed by a highly constitutive ACT-2p actin promoter. γ-ECS gene catalyzes the first step in phytochelatin synthesis pathway. In doing so it increases the concentration of thiol compounds (including phytochelatin, a polypeptide) throughout the plant. Arsenate absorbed from the soil through the roots is transported with transporters via the xylem to the leaves where arc C catalyzes glutathione-coupled electrochemical reduction of arsenate (V) to arsenite (III). Arsenite is then detoxified by the formation of arsenic-protein thiolates with thiol compounds. When grown on arsenic, this plant showed two to three times higher accumulation of arsenic per gram of tissue than its wild counterpart which expressed γ-ECS or arc C alone.

Takeuchi et al. [27] identified a bacteria *Marinomonas communis* out of nine isolated bacterial strains belonging to marine and nonmarine origins. These bacteria could resist up to 510 mg/l of arsenic and could remove 2290 μg of arsenic per gram of dry weight when incubated on a medium containing 5 mg/l of arsenate. This accumulation was the highest accumulation of arsenic reported at that period of time. This makes these bacteria one of the best candidates for removal of arsenic from contaminated water bodies.

Ike et al. [129] acclimatized a mixed culture of heterotrophic bacteria (namely, *Haemophilus*, *Micrococcus*, and *Bacillus*) from soil free of contamination to a high concentration of As (III). Upon addition of arsenic at an initial concentration of up to 1500 mg/l at pH ranging from 7 to 10 and temperatures ranging between 25 and

35 °C, this mixture successfully oxidized As(III) which greatly enhanced the efficiency of removal of arsenic by absorption on activated alumina, an essential pretreatment process from a basal salt medium containing 75 mg/l of As(III).

A study by Tuzen et al. [130] showed that *Ulothrix cylindricum*, green algae, could absorb 67.2 mg/g of As (III) from its aqueous medium which could then be readily desorbed by treatment with 1 M HCl. *Ulothrix cylindricum* also enhanced the process of As (III) absorption and desorption almost by tenfold. A study of the thermodynamic parameters like $\Delta G°$, $\Delta S°$, and $\Delta H°$ of this process revealed that it was an exothermic, spontaneous, and feasible reaction under the studied conditions.

Mirza et al. [131] studied the possibility of phytoextraction of arsenic from synthetic wastewater with *Arundo donax*. The plant showed an increase in its biomass in its root and shoot without showing any symptoms of toxicity when grown in nutrient media supplemented with up to 600 μg/l of arsenic. This finding suggests that *Arundo donax* plants have the potential to be employed in the treatment of water containing arsenic concentration up to 600 μg/l.

Kao et al. [132] isolated a novel As(III)-oxidizing bacteria (As 7325) from the aquifer in the blackfoot disease (BFD) endemic area of Taiwan, which oxidized 2300 μg/l As (III) from in situ As(III)-contaminated groundwater in 1 day under aerobic conditions. They then successfully removed this oxidized As(V) from contaminated groundwater by absorption on As 7325 cell pellets.

Sibi [133] isolated six different microalgae, namely, *Chlorella*, *Oscillatoria*, *Scenedesmus*, *Spirogyra*, and *Pandorina*, which showed a higher rate of absorption of more toxic As (III) than its less toxic As(V) counterpart. The dried biomass of these microalgae showed even higher absorption and much faster kinetics than the living ones.

Jasrotia et al. [134] studied the potential of locally available algal species like *Cladophora* in the possible phycoremediation of arsenic from arsenic-enriched water bodies. Their study revealed the ability of *Cladophora* to survive in arsenic concentrations up to 6 mg/l in water and the plant's ability to absorb arsenic by almost 100% when the arsenic concentration was raised to 80 g/l. HPLC coupled with I CPMS (for arsenic speciation) and electron microscopy studies confirmed the bioabsorption of arsenic in the form of arsenite, arsenate, arsenosugars, MMA, and DMA.

Dey et al. [135] isolated two arsenic-resistant rod-shaped gram-positive bacteria (*Bacillus sp.* and *Aneurinibacillus aneurinilyticus*) from Purbasthali block of Burdwan district, West Bengal, India, which have the ability to oxidize arsenite to less toxic arsenate and can tolerate arsenate concentration up to 4500 ppm and arsenite concentration up to 550 ppm. The isolated *Bacillus sp.* had the ability to remove 51.45% and 51.99% of arsenite and arsenate, respectively, whereas the isolated *Aneurinibacillus aneurinilyticus* had the ability to remove 53.29% and 50.37% of arsenite and arsenate, respectively, from an arsenic-supplemented media after 72 h of incubation.

Guo et al. [119] studied the ability of *Spirulina platensis* in accumulation and biotransformation of arsenic. The dry weight of *Spirulina platensis* could accumulate up to 4.1 mg/kg of arsenic. This species could oxidize toxic As (III) to a lesser toxic As (V) like mono-methyl arsenate (MMA (V)) and dimethyl arsenate with an efficiency of 64% to 86%.

3.11 Problems Associated with Biomethylation As a Means of Detoxification and Bioremediation of Arsenic

Earlier there was this belief that methylation of inorganic arsenic was actually a way of detoxifying it as the acute toxicity of inorganic was found to be much greater than organic arsenicals. However, the process of methylation as a means of detoxification of As (III) may not always hold true. A study by Cullen and Reimer [3] further vindicated this fact as they successfully showed how an MMA(III) derivative was far more toxic than arsenite to *Clavulina humicola*. Methylated forms of As(III) like methyl arsine oxide [$(CH_3AsO)_n$] which showed higher levels of toxicity as compared to arsenite in *Clavulina humicola* were also reported by Cullen and Reimer [3] who also reported that methylated sulfides of arsenic like [$(CH_3AsS)_n$] had far greater toxic effects than oxides of arsenite. Furthermore, the ability of organic arsenicals to generate a wide variety of carcinogenic effects as compared to inorganic arsenic was also reported by Styblo et al. [136]. Human cells responded by showing higher levels of toxicity in the presence of MMA (III) as compared to arsenite [137, 138]. Similar levels of cytotoxicity were reported in dimethyl arsenic acid (DMA) as observed in arsenite in a wide range of human cells [138]. The LD_{50} value of MMA (III) in hamster was found to be much lower than arsenite in hamster [137].These findings go on to reveal that the general notion of detoxification of arsenicals by methylation cannot be accepted universally in all organisms.

3.12 Conclusion

Microalgae are an integral component of our food web and as such play a vital role in the regulation of arsenic in the environment. Hence, the process of arsenic metabolism by microalgae has gradually gained an important position in the scientific community. However, irrespective of the research carried out over many years, there still exists a void in our knowledge about these organisms which need to be filled. This is because there exists a wide variation in the pathways involving biomethylation of arsenic in different microalgae and a sufficient data does not exist regarding the characterization of these enzymes. Moreover, conventional methods in the form of adsorption, ion exchange, coprecipitation, and membrane-based separation are already being used to treat water polluted with arsenic, and the feasibility of such methods are under serious questioning because of the high cost of operations and need of technical manpower. Irrespective of all these bottlenecks, microalgae can emerge as a boon in our fight against the menace of arsenic poisoning. So proper establishment of mechanisms of methylation and docking studies of arsenic on these enzymes are necessary to shed further light on the working of these arsenic methylation enzymes, and with a proper pipelined strategy, remediation of arsenic-contaminated water bodies with microalgae has the potential to grow into a multimillion-dollar industry which is also capable of generating a huge amount of employment.

References

1. Bahar MM, Megharaj M, Naidu R (2013) Toxicity, transformation and accumulation of inorganic arsenic species in a microalga Scenedesmus sp. isolated from soil. J Appl Phycol 25:913–917. https://doi.org/10.1007/s10811-012-9923-0
2. Cullen WR (2008) Is arsenic an aphrodisiac? Royal Society of Chemistry, Cambridge
3. Cullen WR, Reimer KJ (1989) Arsenic speciation in the environment. Chem Rev 89:713–764. https://doi.org/10.1021/cr00094a002
4. Fujiwara S, Kobayashi I, Hoshino S et al (2000) Isolation and characterization of arsenate-sensitive and resistant mutants of Chlamydomonas reinhardtii. Plant Cell Physiol 41:77–83
5. Smedley PL, Kinniburgh DG (2002) A review of the source, behaviour and distribution of arsenic in natural waters. Appl Geochem 17:517–568. https://doi.org/10.1016/S0883-2927(02)00018-5
6. Shen S, Li XF, Cullen WR et al (2013) Arsenic binding to proteins. Chem Rev 113:7769–7792. https://doi.org/10.1021/cr300015c
7. WHO (World Health Organization) (1999) Fact sheet No 210: Arsenic in Drinking Water
8. Mandal B (2002) Arsenic round the world: a review. Talanta 58:201–235. https://doi.org/10.1016/S0039-9140(02)00268-0
9. Paikaray S (2012) Environmental hazards of arsenic associated with black shales: a review on geochemistry, enrichment and leaching mechanism. Rev Environ Sci Biotechnol 11:289–303. https://doi.org/10.1007/s11157-012-9281-z
10. Villaescusa I, Bollinger J-C (2008) Arsenic in drinking water: sources, occurrence and health effects (a review). Rev Environ Sci Biotechnol 7:307–323
11. Aurilio AC, Mason RP, Hemond HF (1994) Speciation and fate of arsenic in three lakes of the Aberjona Watershed. Environ Sci Technol 28:577–585. https://doi.org/10.1021/es00053a008
12. Duncan EG, Maher WA, Foster SD (2015) Contribution of arsenic species in unicellular algae to the cycling of arsenic in marine ecosystems. Environ Sci Technol 49:33–50. https://doi.org/10.1021/es504074z
13. Lee RE (2008) Phycology, 4th edn. Cambridge University Press, New York
14. Maeda S, Mawatari K, Ohki A, Naka K (1993) Arsenic metabolism in a freshwater food chain: blue???Green alga (Nostoc sp.)??? Shrimp (Neocaridina denticulata)??? Carp (Cyprinus carpio). Appl Organomet Chem 7:467–476. https://doi.org/10.1002/aoc.590070705
15. Maeda S, Ohki A, Tokuda T, Ohmine M (1990) Transformation of arsenic compounds in a freshwater food chain. Appl Organomet Chem 4:251–254. https://doi.org/10.1002/aoc.590040312
16. Wang P, Sun G, Jia Y et al (2014) A review on completing arsenic biogeochemical cycle: microbial volatilization of arsines in environment. J Environ Sci (China) 26:371–381. https://doi.org/10.1016/S1001-0742(13)60432-5
17. Smedley PL, Nicolli HB, Macdonald DMJ et al (2002) Hydrogeochemistry of arsenic and other inorganic constituents in groundwaters from La pampa, Argentina. Appl Geochem 17:259–284. https://doi.org/10.1016/S0883-2927(01)00082-8
18. Kinniburgh D, Smedley P (2001) Arsenic contamination of groundwater in Bangladesh. Br Geol Surv Keyworth, U K 1:255. British Geological Survey Report WC/00/19
19. Central Ground Water Board I (1999) Incidence of arsenic in groundwater in West Bengal
20. Smith AH, Lingas EO, Rahman M (2000) Contamination of drinking-water by arsenic in Bangladesh: a public health emergency. Bull World Health Organ 78:1093–1103. https://doi.org/10.1590/S0042-96862000000900005
21. Tseng WP, Chu HM, How SW et al (1968) Prevalence of skin cancer in an endemic area of chronic arsenicism in Taiwan. J Natl Cancer Inst 40:453–463. https://doi.org/10.1093/jnci/40.3.453
22. Hsu K-H, Froines JR, Chen C-J (1997) tudies of arsenic ingestion from drinking water in Northeastern Taiwan: Chemical speciation and urinary metabolites. In: Abernathy CO, Calderson RL, Chappell WR (eds) Arsen Expo Heal Eff. Chapman Hall, London, pp 190–209

23. Kuo T-L (1968) Arsenic content of artesian well water in endemic area of chronic arsenic poisoning. Rep Inst Pathol Natl Taiwan Univ 20:7–13
24. Guo H-R, Chen C-J, Greene HL (1994) Arsenic in drinking water and cancers: A brief descriptive review of Taiwan studies. In: Chappell WR, Abernathy CO, Cothern CR (eds) Arsen Expo Heal Northwood, Sci Technol Lett, pp 129–138
25. Luo ZD, Zhang YM, Ma L et al (1997) Chronic arsenicism and cancer in Inner Mongolia-consequences of well-water arsenic levels greater than 5 ug 1. In: Abernathy CO, Calderon RL, Chappell WR (eds) Arsen Expo Heal Eff. Chapman Hall, London, pp 55–68
26. Ma HZ, Xia YJ, Wu KG et al (1999) Arsenic exposure and health effects in Bayingnormen, Inner Mongolia. In: Chappell WR, Abernathy CO, Calderon RL (eds) Arsen Expo Heal Eff. Elsevier, Amsterdam, pp 127–131
27. Takeuchi M, Kawahata H, Gupta LP et al (2007) Arsenic resistance and removal by marine and non-marine bacteria. J Biotechnol 127:434–442. https://doi.org/10.1016/j.jbiotec.2006.07.018
28. Berg M, Tran HC, Nguyen TC et al (2001) Arsenic contamination of groundwater and drinking water in Vietnam: a human health threat. Environ Sci Technol 35:2621–2626
29. Gurzau E, Gurzau A (2001) Arsenic in drinking water from groundwater in Transylvania, Romania: an overview. In: Chappell WR, Abernathy CO, Calderon RL (eds) Arsen Expo Heal Eff, pp 181–184
30. Del Razo LM, Arellano MA, Cebrián ME (1990) The oxidation states of arsenic in well-water from a chronic arsenicism area of northern Mexico. Environ Pollut 64:143–153. https://doi.org/10.1016/0269-7491(90)90111-O
31. Smith AH (1998) Marked increase in bladder and lung cancer mortality in a region of Northern Chile due to arsenic in drinking water
32. Nicolli HB, Suriano JM, Gomez Peral MA et al (1989) Groundwater contamination with arsenic and other trace elements in an area of the pampa, province of C??Rdoba. Argentina Environ Geol Water Sci 14:3–16. https://doi.org/10.1007/BF01740581
33. Fujii R, Swain WC (1995) Areal distribution of selected trace elements, salinity, and major ions in shallow ground water, Tulare Basin, Southern San Joaquin Valley, California
34. Robertson FN (1989) Arsenic in ground-water under oxidizing conditions, south-West United States. Environ Geochem Health 11:171–185. https://doi.org/10.1007/BF01758668
35. Hindmarsh J, McCurdy R (1986) Clinical and environmental aspects of arsenic toxicity. CRC Crit Rev Clin Lab Sci 23:315–347
36. Nriagu JO, Pacyna JM (1988) Quantitative assessment of worldwide contamination of air, water and soils by trace metals. Nature 333:134–139. https://doi.org/10.1038/333134a0
37. Vallee BL, Ulmer DD, Wacker WEC (1960) Arsenic toxicology and biochemistry. J Occup Environ Med 2:358
38. Borba RP, Figueiredo BR, Matschullat J (2003) Geochemical distribution of arsenic in waters, sediments and weathered gold mineralized rocks from Iron quadrangle. Brazil Environ Geol 44:39–52. https://doi.org/10.1007/s00254-003-0766-5
39. Andreae MO (1980) Arsenic in rain and the atmospheric mass balance of arsenic. J Geophys Res 85:4512. https://doi.org/10.1029/JC085iC08p04512
40. Crecelius EA (1975) The geochemical cycle of arsenic in Lake Washington and its relation to other elements. Limnol Oceanogr 20:441–451. https://doi.org/10.4319/lo.1975.20.3.0441
41. Nimick DA, Moore JN, Dalby CE, Savka MW (1998) The fate of geothermal arsenic in the Madison and Missouri rivers, Montana and Wyoming. Water Resour Res 34:3051–3067. https://doi.org/10.1029/98WR01704
42. Wilkie JA, Hering J (1998) Rapid oxidation of geothermal arsenic (III) in streamwaters of the eastern Sierra Nevada
43. Andreae MO, Andreae TW (1989) Dissolved arsenic species in the Schelde estuary and watershed. Belgium Estuar Coast Shelf Sci 29:421–433. https://doi.org/10.1016/0272-7714(89)90077-2

44. Smedley PL (1996) Arsenic in rural groundwater in GhanaPart special issue: Hydrogeochemical studies in sub-saharan Africa. J Afr Earth Sci 22:459–470. https://doi.org/10.1016/0899-5362(96)00023-1
45. Williams M, Fordyce F, Paijitprapapon A, Charoenchaisri P (1996) Arsenic contamination in surface drainage and groundwater in part of the southeast Asian tin belt, Nakhon si Thammarat Province. southern Thailand Environ Geol 27:16–33. https://doi.org/10.1007/s002540050024
46. Nordstrom DK, Alpers CN, Ptacek CJ, Blowes DW (2000) Negative pH and extremely acidic mine waters from Iron Mountain, California. Environ Sci Technol 34:254–258. https://doi.org/10.1021/es990646v
47. Nordstrom DK, Alpers CN (1999) Negative pH, efflorescent mineralogy, and consequences for environmental restoration at the Iron Mountain superfund site. California Proc Natl Acad Sci 96:3455–3462. https://doi.org/10.1073/pnas.96.7.3455
48. White DE, Hem JD, Waring GA (1963) Chemical composition of subsurface waters. U S Geol Surv Prof Pap 440–F:1–67
49. Lazarević K, Nikolić D, Stosić L et al (2012) Determination of lead and arsenic in tobacco and cigarettes: an important issue of public health. Cent Eur J Public Health 20:62–66
50. Chen C-J, Hsueh Y-M, Lai M-S et al (1995) Increased prevalence of hypertension and long-term arsenic exposure. Hypertension 25:53–60. https://doi.org/10.1161/01.HYP.25.1.53
51. Chen CJ, Chiou HY, Chiang MH et al (1996) Dose-response relationship between ischemic heart disease mortality and long-term arsenic exposure. Arterioscler Thromb Vasc Biol 16:504–510
52. Hansen ES (1990) Shared risk factors for cancer and atherosclerosis-a review of the epidemiological evidence. Mutat Res Genet Toxicol 239:163–179. https://doi.org/10.1016/0165-1110(90)90004-U
53. Barchowsky A, Klei LR, Dudek EJ et al (1999) Stimulation of reactive oxygen, but not reactive nitrogen species, in vascular endothelial cells exposed to low levels of arsenite. Free Radic Biol Med 27:1405–1412. https://doi.org/10.1016/S0891-5849(99)00186-0
54. Smith KR, Klei LR, Barchowsky A (2001) Arsenite stimulates plasma membrane NADPH oxidase in vascular endothelial cells. Am J Phys Lung Cell Mol Phys 280:L442–L449
55. Bunderson M, Coffin JD, Beall HD (2002) Arsenic induces peroxynitrite generation and cyclooxygenase-2 protein expression in aortic endothelial cells: possible role in atherosclerosis. Toxicol Appl Pharmacol 184:11–18. https://doi.org/10.1016/S0041-008X(02)99492-5
56. Lee PC, Ho IC, Lee TC (2005b) Oxidative stress mediates sodium arsenite-induced expression of heme oxygenase-1, monocyte chemoattractant protein-1, and interleukin-6 in vascular smooth muscle cells. Toxicol Sci 85:541–550. https://doi.org/10.1093/toxsci/kfi101
57. Lee MY, Lee YH, Lim KM et al (2005a) Inorganic arsenite potentiates vasoconstriction through calcium sensitization in vascular smooth muscle. Environ Health Perspect 113:1330–1335. https://doi.org/10.1289/ehp.8000
58. Cifuentes F, Bravo J, Norambuena M et al (2009) Chronic exposure to arsenic in tap water reduces acetylcholine-induced relaxation in the aorta and increases oxidative stress in female rats. Int J Toxicol 28:534–541. https://doi.org/10.1177/1091581809345924
59. Wauson EM, Langan AS, Vorce RL (2002) Sodium arsenite inhibits and reverses expression of adipogenic and fat cell-specific genes during in vitro adipogenesis. Toxicol Sci 65:211–219. https://doi.org/10.1093/toxsci/65.2.211
60. Tseng CH (2004) The potential biological mechanisms of arsenic-induced diabetes mellitus. Toxicol Appl Pharmacol 197:67–83
61. Pal S, Chatterjee AK (2005) Prospective protective role of melatonin against arsenic-induced metabolic toxicity in Wistar rats. Toxicology 208:25–33. https://doi.org/10.1016/j.tox.2004.11.005
62. Díaz-Villaseñor A, Sánchez-Soto MC, Cebrián ME et al (2006) Sodium arsenite impairs insulin secretion and transcription in pancreatic β-cells. Toxicol Appl Pharmacol 214:30–34. https://doi.org/10.1016/j.taap.2005.11.015

63. Felix K, Manna SK, Wise K et al (2005) Low levels of arsenite activates nuclear factor-??B and activator protein-1 in immortalized mesencephalic cells. J Biochem Mol Toxicol 19:67–77. https://doi.org/10.1002/jbt.20062
64. Gopalkrishnan A, Rao MV (2006) Amelioration by vitamin a upon arsenic induced metabolic and neurotoxic effects. J Health Sci 52:568–577. https://doi.org/10.1248/jhs.52.568
65. Piao F, Ma N, Hiraku Y et al (2005) Oxidative DNA damage in relation to neurotoxicity in the brain of mice exposed to arsenic at environmentally relevant levels. J Occup Health 47:445–449. https://doi.org/10.1539/joh.47.445
66. Vahidnia A, Romijn F, Tiller M et al (2006) Arsenic-induced toxicity: effect on protein composition in sciatic nerve. Hum Exp Toxicol 25:667–674. https://doi.org/10.1177/0960327106070671
67. Yip SF, Yeung YM, Tsui EYK (2002) Severe neurotoxicity following arsenic therapy for acute promyelocytic leukemia: potentiation by thiamine deficiency [4]. Blood 99:3481–3482. https://doi.org/10.1182/blood-2001-12-0325
68. Dwivedi N, Flora SJS (2011) Concomitant exposure to arsenic and organophosphates on tissue oxidative stress in rats. Food Chem Toxicol 49:1152–1159. https://doi.org/10.1016/j.fct.2011.02.007
69. Yadav RS, Shukla RK, Sankhwar ML et al (2010) Neuroprotective effect of curcumin in arsenic-induced neurotoxicity in rats. Neurotoxicology 31:533–539. https://doi.org/10.1016/j.neuro.2010.05.001
70. Namgung U, Xia Z (2001) Arsenic induces apoptosis in rat cerebellar neurons via activation of JNK3 and p38 MAP kinases. Toxicol Appl Pharmacol 174:130–138. https://doi.org/10.1006/taap.2001.9200
71. Parrish AR, Zheng XH, Turney KD et al (1999) Enhanced transcription factor DNA binding and gene expression induced by arsenite or arsenate in renal slices. Toxicol Sci 50:98–105. https://doi.org/10.1093/toxsci/50.1.98
72. Sasaki A, Oshima Y, Fujimura A (2007) An approach to elucidate potential mechanism of renal toxicity of arsenic trioxide. Exp Hematol 35:252–262. https://doi.org/10.1016/j.exphem.2006.10.004
73. Nandi D, Patra RC, Swarup D (2006) Oxidative stress indices and plasma biochemical parameters during oral exposure to arsenic in rats. Food Chem Toxicol 44:1579–1584. https://doi.org/10.1016/j.fct.2006.04.013
74. Bashir S, Sharma Y, Irshad M et al (2006) Arsenic induced apoptosis in rat liver following repeated 60 days exposure. Toxicology 217:63–70. https://doi.org/10.1016/j.tox.2005.08.023
75. Jain A, Yadav A, Bozhkov AI et al (2011) Therapeutic efficacy of silymarin and naringenin in reducing arsenic-induced hepatic damage in young rats. Ecotoxicol Environ Saf 74:607–614. https://doi.org/10.1016/j.ecoenv.2010.08.002
76. Suzuki T, Tsukamoto I (2006) Arsenite induces apoptosis in hepatocytes through an enhancement of the activation of Jun N-terminal kinase and p38 mitogen-activated protein kinase caused by partial hepatectomy. Toxicol Lett 165:257–264. https://doi.org/10.1016/j.toxlet.2006.05.004
77. Flora SJS, Mehta A, Gupta R (2009) Prevention of arsenic-induced hepatic apoptosis by concomitant administration of garlic extracts in mice. Chem Biol Interact 177:227–233. https://doi.org/10.1016/j.cbi.2008.08.017
78. Li GX, ling PQ, Gao Y et al (2007) Protective effects of hepatocellular canalicular conjugate export pump (Mrp2) on sodium arsenite-induced hepatic dysfunction in rats. Exp Toxicol Pathol 58:447–453. https://doi.org/10.1016/j.etp.2007.02.001
79. TE R, Y H, A T et al (2003) Central organization of mamalian pyruvate dehydrogenase (PD) complex and lipoyl domain-mediated function and control of PD kinases and phosphatase. In: Jordan F, Patel MS (eds) Thiamine: catalytic mechanisms and role in normal and disease states. Marcel Decker, New York, pp 363–386
80. Koike M, Koike K (1975) Structure, assembly and function of mammalian alpha-keto acid dehydrogenase complexes. Adv Biophys 9:187–227

81. Cheng HY, Li P, David M et al (2004) Arsenic inhibition of the JAK-STAT pathway. Oncogene 23:3603–3612. https://doi.org/10.1038/sj.onc.1207466
82. CRANE RK, LIPMANN F (1953) The effect of arsenate on aerobic phosphorylation. J Biol Chem 201:235–243
83. Ratnaike RN (2003) Acute and chronic arsenic toxicity. Postgrad Med J 79:391–396. https://doi.org/10.1136/pmj.79.933.391
84. Dixon HBF (1996) The biochemical action of Arsonic acids especially as phosphate analogues. Adv Inorg Chem 44:191–227. https://doi.org/10.1016/S0898-8838(08)60131-2
85. Lagunas R (1980) Sugar-arsenate esters: thermodynamics and biochemical behavior. Arch Biochem Biophys 205:67–75. https://doi.org/10.1016/0003-9861(80)90084-3
86. Gresser MJ (1981) ADP-arsenate. Formation by submitochondrial particles under phosphorylating conditions. J Biol Chem 256:5981–5983
87. Kenney LJ, Kaplan JH (1988) Arsenate substitutes for phosphate in the human red cell sodium pump and anion exchanger. J Biol Chem 263:7954–7960
88. Aposhian H V (1989) Biochemical toxicology of arsenic. In: Hodgson E, Bend JR, Philpot RM, pp 265–300
89. Delnomdedieu M, Basti MM, Styblo M et al (1994b) Complexation of arsenic species in rabbit erythrocytes. Chem Res Toxicol 7:621–627. https://doi.org/10.1021/tx00041a006
90. Winski SL, Carter DE (1998) Arsenate toxicity in human erythrocytes: characterization of morphologic changes and determination of the mechanism of damage. J Toxicol Env Heal A 53:345–355. https://doi.org/10.1080/009841098159213
91. Delnomdedieu M, Basti MM, Otvos JD, Thomas DJ (1994a) Reduction and binding of arsenate and dimethylarsinate by glutathione: a magnetic resonance study. Chem Biol Interact 90:139–155. https://doi.org/10.1016/0009-2797(94)90099-X
92. Scott N, Hatlelid KM, MacKenzie NE, Carter DE (1993) Reactions of arsenic (III) and arsenic (V) species with glutathione. Chem Res Toxicol 6:102–106
93. Delnomdedieu M, Basti MM, Otvos JD, Thomas DJ (1993) Transfer of arsenite from glutathione to dithiols: a model of interaction. Chem Res Toxicol 6:598–602. https://doi.org/10.1021/tx00035a002
94. Styblo M, Yamauchi H, Thomas DJ (1995) Comparative in vitro methylation of trivalent and pentavalent arsenicals. Toxicol Appl Pharmacol 135:172–178. https://doi.org/10.1006/taap.1995.1220
95. Lin S, Cullen WR, Thomas DJ (1999) Methylarsenicals and arsinothiols are potent inhibitors of mouse liver thioredoxin reductase. Chem Res Toxicol 12:924–930. https://doi.org/10.1021/tx9900775
96. Styblo M, Serves SV, Cullen WR, Thomas DJ (1997) Comparative inhibition of yeast glutathione reductase by arsenicals and arsenothiols. Chem Res Toxicol 10:27–33. https://doi.org/10.1021/tx960139g
97. Das D, Chatterjee A, Mandal BK et al (1995) Arsenic in ground water in six districts of West Bengal, India: the biggest arsenic calamity in the world. Part 2. Arsenic concentration in drinking water, hair, nails, urine, skin-scale and liver tissue (biopsy) of the affected people. Analyst 120:917. https://doi.org/10.1039/an9952000917
98. Benramdane L, Accominotti M, Fanton L et al (1999) Arsenic speciation in human organs following fatal arsenic trioxide poisoning – a case report. Clin Chem 45:301–306
99. Rahman MM, Chowdhury UK, Mukherjee SC et al (2001) Chronic arsenic toxicity in Bangladesh and West Bengal, India—a review and commentary. J Toxicol Clin Toxicol 39:683–700. https://doi.org/10.1081/CLT-100108509
100. Liu J, Zheng B, Aposhian HV et al (2002) Chronic arsenic poisoning from burning high-arsenic-containing coal in Guizhou. China Environ Health Perspect 110:119–122. https://doi.org/10.1289/ehp.02110119
101. Vitayavirasak B, Rakwong K, Chatchawej W (2005) Environmental arsenic exposure of schoolchildren in a former tin mining and smelting community of southern Thailand. Environ Sci 12:195–205

102. Ferreccio C, Sancha AM (2006) Arsenic exposure and its impact on health in Chile. J Health Popul Nutr 24:164–175. https://doi.org/10.2307/23499354
103. Pokhrel D, Bhandari BS, Viraraghavan T (2009) Arsenic contamination of groundwater in the Terai region of Nepal: an overview of health concerns and treatment options. Environ Int 35:157–161 S0160-4120(08)00107-4 [pii]\r10.1016/j.envint.2008.06.003
104. Keshavarzi B, Seradj A, Akbari Z et al (2015) Chronic arsenic toxicity in sheep of Kurdistan Province. Western Iran Arch Environ Contam Toxicol 69:44–53. https://doi.org/10.1007/s00244-015-0157-4
105. Kulasooriya SA (2011) Cyanobacteria : pioneers of planet earth. Ceylon J Sci 40:71–88
106. Singh M, Sharma NK, Prasad SB et al (2013) The freshwater cyanobacterium anabaena doliolum transformed with ApGSMT-DMT exhibited enhanced salt tolerance and protection to nitrogenase activity, but became halophilic. Microbiol (United Kingdom) 159:641–648. https://doi.org/10.1099/mic.0.065078-0
107. Zhang M, Shi X, Yu Y, Kong F (2011) The acclimative changes in photochemistry after colony formation of the cyanobacteria Microcystis aeruginosa. J Phycol 47:524–532. https://doi.org/10.1111/j.1529-8817.2011.00987.x
108. Sabart M, Pobel D, Briand E et al (2010) Spatiotemporal variations in microcystin concentrations and in the proportions of microcystin-producing cells in several Microcystis aeruginosa populations. Appl Environ Microbiol 76:4750–4759. https://doi.org/10.1128/AEM.02531-09
109. Kazmierczak J, Altermann W, Kremer B et al (2009) Mass occurrence of benthic coccoid cyanobacteria and their role in the production of Neoarchean carbonates of South Africa. Precambrian Res 173:79–92. https://doi.org/10.1016/j.precamres.2009.02.002
110. Morin N, Vallaeys T, Hendrickx L et al (2010) An efficient DNA isolation protocol for filamentous cyanobacteria of the genus Arthrospira. J Microbiol Methods 80:148–154. https://doi.org/10.1016/j.mimet.2009.11.012
111. Koksharova OA, Wolk CP (2002) Genetic tools for cyanobacteria. Appl Microbiol Biotechnol 58:123–137. https://doi.org/10.1007/s00253-001-0864-9
112. Wada N, Sakamoto T, Matsugo S (2013) Multiple roles of photosynthetic and sunscreen pigments in cyanobacteria focusing on the oxidative stress. Meta 3:463–483. https://doi.org/10.3390/metabo3020463
113. Costa M, Costa-Rodrigues J, Fernandes MH et al (2012) Marine cyanobacteria compounds with anticancer properties: a review on the implication of apoptosis. Mar Drugs 10:2181–2207. https://doi.org/10.3390/md10102181
114. Rao BD (2015) Antibacterial Activity of Fresh Water Cyanobacteria Antibact Act Fresh Water Cyanobacteria ISSN 6:60–64
115. Mao TK, Van de WJ, Gershwin ME (2005) Effects of a *Spirulina* -based dietary supplement on cytokine production from allergic rhinitis patients. J Med Food 8:27–30. https://doi.org/10.1089/jmf.2005.8.27
116. Colla LM, Oliveira Reinehr C, Reichert C, Costa JAV (2007) Production of biomass and nutraceutical compounds by Spirulina platensis under different temperature and nitrogen regimes. Bioresour Technol 98:1489–1493. https://doi.org/10.1016/j.biortech.2005.09.030
117. Bhattacharya S, Shivaprakash MK (2005) Evaluation of three Spirulina species grown under similar conditions for their growth and biochemicals. J Sci Food Agric 85:333–336. https://doi.org/10.1002/jsfa.1998
118. Cruz-Martinez LC, Jesus CKC, Matsudo MC et al (2015) Growth and composition of arthrospira (spirulina) platensis in a tubular photobioreactor using ammonium nitrate as the nitrogen source in a fed-batch process. Braz J Chem Eng 32:347–356. https://doi.org/10.1590/0104-6632.20150322s00003062
119. Guo Y, Xue X, Yan Y et al (2016) Arsenic methylation by an arsenite S-adenosylmethionine methyltransferase from Spirulina platensis. J Environ Sci (China) 49:162–168. https://doi.org/10.1016/j.jes.2016.06.013
120. Armendariz AL, Talano MA, Wevar Oller AL et al (2015) Effect of arsenic on tolerance mechanisms of two plant growth-promoting bacteria used as biological inoculants. J Environ Sci (China) 33:203–210. https://doi.org/10.1016/j.jes.2014.12.024

121. Paez-Espino D, Tamames J, De Lorenzo V, Cánovas D (2009) Microbial responses to environmental arsenic. Biometals 22:117–130. https://doi.org/10.1007/s10534-008-9195-y
122. Stýblo M, Drobná Z, Jaspers I et al (2002) The role of biomethylation in toxicity and carcinogenicity of arsenic: a research update. Environ Health Perspect 110:767–771. https://doi.org/10.1289/ehp.02110s5767
123. Bentley R, Chasteen TG (2002) Microbial methylation of metalloids: arsenic, antimony, and bismuth. Microbiol Mol Biol Rev 66:250–271. https://doi.org/10.1128/MMBR.66.2.250-271.2002
124. Challenger F (1945) Biological methylation. Chem Rev 36:315–361
125. Hirano SHTKYCX (2005) A new metabolic pathway of arsenite: arsenic-glutathione complexes are substrates for human arsenic methyltransferase Cyt19. Arch Toxicol 79:183–191. https://doi.org/10.1007/s00204-004-0620-x
126. Marapakala K, Qin J, Rosen BP (2012) Identification of catalytic residues in the As(III) S-adenosylmethionine methyltransferase. Biochemistry 51:944–951. https://doi.org/10.1021/bi201500c
127. Katsoyiannis I, Zouboulis A, Althoff H, Bartel H (2002) As(III) removal from ground waters using fixed-bed upflow bioreactors. Chemosphere 47:325–332. https://doi.org/10.1016/S0045-6535(01)00306-X
128. Dhankher OP, Li Y, Rosen BP et al (2002) Engineering tolerance and hyperaccumulation of arsenic in plants by combining arsenate reductase and γ-glutamylcysteine synthetase expression. Nat Biotechnol 20:1140–1145. https://doi.org/10.1038/nbt747
129. Ike M, Miyazaki T, Yamamoto N et al (2008) Removal of arsenic from groundwater by arsenite-oxidizing bacteria. Water Sci Technol 58:1095–1100. https://doi.org/10.2166/wst.2008.462
130. Tuzen M, Sarı A, Mendil D et al (2009) Characterization of biosorption process of as(III) on green algae Ulothrix cylindricum. J Hazard Mater 165:566–572
131. Mirza N, Mahmood Q, Pervez A et al (2010) Phytoremediation potential of Arundo donax in arsenic-contaminated synthetic wastewater. Bioresour Technol 101:5815–5819. https://doi.org/10.1016/j.biortech.2010.03.012
132. Kao A-C, Chu Y-J, Hsu F-L, Liao VH-C (2013) Removal of arsenic from groundwater by using a native isolated arsenite-oxidizing bacterium. J Contam Hydrol 155:1–8. https://doi.org/10.1016/j.jconhyd.2013.09.001
133. Sibi G (2014) Biosorption of arsenic by living and dried biomass of fresh water microalgae – potentials and equilibrium studies. J Bioremed Biodegr https://doi.org/10.4172/2155-6199.1000249
134. Jasrotia S, Kansal A, Kishore VVN (2014) Arsenic phyco-remediation by Cladophora algae and measurement of arsenic speciation and location of active absorption site using electron microscopy. Microchem J 114:197–202. https://doi.org/10.1016/j.microc.2014.01.005
135. Dey U, Chatterjee S, Mondal NK (2016) Isolation and characterization of arsenic-resistant bacteria and possible application in bioremediation. Biotechnol Reports 10:1–7. https://doi.org/10.1016/j.btre.2016.02.002
136. Styblo M, Delnomdedieu M, Thomas DJ (1996) Mono- and dimethylation of arsenic in rat liver cytosol in vitro. Chem Biol Interact 99:147–164
137. Petrick JS, Ayala-Fierro F, Cullen WR et al (2000) Monomethylarsonous acid (MMAIII) is more toxic than Arsenite in Chang human hepatocytes. Toxicol Appl Pharmacol 163:203–207. https://doi.org/10.1006/taap.1999.8872
138. Styblo M, Del Razo LM, Vega L et al (2000) Comparative toxicity of trivalent and pentavalent inorganic and methylated arsenicals in rat and human cells. Arch Toxicol 74:289–299. https://doi.org/10.1007/s002040000134

Algal Biofuel: Still Not a Common Man's Fuel?

4

Aradhana Das, Rajesh Kumar Sahoo, and Enketeswara Subudhi

Abstract

The rapid lifestyle of industrialization and increasing demand of fossil oil that are going to be scarce in future date have led to think the alternative source of renewable energy as fuels to meet our energy demands. Fossil fuel is challenged with increasing price and a decreasing quantity, and burning of the fuels is putting the environment into threat toward pollution and global warming. Various steps toward cultivating oil crops such as *Jatropha*, corn, coconut, soybean, and oil palm have been encouraged, but productivity of oil has been very less, i.e., 5% of total biomass, and it needs vast acres of cultivated land. Therefore, to overcome the problem, today's world is moving toward microalgae cultivation, which in comparison can grow faster in wastelands/uncultivated lands and can produce up to 80% of the dry weight of algae biomass. Microalgae are phototrophic and are able to transform carbon dioxide into biofuels, valuable bioactive compounds, foods, and feeds. In spite of all positivity, microalgae biofuel is still not common man's fuel due to various hurdles. Overhead harvesting cost is 20–30% higher to the cultivation cost of algae; it can reduce the nonrenewable resources (nitrogen, phosphorus) for which still date it cannot reach to common man. However, limited supply of these renewable oils and high cost stop it to be a potential challenger in the face of other petroleum-based fuels. Overall, economic feasibility and environmental suitability cannot be forgotten when venturing into scaling up for future commercialization.

A. Das · R. K. Sahoo · E. Subudhi (✉)
Biomics & Biodiversity Laboratory, Centre for Biotechnology,
Siksha 'O' Anusandhan (Deemed to be University), Bhubaneswar, Odisha, India
e-mail: aradhanadas@soa.ac.in; enketeswarasubudhi@soa.ac.in

L. B. Sukla et al. (eds.), *The Role of Microalgae in Wastewater Treatment*,
https://doi.org/10.1007/978-981-13-1586-2_4

4.1 Introduction

The social and economic development of a nation is mainly propelled by the amount and type of energy it can generate. As constant progress of the world to expand its economy in modernization, industrialization of lifestyle with growing population and increasing demand for vehicles for transportation has directed to substantial intensification in fossil fuel, i.e., petroleum-based fuel use [1, 2]. The world energy consumption has been projected to grow to an extent of 56% during 2010–2040 as reported by IEO2013 [3]. The continuous overexploitation of petroleum fuel has increased the atmospheric carbon dioxide concentration, which possibly can upset across the globe due to change in climate and global warming [4]. In addition, increased fuel requirement has also led to the fast exhaustion of the sources of such energy. About 80% of global energy fuels are supplied by fossil fuels; natural gas, coal, and crude oil but unless resources of renewable energy are exploited; such huge demand of energy at global scale can never be reached. Therefore, worldwide demands of motor and power generation fuels, with environmental consequences, have headed for finding novel substitutes which are not only sustainable and renewable but also eco-friendly and economic [5, 6]. Efforts are being taken to replace fossil fuels [7–13] from oil palm, rapeseed, *Jatropha*, and soybean but are subjected to foremost disagreement as they require land for their production which has been used for raising food crops [2]. As compared to traditional crop, algae can be cultivated in submerged area as well as in seawater [13] to produce 300-fold oil per unit area.

Efforts are being taken in the present years to explore the potentials of utilizing algae as a resource of oil and gas of biological origin aiming at applications for energy generation. Algae are not only the fastest-growing plants but also are the huge diverse group of both unicellular and multicellular autotrophs. They have the ability to thrive under varied environmental conditions like freshwater as well as saline and seawater [14, 15]. They have phototrophic-driven cell factories that convert sunlight, water, and carbon dioxide to potential biofuels, foods, and feeds, and high-value bioactive compounds of triacylglycerides (fats) and polysaccharides (sugars) produced by them are proved to be the principal starting material for production of biodiesel. As compared to different energy crops, microalgae pose enormous amount of treasured properties, higher efficiency in photosynthesis, higher rate of product accumulation [16], higher rate of production of biomass, and higher ratio of carbon to nitrogen (C/N) as well as carbon to phosphorous (C/P) [17, 18]. Therefore, biodiesel of microalgal origin has the potential to replace the fuels derived from petroleum, such as gasoline, jet fuel, and diesel. Even though algae act to be a hopeful resource for production of biofuel, several environmental and economic bottlenecks lie in its large-scale application. The present cost of production of algal biofuels is substantially higher than fossil fuel, but in the near future, it would be likely to come down. Still of many efforts, it cannot reach to common people because of many hurdles. Many researchers are putting efforts, focusing on the biosynthetic pathways, and improving its economics [19].

4.2 Different Problems to Commercialization

4.2.1 Demand for Land

As a consequence of rapid growth of industrialization, there are huge demands for fossil fuel in substitute of petroleum which are going to be scarce in future era. In the report of US Energy Information Administration, in the year 2008, the USA required 19,497,950 barrels of oil per day. Growing terrestrial plants for biodiesel does not seem to compete with the oil prospective in the future, because it needs land not only to grow food but also needs to feed the animals. However, for growing algae only about 1% of the wasteland/feeding land (limited productivity land) is enough to produce oil to replace the entire diesel used. According to Chisti [23] replacement of 50% US transport fuel, algal biodiesel would only require 1.1–2.5% of farming land to yield up to 30–70% of oil content which is relatively very high in land area coverage to other feedstock cultivation (see Fig. 4.1).

It is evident from recent survey that present technology has the prospective to generate 220 × 109Lyr-1 of oil from microalgae which is equivalent to 48% of existing import of US petroleum used for transportation basing on fuel utilization in 2011 [20]. As reported by Gerbens-Leenes et al. [21], 17,000 km2, i.e 1% of the total farming area will be required for microalgal biofuel to meet 3.5% of fuel demand of European Union in 2030. Therefore, 28% of the existing area will be needed in the entire world for algal biodiesel production. Evidently, in comparison algae can yield more oil than other biomass feedstocks. In open pond, the amount of land needed to supply fuel demands depends directly on the type of biomass

Fig. 4.1 Land area needed under cultivation to meet 50% requirement of transport fuels (liquid) in the USA

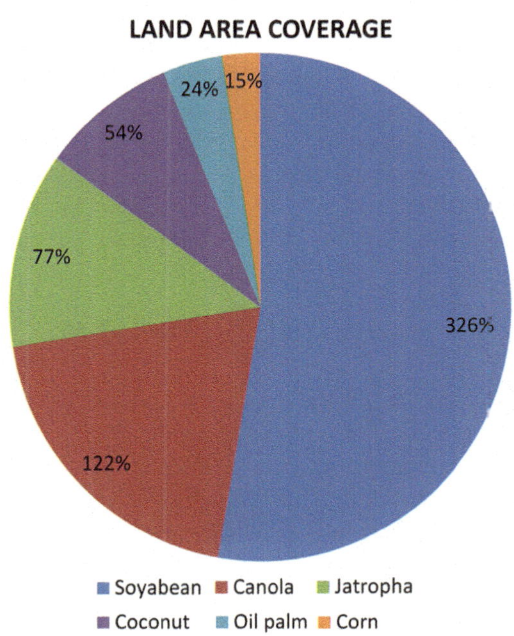

used. Application of photobioreactors, is encouraged as it needs less amount of water as well as land compared to open ponds [22] but still is not economical as yet because it needs more fossil fuel energy for their operation even at a small scale than the energy it can generate. Therefore, locating open pond or photobioreactor for microalgae cultivation, a proper site selection should be one of the main criteria for cultivation as it needs a suitable land topography, i.e., slope, appropriate water supply with proper salinity, good climatic conditions (temperature, proper nutrients, and carbon supply), and geological features (porosity, permeability, compactness). These are the important factors to be taken into consideration for providing optimal growth conditions.

4.2.2 Carbon Dioxide Availability

Microalgae are the unicellular organisms and capable to change solar energy to chemical energy. During the conversion process, CO_2 is essential requirement for biodiesel production. Chisti [28] reported that for production of 1 kg microalgae biomass, they require 1.8 kg CO_2. For pilot-scale production of microalgal biomass which depends on availability of CO_2, it contributes almost half of the cost of biomass production [28]. Many of the researchers prove that microalgae have ability to concentrate the CO_2 from the culture medium because absorption of CO_2 from atmosphere is not sufficient for large biomass production [23–25]. In addition, there are several non-biological technologies which are developed for capturing CO_2 from atmosphere but are subjected to economic challenges. Some researchers also developed biological approaches, i.e., genetic modification of microalgae for capturing CO_2, but it has been less effective for rapid growth of microalgae. Carbon dioxide is necessary for algal biofuel production, but absorption of CO_2 from atmosphere to culture is difficult because its concentration in the atmosphere is very low. Microalgae are generally growing in alkaline condition, and bicarbonate can be used as alternative source for supplying CO_2 [26], but this strategy may not work in marine algae because salts tend to precipitate once the pH exceeds to 8.0. From economic point of view, the regular supply of CO_2 is very expensive and also increasing the total cost of the biofuel.

4.2.3 Nutrient Requirement

Nitrogen and phosphorus are two essential elements for production of biofuel from microalgae. But, microalgae are producing biofuel in deficiency of these elements. Nitrogen fertilizer can be produced from atmospheric nitrogen by Haber-Bosch process for producing but it involves the use of fossil energy. Metz et al. [27] reported that global energy up to 1.2% is consumed to produce N fertilizer for agriculture. Chisti [28] reported that the US Agriculture Department consumed 5.4 million ton of N and 1.1 million ton of P for production of 82 million ton algal biomass. If huge amount of N and P is diverted toward the utilization of algae growth, how would we provide the same for food production? However, efficiencies of uptake of nutrient and their content in algae may vary with growth conditions, time of exposure, the species, and the percentage of carbohydrates, proteins, and lipids in the harvested biomass.

4.2.4 Water Supply

Recently much work has been done in the field of enhancement of lipid accumulation in single microalga cell. High lipid productivity not only depends upon nutrition limitation but also some other factors such as light, temperature, and water quality. Among these, water is very essential for biomass production. Use of freshwater for production of any goods and services in the entire process is called water footprint. For algal fuel production, the water footprint is dependent upon (1) evaporation or loss of water during culture and (2) transformation of freshwater to wastewater. There are various factors which are involved in evaporative loss such as local climatic conditions, irradiance level, the air temperature, the wind velocity, and the absolute humidity. Becker [29] reported that freshwater evaporation in tropical areas was 10 liter/m which is equivalent to 10 mm d^{-1}. Freshwater is also necessary for washing the biomass and lipid extraction. Regular supply of freshwater is needed for substantial production of biofuel. For 1 kg biofuel production by freshwater alga *C. vulgaris*, 3727 kg freshwater was needed [30]. During algal culturing in large open pond, the maintenance of freshwater is required due to evaporative loss [28]. Utilization of huge amount of freshwater for biofuel production does not appear to be a practical option. The consumption of freshwater is reduced by recycling the water, but it loses the nutrients and risk of bacterial, fungal, or virus infection as well as other inorganic and organic compound/metabolites as inhibitors. Brackish water may be recycled, as substitute of freshwater, but pretreatment of water is required to remove inhibitors. In contrast, employment of photobioreactors for production of microalgal biofuel has significantly lower water demand. Yang et al. [30] stated the reduction of water trail in bioreactors can expedite superior regulation of culture and can generate thickly grown algal culture per unit volume of used water. Regular water spray is required to prevent the overheating of bioreactor during operation. The production of biofuel in photobioreactor is very cost intensive due to its maintenance, installation, and operation. Dependable supply of water at a low cost is very difficult to make available for the successful production of biofuel from algae.

4.3 Present Static Economic Status of Liquid Fuels

There are significantly various challenges and hurdles in making fossil fuels capable of competing with petroleum to reach common people as it can touch our routine life in many forms. At present, the evaluation price is US$300–2600 of a barrel of algae-based fuel as compared with $40–80 for petroleum which is very high related to crude oil [31–33]. As the petroleum requirement increases considerably to meet the demand, its price also affects the economic feasibility [34, 35]. Production of fuel from algae in a commercial venture is still subjected to questions as compared to other renewable energy resources, unless otherwise it is superior or nearer to prevailing fuels of petroleum origin from economic point of view. However, the challenges it has been subjected to and the strategies it has to follow to come over of this economic incongruity are very much highlighted in the present article.

4.4 Conclusion

The entire facts of using microalgae as fuel in substitute of fossil fuels are unsuccessful except the production at an industrial scale which is eco-friendly and economically viable. No industrial scale setup till date has a large-scale algal biofuel production that can support both the environment and economic. Both, the costs incurred in investment capital and operation make the algal fuel, making highly expensive so as to make it commercially suitable. The time is waited when production costs of algae will come down and relatively that of fossil fuels will rise high. With the advent of real game-changing breakthrough of start-ups on algal fuel production, hopefully the age of green fuel may bring into being. Probably, more financial support on research and development is desirable to crack this boundless prospective into economically viable, and thus, it is too premature to declare if the future biofuels need can bank upon algae.

References

1. Agarwal AK (2007) Biofuels (alcohol and biodiesel) applications as fuels for internal combustion engines. Prog Energy Combust Sci 33:233–271
2. Nigam PS, Singh A (2011) Production of liquid biofuels from renewable resources. Prog Energy Combust Sci 37:52–68. https://doi.org/10.1016/j.pecs.2010.01.003
3. International Energy Outlook 2013 (2013) International Energy Outlook 2013 (IEO2013), released by the U.S. Energy Information Administration (EIA). Published: July 25, 2013. [Internet] http://susris.com/2013/07/25/international-energy-outlook-2013/
4. Parry ML (2007) Intergovernmental Panel on Climate Change, Working Group II, World Meteorological Organization, United Nations Environment Programme. Summary for Policymakers. In: Parry ML, Canziani OF, Palutikof JP, van der Linden PJ, Hanson CE (eds) Climate change 2007: impacts, adaptation and vulnerability. Contribution of Working Group II to the Fourth Assessment Report of the Intergovernmental Panel on Climate Change. Cambridge University Press, Cambridge
5. Alemán-Nava GS, Casiano-Flores VH, Cárdenas-Chávez DL, Díaz-Chavez R, Scarlat N, Mahlknecht J, Dallemand JF, Parra R (2014) Renewable energy research progress in Mexico: a review. Renew Sust Energ Rev 32:140–153. https://doi.org/10.1016/j.rser.2014.01.004
6. Mata TM, Martins AAA, Caetano NS (2010) Microalgae for biodiesel production and other applications: a review. Renew Sust Energ Rev 14:217–232. https://doi.org/10.1016/j.rser.2009.07.020

7. Korres NE, Singh A, Nizami AS, Murphy JD (2010) Is grass biomethane a sustainable transport biofuel? Biofuels Bioprod Biorefin 4:310–325. https://doi.org/10.1002/bbb.228
8. Pant D, Van Bogaert G, Diels L, Vanbroekhoven K (2010) A review of the substrates used in microbial fuel cells (MFCs) for sustainable energy production. Bioresour Technol 101:1533–1543. https://doi.org/10.1016/j.biortech.2009.10.017
9. Prasad S, Singh A, Joshi HC (2007a) Ethanol as an alternative fuel from agricultural, industrial and urban residues. Resour Conserv Recycl 50:1–39. https://doi.org/10.1016/j.resconrec.2006.05.007
10. Prasad S, Singh A, Joshi HC (2007b) Ethanol production from sweet sorghum syrup for utilization as automotive fuel in India. Energy Fuel 21:2415–2420. https://doi.org/10.1021/ef060328z
11. Singh A, Nigam PS, Murphy JD (2010a) Renewable fuels from algae: an answer to debatable land based fuels. Bioresour Technol. https://doi.org/10.1016/j.biortech.2010.06.032
12. Singh A, Pant D, Korres NE, Nizami AS, Prasad S, Murphy JD (2010b) Key issues in life cycle assessment of ethanol production from lignocellulosic biomass: challenges and perspectives. Bioresour Technol 101:5003–5012
13. Singh A, Smyth BM, Murphy JD (2010c) A biofuel strategy for Ireland with an emphasis on production of biomethane and minimization of land-take. Renew Sust Energ Rev 14:277–288
14. Carlsson A, Beilen van J, Möller R, Clayton D, Bowles DE (2007) Micro-and macroalgae – utility for industrial applications Bioproducts, E. R. t. E. P. o. S. R.-. and Crops, f. N.-f., CNAP, University of York: 86
15. Schenk PM, Thomas-Hall SR, Stephens E, Marx UC, Mussgnug JH, Posten C, Kruse O, Hankamer B (2008) Second generation biofuels: high-efficiency microalgae for biodiesel production. Bioenergy Res 1:20–43. https://doi.org/10.1007/s12155-008-9008-8
16. Minowa T, Yokoyama SYA, Kishimoto M, Okakura T (1995) Oil production from algal cells of Dunaliella tertiolecta by direct thermochemical liquefaction. Fuel 74:1735–1738. https://doi.org/10.1016/0016-2361(95)80001-X
17. Haag AL (2007) Algae bloom again. Nature 447:520–521. https://doi.org/10.1038/447520a
18. Shurin JB, Mandal S, Abbott RL (2014) Trait diversity enhances yield in algal biofuel assemblages. J Appl Ecol 51:603–611. https://doi.org/10.1111/1365-2664.12242
19. Singh A, Nigam PS, Murphy JD (2011) Mechanism and challenges in commercialisation of algal biofuels. Bioresour Technol 102(1):26–34
20. Wigmosta MS, Coleman AM, Skaggs RJ, Huesemann MH, Lane LJ (2011) National microalgae biofuel production potential and resource demand. Water Resour Res 47. https://doi.org/10.1029/2010WR009966
21. Gerbens-Leenes PW, Xu L, De Vries GJ, Hoekstra AY (2014) The blue water footprint and land use of biofuels from algae. Water Resour Res 50:8549–8563. https://doi.org/10.1002/2014WR015710
22. Pulz O (2001) Photobioreactors: production systems for phototrophic microorganisms. Appl Microbiol Biotechnol 57:287–293. https://doi.org/10.1007/s002530100702
23. Badger MR, Hanson D, Price GD (2002) Evolution and diversity of CO2 concentrating mechanisms in cyanobacteria. Funct Plant Biol 29:161–173. https://doi.org/10.1071/PP01213
24. Giordano M, Beardall J, Raven JA (2005) CO_2 concentrating mechanisms in algae: mechanisms, environmental modulation, and evolution. Annu Rev Plant Biol 56:99–131
25. Meyer M, Griffiths H (2013) Origins and diversity of eukaryotic CO2-concentrating mechanisms: lessons for the future. J Exp Bot 64:769–786. https://doi.org/10.1093/jxb/ers390S
26. Chi Z, O'Fallon JV, Chen S (2011) Bicarbonate produced from carbon capture for algae culture. Trends Biotechnol 29:537–541. https://doi.org/10.1016/j.tibtech.2011.06.006
27. Metz B, Davidson O, De Coninck H, Loos M, Meyer L (2005) Carbon capture and storage. Cambridge University Press, Cambridge
28. Chisti Y (2013) Constraints to commercialization of algal fuels. J Biotechnol 167:201–214. https://doi.org/10.1016/j.jbiotec.2013.07.020
29. Becker EW (1994) Microalgae: biotechnology and microbiology. Cambridge University Press, Cambridge

30. Yang J, Xu M, Zhang X, Hu Q, Sommerfeld M, Chen Y (2011) Life-cycle analysis on biodiesel production from microalgae: water footprint and nutrients balance. Bioresour Technol 102:159–165. https://doi.org/10.1016/j.biortech.2010.07.017
31. Alabi AO, Tampier M, Bibeau E (2009) Microalgae technologies and processes for biofuels/bioenergy production in British Columbia. BC Innovation Council, Vancouver
32. Dismukes GC, Carrieri D, Bennette N, Ananyev GM, Posewitz MC (2008) Aquatic phototrophs: efficient alternatives to land-based crops for biofuels. Curr Opin Biotechnol 19:235–240. https://doi.org/10.1016/j.copbio.2008.05.007
33. U.S. Energy Information Administration, Department of Energy (DOE) crude oil estimates/statistics: http://www.eia.doe.gov/emeu/international/reserves.html; http://www.eia.doe.gov/energyexplained/index.cfm?page=oil_home#tab2; http://tonto.eia.doe.gov/dnav/pet/pet_cons_psup_dc_nus_mbbl_a.htm
34. Xu H, Miao X, Wu Q (2006) High quality biodiesel production from a microalgae chlorella photothecoides by heterotrophic growth in fermenters. J Biotechnol 126:499–507
35. Yusuf C (2007) Biodiesel from microalgae. Biotechnol Adv 25:294–306. https://doi.org/10.1016/j.biotechadv.2007.02.001

Bioreduction of Hexavalent Chromium Using Microalgae

5

Debabrata Pradhan and Lala Behari Sukla

Abstract

Hexavalent chromium (Cr(VI)) is enriched in the water system of environment above the regulatory level due to different human activities. Cr(VI) has the chemical properties favorable for its dissolution in the water environment at an elevated concentration level. This concerns the environmentalists as Cr(VI) in water is carcinogenic to different organs of the living organisms. Different techniques like chemical, biological, and combination of both have been undertaken using various methods to remove Cr(VI) from the water. Primarily bioremediation including bioreduction and biosorption has potential to remove Cr(VI) from water. Also some other processes like microbial fuel cells and biostimulation sideline the Cr(VI) removal from water along with different primary objectives. Among the living organism, microalgae have great potential to remove Cr(VI) from water. They have the unique photosynthesis and cellular metabolisms compensating the Cr(VI) removal. The use of microalgae in bioremediation for removal of pollutants from the contaminated water is a practical interest due to different advantages as it requires low energy with reduced sludge formation and carbon dioxide sequestration. Most of the algae follow an adsorption followed by reduction of Cr(VI) during the bioremediation process.

D. Pradhan (✉) · L. B. Sukla
Biofuels and Bioprocessing Research Center, Institute of Technical Education and Research, Siksha 'O' Anusandhan (Deemed to be University), Bhubaneswar, Odisha, India
e-mail: debabratapradhan@soa.ac.in; lalabeharisukla@soa.ac.in

L. B. Sukla et al. (eds.), *The Role of Microalgae in Wastewater Treatment*,
https://doi.org/10.1007/978-981-13-1586-2_5

5.1 Introduction

Water pollution is one of the major environmental pollutions resulted from the biogeochemical water cycle. Many poisonous and harmful chemicals produced from different manufacturing industries contaminate water supplies [1]. Worldwide different countries are in mission for control and abatement of the water pollutions [2, 3]. Heavy and toxic metals are one of the major reasons for pollution of water. The standard levels of different toxic metals in the drinking water are summarized by the regulatory authorities of different countries [4]. Hexavalent chromium or Cr(VI) is an important water pollutant proved to be carcinogenic even at low levels of parts per billion (ppb) [5]. The solubility and thermodynamic factors of Cr(VI) favor its transportation within the ecosystems [6].

The enrichment of Cr(VI) in water is related to the different industrial activities such as alloying industries, cement, chemical plants, electroplating, asbestos, contaminated landfill, glassmaking, industrial effluents, tobacco, rocks, and leather tanning [4, 7, 8]. Most of the chromium is deposited in the Cr_2O_3 and $FeO.Cr_2O_3$. Chromium is extracted as metallic chromium through different industrial oxidation and reduction process [6]. About 90% of total chromium is used as components of alloys and 10% is used in cement, leather tanning, refractory, electroplating, pigment, ceramics, glass, machinery, and wood preservation industries [8]. Refractory bricks used in ceramic industries are produced from chromite ores [9].

5.2 Chemistry of Cr(VI)

The trivalent chromium or Cr(III) and hexavalent chromium or Cr(VI) are the most stable chromium ions present in the environment [10]. Cr(VI) exists as $Cr_2O_7^{2-}$, $HCrO_4^-$, H_2CrO_4, and CrO_4^{2-}, and Cr(III) exists as Cr^{3+}, CrO^+, $Cr(OH)_2^+$, $HCrO_2$, and CrO_2^- in the solution state [11]. Na_2CrO_4 is a primary product of the lime treatment of minerals and further transformed to other compounds and ions such as CrF_6, $Cr_2O_7^{2-}$, CrO_3, $CrOCl_4$, $H_2Cr_2O_7$, CrO_4^{2-}, $HCrO_4^-$, and H_2CrO_4 [12]. The Cr(VI) species are strong oxidizing agent and they are used in different manufacturing industries. The industries release wastewater to the environment containing high Cr(VI) level [5]. The stability of Cr(III) and Cr(VI) at different pH and Eh are shown in Fig. 5.1. Pradhan et al. [4] described the redox chemistry of chromium. The best mechanism of Cr(VI) removal from water is the reduction of highly soluble Cr(VI) species to less soluble Cr(III). The less soluble Cr(III) can be removed by immobilizing them in the physical and chemical processes. Adsorption process, especially physical adsorption, is a suitable to remove less soluble Cr(III) [14].

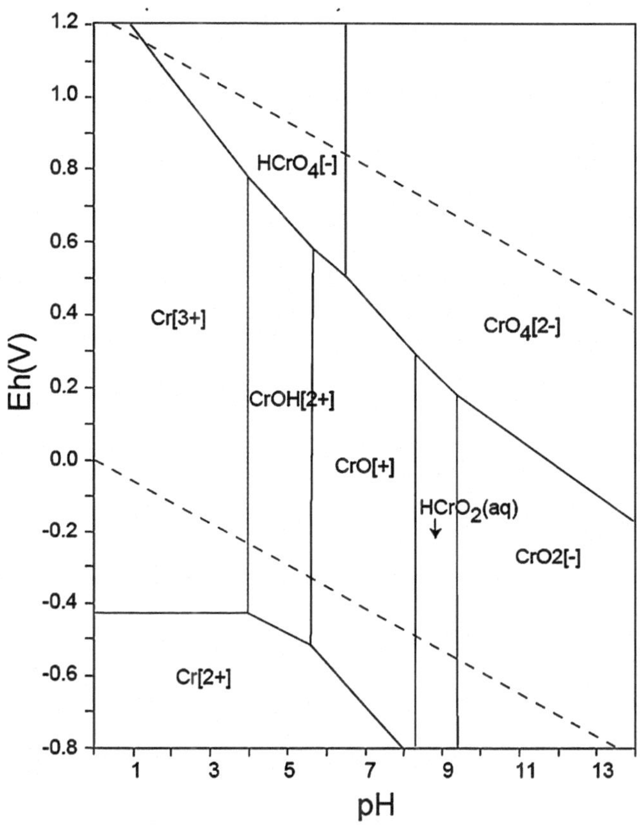

Fig. 5.1 Eh-pH phase diagram for chromium [13]

5.3 Biological Toxicity Mechanism of Cr(VI)

The anionic species of Cr(VI) are stable less than pH 6.0 (Fig. 5.1). The anionic species CrO_4^{2-} has similarity with SO_4^{2-} ion. This similarity of CrO_4^{2-} with SO_4^{2-} makes them easy to enter into the biological cells through the sulfate transport system [15]. The CrO_4^{2-} oxidizes different components of cells forming different reactive species inside the cells and itself reduced to Cr(III) with intermediates of Cr(V) and Cr(IV). It forms different intracellular reactive species like O_2^-, O, O_2H^-, OH^-, and free radicals during reduction of CrO_4^{2-} [11]. The higher the Cr(VI) concentration, the more the reactive species generation [16]. The intracellular reactive species cause both oxidative and nonoxidative forms of DNA damage, i.e., Cr-DNA binding (adducts) [15]. Also metabolism of Cr(VI) is associated with the single-strand DNA breaks [17]. Different diseases like liver cancer, kidney cancer, lung cancer, dermatitis, dermal necrosis, sperm damage, and dermal corrosion are developed due to the alteration of DNA profile [5, 18–22, 51].

5.4 Conventional Cr(VI) Treatment

Reduction of Cr(VI) is an emerging field of research due to its numerous toxicity effects on the human body. The convention treatment of Cr(VI) emphasizes on direct reduction [23], composite ceramic adsorption [24], resin adsorption [25], thermal treatment [26], nanomaterial-catalyzed reduction [27], electrolysis and electrocoagulation [28, 29], desalination [52], and catalytic reduction [30]. The conventional methods involve different intensive subprocesses and utilize substantial chemical reagents and generate toxic sludge, making the overall process complicated [10]. The conventional methods comprise of subprocesses that are not suitable for the in situ operation of Cr(VI) removal. Overall the conventional methods are expensive and not eco-friendly. Thus an alternative treatment method for detoxification of Cr(VI) is preferable. Bioremediation is a potential alternative treatment method conquering the drawbacks of conventional treatment methods [31]. The Cr(VI) acts as an electron acceptor stimulating the oxidation required for the microbial growth in the bioremediation process. Bioremediation uses different bacteria, plants, algae, fungi, and some derivative of organisms for the detoxification of Cr(VI) [10].

5.5 Algae for Cr(VI) Reduction

Romanenko and Koren'kov [32] first reported the microbial bioreduction of Cr(VI) in the late 1970s. They observed *Pseudomonas* species grown could reduce Cr(VI) in an anaerobic condition. Later Cr(VI) reduction using different microorganisms has been a hot bioremediation topic for the environmental researcher. For the bioreduction of Cr(VI), parameters like agitation, initial Cr(VI) concentration, temperature, nutrient supplements, pH, cell immobilizers, and reactor design have been proved as influential parameters [8]. Photosynthetic microalgae absorb metal ions dissolved in water during the autotrophic primary production [33]. Microalgae detoxify different metal ions present in the water, either by changing the metal ions thermodynamic stability in water or complex formation [34]. The advantage of using microalgae as Cr(VI) reducers is they are available less expensively and conveniently.

A microalgae species *Chlorella vulgaris* has ability to reduce Cr(VI) photochemically. Deng et al. [35] used *C. vulgaris* for the Cr(VI) reduction. For the photochemical influence, the *C. vulgaris was* irradiated with a light of wavelength 365 nm using a metal halide lamp having output power of 250 W. Different parameters were varied during the photochemical reduction. They observed the Cr(VI) was increasingly reduced by the *C. vulgaris* with increase of exposure time and algae concentration, and decrease of pH and initial Cr(VI) concentration. In another study, a laboratory stock microalgae culture *C. vulgaris* was used to remove Cr(VI) from a wastewater sample. The wastewater sample was collected from a tannery industry and diluted with tap water in a ratio of 1:1 before starting the experiments. The Cr(VI) concentration in the wastewater sample was 3.22 mg/L. Under irradiation of fluorescent lights of 149–299 μmol photons/m^2/s, the salt-tolerant microalgae *C. vulgaris* removed Cr(VI) completely from the diluted water sample at 28 °C in 12 days [36]. They did not study effectively the organelle activities of the *C. vulgaris* cell during the reduction. Chen et al. [37] reported mechanism of the Cr (VI) reduction is due to the organelles of the *C. vulgaris*.

The microalgal biomass of *C. vulgaris* was used as an adsorbent for chromium-removing process. It was observed that Cr removal though *C. vulgaris* was occurring not only through the adsorption mechanism but also through reduction mechanism. The results of X-ray absorption near-edge spectroscopy showed the Cr(VI) reduction by *C. vulgaris*. The enzymatic chromium reductase is the potential way to reduce Cr(VI). The Cr(VI) reduction was observed even in the dead cells revealing that a nonenzymatic reduction route is possibly involved in this process. Furthermore glutathione released from the broken cells played an auxiliary role in the Cr(VI) reduction [38]. The dry biomass of an isolated microalgal *Chlorella miniata* was used for the Cr(VI) adsorption. The biosorption coupled with bioreduction was found to be involved in the Cr(VI) removal. This biosorption coupled with bioreduction was confirmed in the desorption studies as the Cr(III) engaged in most of the adsorption spots. The Cr(VI) uptake increased with decrease of the initial pH. The total Cr uptake was maximum at an initial pH -3. The FTIR analysis revealed the amino and carboxylate groups were the adsorption sites of Cr(VI) and Cr(III) [39]. Dried biomass of *C. vulgaris* was used as absorbent for Cr(VI) removal from electroplating and galvanizing industry effluents. The pH was a limiting factor for Cr(VI) sorption. The highest Cr removal was observed at a concentration of 81 3 mg/L was observed at. Freundlich isotherm model fitted the experimental data well [40]. Waste microalgae biomass residues of *C. vulgaris* from a biodiesel production unit were used as adsorbent for Cr(VI) uptake [41]. The removal of Cr(VI) increased as the pH decreased and temperature increased. The Sips isotherm was well fitted to the experimental data. At an optimum condition of pH 1.5 and temperature of 25 °C, the total chromium uptake was found to be 43.3 mg/g. X-ray photoelectron spectroscopy revealed an adsorption of Cr(VI) followed by reduction to Cr(III) was the mechanism for the adsorption. The FTIR study indicated a similar result that was reported by Han et al. [39] above.

Cr(VI) biosorption by raw algae is associated with a high organic compound leaching. A raw *Sargassum* sp. seaweed was modified with HCl, NaOH, $CaCl_2$, HCOH, and $C_5H_8O_2$ in order minimize organic leaching during Cr(VI) biosorption. The chemical modification by 0.2% HCOH was found to be enough for seizing the organic leaching. The maximum adsorption achieved in both modified seaweed and raw seaweed at pH 2.0 with the uptake of 1.1 and 0.6 mmol/g, respectively [42], A macroalga *Sargassum cymosum* was used for Cr(VI) removal which followed a reduction followed by adsorption process [43]. The alginate compound was first extracted from the Brazilian brown seaweed *Sargassum filipendula*. Then the residue was used for Cr(III) and Cr(VI) removal from aqueous solutions. The biosorption followed by the reduction was observed for the bioremediation process. The Langmuir adsorption isotherm model was fitted to the Cr remediation with maximum biosorption capacities of 0.819 and 0.635 mmol/g for Cr(total) and Cr(III), respectively [44]. A brown macroalga *Pelvetia canaliculata* was used as Cr(VI) remover from an acidic electroplating wastewater varying different adsorption parameters such as pH, initial Cr(VI) concentration, temperature, and biomass concentration. The acidified *P. canaliculata* biomass reduced 2.4 mmol/g Cr(VI) and adsorbed 1.9 mmol/g Cr(III). However, the raw *P. canaliculata* biomass packed in a column reactor showed the uptake of 2.1 mmol/g Cr(VI) [45].

Dry biomass of marine green algae *Halimeda gracilis* was used to remove Cr(VI). The optimized condition for Cr(VI) removal was pH 4.9, sorbent dosage 2.2 g/L, agitation speed 136 rpm, and contact time 47 min with adsorption capacity 55.55 mg/g. The adsorption was a pseudo-second-order intraparticle diffusion, pseudo-first-order kinetics, power function, and Elovich kinetic models. About 80% of the sorbent was regenerated in 0.2 M HCl solution [46]. Chemical modification of marine brown algae *Cystoseira indica was performed* for Cr(VI) removal. Four specified biomass specimen were prepared by cross-linking with epichlorohydrin (CB1, CB2), oxidized by potassium permanganate (CB3), and only washed by distilled water (RB). The maximum Cr(VI) removal was observed at a pH 3.0 in 2 h only for all four modified biomass. The maximum adsorption capacity of Cr(VI) were found to be 22.7, 24.2, 20.1, and 17.8 mg/g, respectively, for CB1, CB2, CB3, and RB [47]. Valorization with acid of seaweed *Laminaria digitata was done Cr(VI)* remediation. The protonated *L. digitata biomass could remove* 2.2 mmol/g. Cr(VI) removal followed a reduction-cum-adsorption mechanism. The adsorption capacity of the biomass increased with the extend of Cr(VI) reduction as more active sites were formed due to oxidation of biomass during the reduction process [48]. Raw and acidified biomass of alga *Oedogonium hatei* were used for Cr(VI) removal from an aqueous solution. The raw and acid-treated alga, respectively, adsorbed 30 and 35 mg/g of Cr(VI) at an optimum condition of pH 2.0, contact time 2 h., adsorbent dose 0.8 g/L, and temperature 45 °C. The Cr(VI) adsorption process followed a pseudo-first-order kinetic model and was found to be spontaneous and endothermic in nature. The experimental data were well fitted to both the Freundlich and Langmuir adsorption isotherm. The FTIR analysis revealed functional group like -COOH, -OH, and $-NH_2$ groups played major role in the biosorption process. Up to 75% of the loaded adsorbent was regenerated in 0.1 M NaOH solution [49]. The activated carbon from blue-green algal bloom showed maximum adsorption capacity of 155.52 mg/g of Cr(VI) from aqueous solutions. The adsorption process followed the pseudo-second-order kinetic model. The experimental data well fitted to Freundlich adsorption isotherm [50].

5.6 Conclusions

The Cr(VI) level in the water increases above the standard regulatory concentration due to different anthropogenic activities such as pigment synthesis, metal finishing, leather tanning, chromite mining, and electroplating. This is started with the mining of chromite or ferrous chromite ores followed by different industrial activities resulting in Cr(VI) concentration in water. The Cr(VI) toxicity in the living organisms is developed due to cellular Cr(VI) reduction forming hyperreactive species inside the cells, Cr(III)-protein complex, and altered DNA sequence. The Cr(VI) in the water can be removed by either the chemical or biological or combination of both processes. Bioremediation including bioreduction and biosorption has potential to remove Cr(VI) from water. A variety of algal species have shown positive correlation toward Cr(VI) uptake. Some microalgae directly reduce Cr(VI) during the photosynthesis. But, the biosorption process using biomass of microalgae has a wide potential for the purpose of Cr(VI) removal. The Cr(VI) acts as an electron acceptor forming active sites on the surface of biomass in the redox reactions. The active sites on the biomass easily adsorb the reduced Cr(III) in the cation-anion interactions. The major active sites are carboxyl, hydroxyl, and amine group present in the biomass or formed during the reduction of Cr(VI). Overall the biosorption process of Cr(VI) follows a reduction coupled with adsorption technique for its removal from water.

Acknowledgments The authors are grateful to Prof. (Dr.) M. R. Nayak, President, SOA (Deemed to be University), for providing the infrastructures and encouragement throughout.

References

1. United States Geological Survey (USGS) (1998) Ground water and surface water: a single resource. Circular 1139 by Denver, C
2. Bureau of Indian Standards (2012) Indian standard drinking water specification (second revision), ICS 13.060.20, IS 10500:2012
3. United States Environmental Protection Agency (2009) National primary drinking water regulations. EPA 816-F-09-004
4. Pradhan D, Sukla LB, Sawyer M, Rahman PKSM (2017) Recent bioreduction of hexavalent chromium in wastewater treatment: a review. J Ind Eng Chem. https://doi.org/10.1016/j.jiec.2017.06.040 in press
5. Kimbrough DE, Cohen Y, Winer AM, Creelman L, Mabuni C (1999) A critical assessment of chromium in the environment. Crit Rev Environ Sci Technol 29(1):1–46
6. Gheju M (2011) Hexavalent chromium reduction with zero-valent iron (ZVI) in aquatic systems. Water Air Soil Pollut 222:103–148
7. Bielicka A, Bojanowska I, Wiśniewski A (2005) Two faces of chromium-pollutant and bioelement. Pol J Environ Stud 14(1):5–10
8. Dhal B, Thatoi HN, Das NN, Pandey BD (2013) Chemical and microbial remediation of hexavalent chromium from contaminated soil and mining/metallurgical solid waste: a review. J Hazard Mater 250-251:272–291
9. Westbrook J (1983) Chromium and chromium alloys. In: Grayson M (ed) Kirk-Othmer encyclopedia of chemical technology, vol 6, 3rd edn. Wiley-Interscience, New York, pp 54–82
10. Malaviya P, Singh A (2016) Bioremediation of chromium solutions and chromium containing wastewaters. Crit Rev Microbiol 42(4):607–633
11. Barrera-Díaz CE, Lugo-Lugo V, Bilyeu B (2012) A review of chemical, electrochemical and biological methods for aqueous Cr(VI) reduction. J Hazard Mater 223-224:1–12
12. Langard S (1990) One hundred years of chromium and cancer: a review of epidemiological evidence and selected case reports. Am J Ind Med 17:189–215
13. Geological Survey of Japan (2005) Atlas of Eh-pH diagrams, intercomparison of thermodynamic databases. Open File Report No.419. Prepared by National Institute of Advanced Industrial Science and Technology, p 78

14. Rai D, Eary LE, Zachara JM (1989) Environmental chemistry of chromium. Sci Total Environ 86:15–23
15. Zhitkovich A (2011) Chromium in drinking water: sources, metabolism, and Cancer risks. Chem Res Toxicol 24:1617–1629
16. Lamb A, Evans G, King JR (2013) Mathematical modelling of toxicity associated with intracellular chromium reduction. Bull Math Biol 75(9):1472–1500
17. Messer J, Reynolds M, Stoddard L, Zhitkovich A (2006) Causes of DNA single-strand breaks during reduction of chromate by glutathione in vitro and in cells. Free Radic Biol Med 40:1981–1992
18. Gibb HJ, Lees PS, Pinsky PF, Rooneym BC (2000) Lung cancer among workers in chromium chemical production. Am J Ind Med 38:115–126
19. Kim SH, Lee IC, Baek HS, Moon C, Kang SS, Bae CS, Kim SH, Shin DH, Kim J (2012) C. Pycnogenol® prevents hexavalent chromium-induced spermatotoxicity in rats. Mol Cel Toxicol 8:249–256
20. Singh P, Chowdhuri DK (2016) Environmental presence of hexavalent but not trivalent chromium causes neurotoxicity in exposed drosophila melanogaster. Mol Neurobiol. Available at: https://doi.org/10.1007/s12035-016-9909-z
21. U.S. Environmental Protection Agency (1998) Toxicological review of trivalent chromium. CAS No. 16065-83-1. In support of summary information on the Integrated Risk Information System (IRIS). USEPA, Washington, DC
22. Zhang J, Li X (1987) Chromium pollution of soil and water in Jinzhou. Chin J Prev Med 21:262–264
23. Yuan Y, Yang S, Zhou D, Wu F (2016) A simple Cr(VI)–S(IV)–O_2 system for rapid and simultaneous reduction of Cr(VI) and oxidative degradation of organic pollutants. J Hazard Mater 307:294–301
24. Ai L, He J, Wang Y, Wei C, Zhan J (2016) Aerosol-assisted in situ synthesis of iron–carbon composites for the synergistic adsorption and reduction of Cr(VI). RSC Adv 6:56108–56115
25. Chang X, Li M, Liu Q, Liu Q, Yao J (2016) Adsorption–reduction of chromium(VI) from aqueous solution by phenol–formaldehyde resin microspheres. RSC Adv 6:46879–46888
26. Meegoda JN, Kamolpornwijit W, Batagoda JH (2016) A detailed laboratory scale feasibility study of recovering metallic iron and chromium from chromium contaminated soils. Indian Geotech J. Available at: https://doi.org/10.1007/s40098-016-0208-4
27. Celebi M, Yurderi M, Bulut A, Kaya M, Zahmakiran M (2016) Palladium nanoparticles supported on amine-functionalized SiO_2 for the catalytic hexavalent chromium reduction. Appl Catal B Environ 180:53–64
28. Hossini H, Rezaee A, Masoumbeigi H (2014) Optimization of chromium reduction and sludge production by bipolar electrocoagulation using response surface methodology. J Health Policy Sustain Health 1(1):1–7
29. Weng W, Wang M, Gong X, Wang Z, Wang D, Guo Z (2016) Direct electro-deposition of metallic chromium from K_2CrO_4 in the equimolar $CaCl_2$-KCl molten salt and its reduction mechanism. Electrochim Acta 212:162–170
30. Fan Z, Zhao Y, Zhai W, Qiu L, Li H, Hoffmann MR (2016) Facet-dependent performance of BiOBr for photocatalytic reduction of Cr(VI). RSC Adv 6:2028–2031
31. Lovley DR (1995) Microbial reduction of iron, manganese, and other metals. Adv Agron 54:175–231
32. Romanenko VI, Koren'kov VN (1977) A pure culture of bacteria utilizing chromates and bichromates as hydrogen acceptors in growth under anaerobic conditions. Mikrobiologiya 43:414–417
33. Salt D, Pickering I, Prince R, Gleba D, Dushenkov S, Smith R, Raskin I (1997) Metal accumulation by aquacultured seedlings of Indian mustard. Environ Technol 31:1636–1644
34. Duarte B, Delgado M, Cac-ador I (2007) The role of citric acid in cadmium and nickel uptake and translocation, in *Halimione portulacoides*. Chemosphere 69:836–840
35. Deng L, Wang HL, Deng NS (2006) Photoreduction of chromium(VI) in the presence of algae, *Chlorella vulgaris*. J Hazard Mater 138:288–292

36. Das C, Naseera K, Ram A, Meena RM, Ramaiah N (2016) Bioremediation of tannery wastewater by a salt-tolerant strain of *Chlorella vulgaris*. J Appl Phycol. Available at: https://doi.org/10.1007/s10811-016-0910-8
37. Chen Z, Song S, Wen Y (2016) Reduction of Cr (VI) into Cr (III) by organelles of *Chlorella vulgaris* in aqueous solution: an organelle-level attempt. Sci Total Environ 572:361–368
38. Yen HW, Chen PW, Hsu CY, Lee L (2017) The use of autotrophic *Chlorella vulgaris* in chromium (VI) reduction under different reduction conditions. J Taiwan Inst Chem Eng 74:1–6
39. Han X, Wong YS, Wong MH, Tam NFY (2007) Biosorption and bioreduction of Cr(VI) by a microalgal isolate, *Chlorella miniata*. J Hazard Mater 146:65–72
40. Sibi G (2016) Biosorption of chromium from electroplating and galvanizing industrial effluents under extreme conditions using *Chlorella vulgaris*. Green Energy Environ 1:172–177
41. Xie Y, Li H, Wang X, Ng IS, Lu Y, Jing K (2014) Kinetic simulating of Cr(VI) removal by the waste *Chlorella vulgaris* biomass. J Taiwan Inst Chem Eng 45:1773–1782
42. Yang L, Chen JP (2008) Biosorption of hexavalent chromium onto raw and chemically modified *Sargassum* sp. Bioresour Technol 99:297–307
43. de Souza FB, Brandão HDL, Hackbarth FV, de Souza AAU, Boaventura RAR, de Souza SMAGU, Vilar VJP (2016) Marine macro-alga *Sargassum cymosum* as electron donor for hexavalent chromium reduction to trivalent state in aqueous solutions. Chem Eng J 283:903–910
44. Bertagnolli C, da Silva MGC, Guibal E (2014) Chromium biosorption using the residue of alginate extraction from *Sargassum filipendula*. Chem Eng J 237:362–371
45. Hackbarth FV, Maass D, de Souza AAU, Vilar VJP, de Souza SMAGU (2016) Removal of hexavalent chromium from electroplating wastewaters using marine macroalga *Pelvetia canaliculata* as natural electron donor. Chem Eng J 290:477–489
46. Jayakumar R, Rajasimman M, Karthikeyan C (2014) Sorption of hexavalent chromium from aqueous solution using marine green algae *Halimeda gracilis*: optimization, equilibrium, kinetic, thermodynamic and desorption studies. J Environ Chem Eng 2:1261–1274
47. Basha S, Murthy ZVP, Jha B (2008) Biosorption of hexavalent chromium by chemically modified seaweed, *Cystoseira indica*. Chem Eng J 137:480–488
48. Dittert IM, Brandão HL, Pina F, da Silva EAB et al (2014) Integrated reduction/oxidation reactions and sorption processes for Cr(VI) removal from aqueous solutions using *Laminaria digitata* macro-algae. Chem Eng J 237:443–454
49. Gupta V, Rastogi A (2009) Biosorption of hexavalent chromium by raw and acid-treated green alga *Oedogonium hatei* from aqueous solutions. J Hazard Mater 163:396–402
50. Zhang H, Tang Y, Cai D, Liu X et al (2010) Hexavalent chromium removal from aqueous solution by algal bloom residue derived activated carbon: equilibrium and kinetic studies. J Hazard Mater 181:801–808
51. Headlam HA, Lay PA (2016) Spectroscopic characterization of genotoxic chromium (V) peptide complexes: Oxidation of Chromium (III) triglycine, tetraglycine and pentaglycine complexes. J Inorg Biochem 162:227–237
52. An Z, Zhang H, Wen Q, Chen Z, Du M (2014) Desalination combined with hexavalent chromium reduction in a microbial desalination cell. Desalination 354:181–188

Removal of Radon from Radionuclide-Contaminated Water Using Microalgae

6

Debabrata Pradhan and Lala Behari Sukla

Abstract

Bioremediation using microalgae is a potential alternative way to different conventional water treatment processes. The advantages of using microalgae are as follows: it requires low energy with reduced sludge formation and carbon dioxide sequestration. This review summarizes the possible bioremediation of radioactive radon from water system as radon gas is α- and β- radiation emitter. The major health hazard occurs due to indoor air radon which is usually released from the potable water contaminated with dissolved radon supplied through different distribution systems. Standard techniques like aeration and activated carbon filtration are the conventional techniques applied to remove radon from drinking water. However, both the processes face different technological drawbacks from complex designing, maintaining uniform concentration flow, short-lived decay products, and risk of sewer recontamination. The activated intercellular polysaccharides of the microalgae cell can be a potential accumulator of radon from drinking water as the activated intercellular polysaccharides mimic the activated carbon. The microalgae do not grow under the influence of high energy radiation; but, due to the evolution of this kind of microorganisms under prolonged influence of high energy radiation, they overcome the physiological stress in the extreme environment for their growth. A number of algal species grow in the highly radioactive sites. They accumulate different radioactive elements as well as α-, β-, and γ-radiations.

D. Pradhan (✉) · L. B. Sukla
Biofuels and Bioprocessing Research Center, Institute of Technical Education and Research, Siksha 'O' Anusandhan (Deemed to be University), Bhubaneswar, Odisha, India
e-mail: debabratapradhan@soa.ac.in; lalabeharisukla@soa.ac.in

L. B. Sukla et al. (eds.), *The Role of Microalgae in Wastewater Treatment*,
https://doi.org/10.1007/978-981-13-1586-2_6

Abbreviations

μSv/y	10^{-6} × Sievert per year
BARC	Bhaba Atomic Research Center, Mumbai, India
BIS	Bureau of Indian Standard
BOD	Biochemical oxygen demand
Bq/L:	Becquerel per liter, 1 Bq/m^3 = 0.027 pCi/L
Bq	Becquerel. Unit of radioactivity, equal to one nuclear transformation per second
COD	Chemical oxygen demand
FOG	Fat, oil, and grease
GAC	Granular activated carbon
Gy	Gray, an SI unit of ionization radiation, 1 Gy = 1 J/kg
mBq/L	Millibecquerel per liter
Rn-222	Radon with atomic mass 222 (most stable isotope of radon)
Sv	Sievert. Unit of collective effective dose, equal to the sum of effective doses of individuals among a group. 1 Sv = 1 J/kg
Sv/y	Sievert per year
UNSCEAR	United Nations Scientific Committee on the effects of atomic radiation
USDWR	United States Drinking Water Regulation
USEPA	United States Environmental Protection Agency
WHO	World Health Organization
α	Alfa ionization radiation
β	Beta ionization radiation
mrem/yr.	milliremper year. (rem: Roentgen Equivalent Man is a measurement that correlates the dose of any radiation to the biological effect of that radiation), 1 Sv = 100 rem

6.1 Introduction

Radon (Rn) radiates high energy α- and β-radiations. Radioactive Rn has 36 isotopes, but only Rn-222 is stable [1]. Due to the specific chemical properties, potable water supplied to household from different resources are contaminated with Rn-222 [2]. Most of the waterborne radon entering a home is released into the indoor air through normal water use activities; later, this indoor air radon is inhaled by the human being. According to the UNSCEAR report, the average effective radon dose caused due to inhalation is directly proportional to the concentration of radon present in the drinking water [3, 4]. The acceptable limits for gross α- and β-radiation are listed in Table 6.1. Radon is an established human lung carcinogen second to that of tobacco smoke. About 10–15% of total death due to lung cancer in the United States is due to indoor Rn-222 exposure [2]. Besides lung cancer other diseases such as acute leukemia, kidney cancer, prostate cancer, malignant melanoma, and

Table 6.1 Acceptable limit of radiation in drinking water

Sl. no.	Radiations	NPDWR, USEPA [5]	BIS [6]	WHO [7]	EU [8]
1.	Gross α emitters	15.0 pCi/L	0.1 Bq/L	0.5 Bq/L	0.04 Bq/L
2.	Gross β emitters	4.0 mrem/yr	1.0 Bq/L	1.0 Bq/L	0.4 Bq/L

childhood cancer are developed due to indoor Rn-222 inhalation [2]. In vitro mutagenicity studies revealed radon alters the gene sequence resulting propagation and separation of cancer cells. Kjellberg and Wiseman [9] reported the dissolved Rn-222 in drinking water absorbed in the stomach cell has positive correlation with stomach cancer.

Different water treatment techniques like aeration, lime softening, GAC filtration, reverse osmosis, coagulation and flocculation, sand filtration, and hydrous manganese filtration have been used for treatment of radionuclides and other pollutants present in water [10]. The GAC adsorption and diffused aeration techniques have been reported as potentially cost-effective treatment processes to remove Rn-222 [11]. But, the major concern is the accumulation of Rn-222 in different tools of the above two processes which ultimately release back to the water reservoirs resulting in an elevated radiation levels [10, 12, 13]. Bioremediation emerges as a potential alternative to the above technologies as it is simple to scale up for in situ operation, requires low-cost equipments, is eco-friendly and cost-effective, has no emission, and gives almost a closed-loop process [14–18]. In the bioremediation process, different indigenous microorganisms or microphytes play important role in switching the contaminants through the metabolic process. The use of microalgae in bioremediation is a practical interest in wastewater treatment due to different advantages as it requires low energy with reduced sludge formation and carbon dioxide sequestration. Extremophilic microalgae can grow under variable conditions of acid, alkali, pH, temperature, light, CO_2, and metal ions [19]. Due to their sustainable evolution mechanism in the extreme environments, they can be potential radiation accumulator in the bioremediation process. But adequate reports have not been presented to predict the mechanism of accumulation of radionuclides by microalgae. Only few authors have shown the tolerance of microalgae to radiation without any genome defects [20–28].

6.2 Radon in Water

Radon is an inert gas and always tends to be in the elemental state. It does not form any chemical compounds or complexes; its concentration in water is unlikely changed by natural geochemical process except dilution and seasonal variation. Sometimes Rn-222 concentration in some geological water is above 10,000 Bq/L which is much higher than the acceptable level as shown in Table 6.1. Rn-222 is the only inert gas in the decay series of uranium (U-238) and thorium (Th-232) at standard temperature and pressure. The parents of Rn-222 in the decay series, i.e., radium and uranium, are highly soluble in water in the form of Ra^{2+} and U^{6+}.

During the decay Rn-222 reaches 99% secular equilibrium with its immediate parent Ra-226 in 25.4 days in a closed system [29]. But in an open system like groundwater, transport of daughter and parent causes non-prevailing of secular equilibrium causing continuous α decay. Ra-226 decays continuously produce Rn-222 to achieve secular equilibrium enriching its concentration. In some location its concentration is much higher than the regulatory level. It produces stable daughter isotopes, and those are Pb-210 ($t_{1/2}$ = 22.3 year), Bi-210 ($t_{1/2}$ = 5.013 days), Po-210 ($t_{1/2}$ = 138.376 days), and stable Pb-206. The density of radon gas is 9.074 g/L [30]. The solubility of radon gas in water is 1.67×10^{-4} (mole fraction) at 25 °C, which is more than oxygen (2.29×10^{-5}) [31]. Due to the adequate solubility, Rn-222 is mechanically transported to water bodies from different sources by means of direct contamination of Rn-222 or decay product of Ra-226 resulting in elevated Rn-222 concentration [32]. On the basis of literature, a biogeochemical radon cycle is proposed which is shown in Fig. 6.1. Due to hydrological cycle, a variety of contaminants including Rn-222 are mixed with water in the form of dissolved or suspended state. In the environment the sources of Rn are (1) terrestrial crusts like U-ore, phosphate rock, granite, and lime stone; (2) building materials and their conjunctions; (3) plumbing wastes; (4) native radioactive decay; (5) continental temperature inversion due to seasonal variations; (6) meteorological variation like wind speed, humidity, and temperature; (7) accumulation due to low atmospheric pressure; (8) basement soil to indoor due to structural defects of floors; (9) carrier gas like water vapor and CO_2; (10) uranium mining wastes; (11) cereal and pulses; and (12) by-product of uranium ore processing in HCl or HBr [2, 10, 33, 34]. Rn-222 is introduced to water from the above sources by different natural and artificial activities.

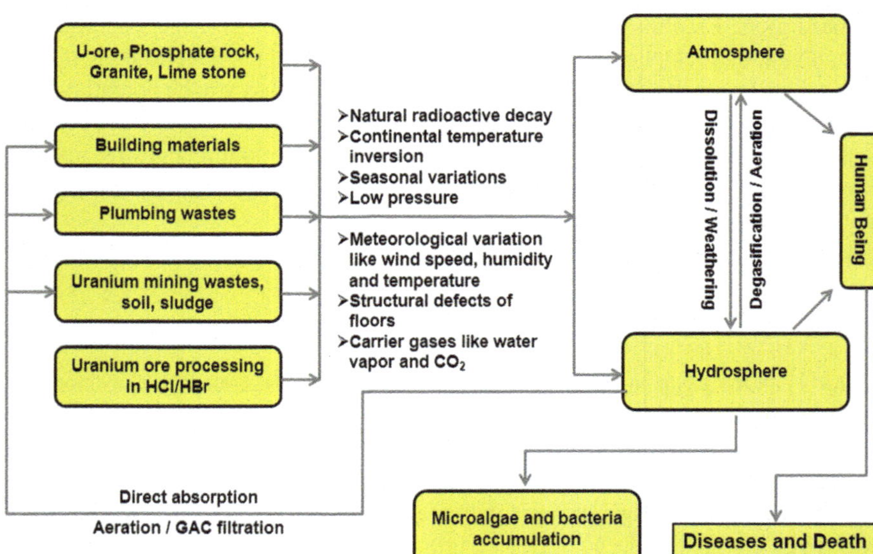

Fig. 6.1 Proposed geochemical radon cycle [2, 10, 33]

Several articles have reported the elevated Rn concentration in drinking water in the Indian province such as Tamil Nadu, Karnataka, Punjab, Haryana, Rajasthan, and Himalaya regions [13, 35–37]. The occurrence of radioactive Th-232, U-238, Ra-226, K-40, and Cs-137 has been reported in the daily consumables like rice, wheat, pulses, and drinking water of Odisha's Chhatarpur area in India [33]. The ingestion doses were estimated in the range of 110–937, 11–308, and 0.6–3.0 μSv/year from the cereals, pulses, and drinking water, respectively. Water samples collected from the borewells of Mandya region of Karnataka, India, have been reported to be contaminated with Rn-222 ranging from 6.5 to 44.9 Bq/L. The ingestion dose due to the borewell contamination varies from 26.4 to 178.6 μSv/year [36]. In the Amritsar city of Punjab, India, Rn-222 concentration ranging from 0.5 to 11.1 Bq/L was found in the drinking water samples [38]. The mineral water used for drinking purpose in two states of Brazil, i.e., São Paulo and Minas Gerais, has been reported to be contaminated with Rn-222 and Rn-220 [39]. The water utilities used for the distribution of water have been reported to be contaminated with radium, thorium, and uranium isotopes, and their gross alpha and beta activity were 255 and 181 pCi/g, respectively [10]. The water samples collected from Rajasthan, India, were contaminated with Rn-222 ranging from 0.6 to 21.9 Bq/L, and the total annual effective dose ranged from 1.4 to 60.0 μSv/year [3]. The drinking water samples collected from Shimoga district of Karnataka, India, were estimated to be ranged from 3.0 to 39.0 Bq/L. About 45% of total collected drinking water samples were contaminated with Rn more than the recommended MCL [40]. The uranium concentration ranging from 2.0 to 3.5 ppm was found to be contaminated in the soil samples collected from Ganderbal district of Jammu and Kashmir, India. The Rn exhalation rate due to the uranium in the soil was found to vary between 5.0 and 21.9 mBq/kg/h [35]. A raw water sample from the inlet of a WTP showing total radium activity of 0.25 Bq/L was introduced to the standard water treatment processes such as flocculation/sedimentation, sand filtration, and reverse osmosis to evaluate the radium removal efficiency [41]. It was found that the flocculation/sedimentation, sand filtration, and reverse osmosis, respectively, removed 33, 22, and 98% of radium from the water in the individual step. The solid waste generated in the sedimentation and sand filtrations stages were found to contain 4500 and 6752 Bq/kg of radium, respectively. In another study the air inside a WTP was found to be ranged from 2 to 18 Bq/m^3 [42].

6.3 Conventional Radon Removal

6.3.1 Aeration

Radon from drinking water can be removed through aeration. In the aeration process, clean air is injected to water in order to remove certain dissolve gases and metals [43]. Due to its high Henry constant of radon solubility in water aeration, it is a suitable process to remove it from water. The aeration techniques involved storage and/or storage with minimal aeration. The diffused bubble aeration technique was used to efficiently remove radon from a moderately contaminated groundwater supplies [44]. The important parameters are air-to-water ratios and detention times for the diffused bubble aeration technique. In an experimental study of the diffused bubble aeration technique, about 97% of radon could be removed at an optimized condition of detention time and air-to-water ratio respectively 19 min and 12 [45]. The efficiency of the aeration solely depends on the air-to-water ratio which determines the cost of the process. Also there are so many factors to be considered to design an aerator especially for treatment of Rn-222 making the process too complex [12]. Several technical failures occur during operation of the aerators. Moreover the air ventilated to atmosphere cause high level of Rn-222 airborne. Raucher and Drago [13] estimated the annual installation and operation cost of aeration technique for radon removal across the United States was $2.6 billion.

6.3.2 Activated Carbon Adsorption

Rn is an inert gas found in the elemental form instead of forming any chemical compound. Instead of chemisorptions, the physical adsorption on the GAC surface is a suitable process to immobilize Rn from water. Annanmaki and Turtiainen [12] reported about 99% Rn removal applying a GAC adsorption process. But the major problem is production of the short-lived decay products of radon retained on the GAC which emits gamma rays as a consequence. Also the radioactivity retained in the GAC returned to the sewer again on its disposal. Karunakara et al. [46] used coconut-based granular activated charcoal as cylindrical adsorbent beds to remove Rn-222 and Rn-220 from air. The adsorption coefficient varied from 2.2 to 4.1 m^3/kg. The adsorption coefficient had a positive linear relation with the flow rate. The adsorbent could be regenerated on heating at 100 °C. In another study different GAC adsorbents successfully adsorbed radon from water supplies containing Rn 1500 to 750,000 pCi/L [47]. The removal of Rn from drinking water using GAC is a positive benefit for the water users. However, the accumulation of radon on activated carbon causes radiologic hazards for the water treatment plant operators, and the spent carbon may be considered a low-level radioactive waste.

6.3.3 Reverse Osmosis

Reverse osmosis is an advanced water treatment technique used to remove different trace elements and ions present in the water. A two-pass reverse osmosis system performed well to remove radionuclides from geothermal water [48]. The removal efficiency was found to be ranging from 70.7 to 77.2% for both radium and uranium with the gross α and β activities reduced to undetermined level. Commercial reverse osmosis equipment was examined for removing radioactivity as well as salinity from a domestic water. The reverse osmosis equipment removed most of the elements up to 94%; however, it failed to remove gaseous radon in the water [49].

6.4 Algae as Radon Remover

The activated intercellular polysaccharides of the biological cell mimic the properties of activated carbon. Since the activated carbon has ability to absorb Rn-222 from drinking water, the activated intercellular polysaccharides of the biological cell of radio-tolerant organism can show a potential accumulator of Rn from drinking water [12, 27]. Microalgae are found to contain up to 61% carbohydrates, 45% proteins, and 4% lipids which can be utilized as radionuclide accumulator [14, 18, 21, 27]. But the major issue is α- and β-tolerant organism finding due to lack of adequate research on this field. Secondly, some microbes have shown extraordinary tolerance to the high energy radiations above the standard regulation level. Some algae species have shown tolerance and accumulation power toward different radioactive isotopes, and those are listed in Table 6.2. The extremophilic species like *Thermococcus gammatolerans*, *Pyrococcus furiosus*, *Halobacterium* sp., *Chroococcidiopsis* sp., and *Alternaria alternata* are extremely radio resistant similar to the *Deinococcus radiodurans* [28, 50]. The evolution of the radioresistance properties in *D. radiodurans* has been studied, far less in other organisms. The evolution is mainly due to protection against oxidative stress, an efficient conventional DNA repair tool box, original DNA repair mechanisms, and a condensed nucleoid [50].

Few bacteria and microalgae showed their potential to accumulated radioactive elements from water [20–28]. The microorganisms have developed an adaptation method to overcome the physiological stress in radioactive environments. The most studied radio-tolerant microorganism is *Deinococcus radiodurans* [21]. The microalgae can sustain irradiation stress up to 20,000 Gy. The analysis revealed the resistance to radiation dose by the microalgae due to the genome multiplication and speedy DNA repair mechanisms [27]. Another group of scientist in Australia investigated the tolerance ability of *Chlorophyceae* green microalgae to ionizing

Table 6.2 Algae tolerance toward radionuclides

Algae	Tolerance/treatment to different radionuclides	References
Closterium moniliferum	Strontium-90	[25, 26]
Coccomyxa actinabiotis	Ag-110, Co-60, Co-58, Sb-124, Cr-51, Zn-65, Mn-54, Cs-137, U-238, C-14	[27, 28, 50]
Chara sp., *Nitella* sp., *Pistia* sp., *Jussia* sp., *Eichornia* sp., *Hydrilla*	Ra-226	[51]
Polysiphonia fucoides, *Furcellaria lumbricalis*	Cr-51, Mn-54, Co-57,Co- 60, Zn-65, Sr-85,Cd-109, Ag-110, Sn-113, Cs-137, Am-241	[52]
Ulva sp., *Ecklonia radiata*	I-131	[22]
Deinococcus radiodurans	Radioactive waste mixed with Hg(II)	[21]
Chroococcidiopsis	Artificial ionization radiation from tungsten disk	[20]
Jania longifurca, *Cystoseira*, *Sargassum vulgare*	Cs-137, Po-210, Pb-210	[53]
Chlorophyceae	Artificial γ-rays and fast neutrons	[23]

radiations [23]. Radioactive iodine (I-131) as source of radiation has been detected in macroalgae. Different I-131 contaminated sample from liquid effluent and sludge of STPs were evaluated in terms of concentration factors (ratio of concentration in algae to the surrounding media). The I-131 level in other macroalgae such as *Ulva sp.* and *Ecklonia radiata* was also estimated. The radioactive I-131 was accumulated by macroalgae *Ulva* sp. and *E. radiata* in the shoreline outfall of a coastal environment in Australia. The concentration factors of *Ulva* sp. and *E. radiata* were found to be 177 and 528, respectively [22]. The advantage of using microalgae for bioremediation is due to the capacity of rapid biomass generation intern which increased potential adsorption sites for contaminants in water.

In the Nature news, it has been reported that algae hold a key role in remediation of radiations following the Fukushima nuclear accident in Japan. This novel conclusion was claimed at a meeting of the American Chemical Society [25, 26]. The alga was characterized as *Closterium moniliferum*. The *C. moniliferum* removed radioactive Sr-90 from the surrounding water containing radioactive contaminants. The mechanism of remediation was described as the formation of vacuoles, a subcellular structure, followed by deposition in the crystal [25]. In another study the newly discovered unicellular microalgae *Coccomyxa actinabiotis* is isolated from a nuclear facility displaying exceptional resistance to ionizing radiation, accumulation, and detoxification of radioactive isotopes [27, 28, 50]. Bioremediation is an interesting alternative for mitigation of radiations released from aqueous effluents of nuclear industries. The microalgae can fight against the radiations during unexpected accidents in the nuclear industries. A new isolated microalgae *C. actinabiotis* withstand at a radiation doses up to 20,000 Gy. The isolated *C. actinabiotis* rapidly decontaminated the major radioactive metals from the wastewater of nuclear industries and 85% of carbon-14 in few hours only. This autotrophic eukaryote *C. actinabiotis* showed tolerance similar to that of the famous radioresistant prokaryote *D. radiodurans*. The nuclear magnetic resonance showed the *D. radiodurans* is hardly affected by radiation doses of up to 10,000 Gy with easy recovery of the cellular functions. Another report suggested the radiation resistance of *Chroococcidiopsis* strains can survive prolonged radiation through DNA repair mechanism [20].

Highly adapted filamentous algae grow in the wastewater at the uranium industries at Jaduguda, India. The algae accumulated Ra-226 from the wastewater of tailing pond, and its concentration was found to be 9850 Bq/kg in free-floating algal species with the maximum concentration factor of 8.6 [51]. Bioaccumulation ability of radionuclides such as Cr-51, Mn-54, Co-57, Co-60, Zn-65, Sr-85, Cd-109, Ag-110, Sn-113, Cs-137, and Am-241 by red algae species *Polysiphonia fucoides* and *Furcellaria lumbricalis* isolated from the southern Baltic Sea was determined under laboratory conditions. Compared to the *F. lumbricalis*, *P. fucoides* demonstrated better bioaccumulative properties toward most of the investigated radionuclides [52].

The study of accumulation of radionuclides using different microalgae is in the elementary stage though numerous researches have been done for application of the rapid generating algae biomass for wastewater treatment process. It is too early to speak about the mechanism of the accumulation of radionuclides by the microalgae.

The phycologists have a lot of scopes in the bioremediation of radionuclides using different naturally occurring microalgae. So the future research may be focused on isolation, characterization, taxonomy, selection of strain, tolerance study, genomic and metagenomics study for radio ecosystem, culturing of microalgae, biomass generation and harvesting, adaptability, accumulation of radionuclide, metagenomics study post accumulation, optimization of growth, degrading parameter correlation with metagenomic study with spectroscopic analysis, kinetic parameters for a low cost, and less complex bioremediation process for the radionuclides including the Rn-222.

6.5 Conclusions

The potable water supplied to household from either groundwater or water reservoir containing Rn-222 is the major reason for indoor radon contamination. Prolong inhalation or ingestion of the gas has several health effects. Two processes such as liquid phase granular activated carbon adsorption and air stripping with vapor phase carbon have been performed commercially to remove the radon from water despite different technological drawbacks like complex designing, uniform concentration flow, short-lived decay products, and risk of sewer recontamination associated with the processes. So time has come to search for a new technology for complete removal of carcinogenic radon from water environment. Algae holds promising in this scenario. Bioremediation is an interesting alternative for mitigation of radiations released from aqueous effluents of nuclear industries. The microalgae can fight against the radiations during unexpected accidents in the nuclear industries. The evolution of the radioresistance properties in different algae species is mainly due to protection against oxidative stress, original DNA repair mechanisms, and condensed nucleoid. Therefore phycoremediation may be a promising opportunity for the environmentalist to remove radon from the water system.

Acknowledgments The authors are grateful to Prof. (Dr.) M. R. Nayak, President, SOA (Deemed to be University), for providing the infrastructures and encouragement throughout.

References

1. Chauhan V, Howland M, Kutzner B, McNamee JP, Bellier PV, Wilkins RC (2012) Biological effects of alpha particle radiation exposure on human monocytic cells. Int J Hyg Environ Health 215:339–344
2. Zoughool MA, Krewski D (2009) Health effects of radon: a review of the literature. Int J Radiat Biol 85:57–69
3. Mittal S, Rani A, Mehra R (2016) Estimation of radon concentration in soil and groundwater samples of northern Rajasthan, India. Journal of Radiation Research and Applied Sciences 9:125–130
4. UNSCEAR (2000) United Nations scientific committee on the effects of atomic radiation. Sources and effects of ionizing radiation. Report to the general assembly with scientific annexes. New York, United Nations; 2008

5. NPDWR (2009) National Primary Drinking Water Regulations, USEPA, EPA 816-F-09-004, May, 2009
6. IS (2012) Indian standard drinking water-speciation (second revision), Bureau of Indian Standards, IS 10500, May, 2012
7. WHO (2004) World health organization, guidelines for drinking water quality, vol 1, 3rd edn
8. EU (2013) Council Directive 2013/51/Euratom of 22 October 2013 laying down requirements for the protection of the health of the general public with regard to radioactive substances in water intended for human consumption Web: http://eur-lex.europa.eu/legal-content/EN/TXT/?uri=CELEX:32013L0051
9. Kjellberg S, Wiseman JS (1995) The relationship of radon to gastrointestinal malignancies. Am Surg 61:822–825
10. Lytle DA, Sorg T, Wang L, Chen A (2014) The accumulation of radioactive contaminants in drinking water distribution systems. Water Res 50:396–407
11. Lowry JD, Brutsaert WF, McEnerney T, Molk C (1984) GAC adsorption and diffused aeration for the removal of radon from water supplies. Proceedings AWWA annual conference, Dallas, TX, USA; code 7121, 315–338
12. Annanmaki M, Turtiainen T (2000) Treatment techniques for removing natural radionuclides from drinking water, final report of the treatment techniques for removing natural radionuclides from drinking water (TENAWA) project. ISBN:951-712-353-1
13. Raucher RS, Drago JA (1992) Estimating the cost of compliance with the drinking water standard for radon. American Water Works Association 84:51–65
14. Aishvarya V, Jena J, Pradhan N, Panda PK, Sukla LB (2015) Microalgae: cultivation and application. In: Sukla LB, Pradhan N, PandaS, Mishra BK (eds) Environmental microbial biotechnology. Springer, pp 289–311. ISBN:978-3-319-19017-4
15. Kim DJ, Pradhan D, Ahn JG, Lee SW (2010) Enhancement of metals dissolution from spent refinery catalysts using adapted bacteria culture – effects of pH and Fe(II). Hydrometallurgy 103:136–143
16. Moheimani NR, Webb JP, Borowitzka MA (2012) Bioremediation and other potential applications of coccolithophorid algae: a review. Algal Res 1:120–133
17. Pradhan D, Mishra D, Kim DJ, Ahn JG, Roy Chaudhury G, Lee SW (2010) Bioleaching kinetics and multivariate analysis of spent petroleum catalyst dissolution using two acidophiles. J Hazard Mater 175:267–273
18. Sukla LB, Nayak M, Jena J, Panda HS et al (2011) Large scale cultivation of brackish water isolates *Scenedesmus* sp. in raceway pond for biodiesel production. World congress for man and nature(WCMANU_2011)
19. Varshney P, Mikulic P, Vonshak A, Beardall J, Wangikar PP (2015) Extremophilic micro-algae and their potential contribution in biotechnology. Bioresour Technol 184:363–372
20. Billi D, Friedmann EI, Hoffer KG, Caiola MG, Friedmann RO (2000) Ionizing-radiation resistance in the desiccation-tolerant Cyanobacterium *Chroococcidiopsis*. Appl Env Micro 66:1489–1492
21. Brim H, McFarlan SC, Fredrickson JK, Minton KW, Zhai M, Wackett LP, Daly MJ (2000) Engineering *Deinococcus radiodurans* for metal remediation in radioactive mixed waste environments. Nat Biotechnol 18:85–90
22. Carolan JV, Hughes CE, Hoffmann EL (2011) Dose assessment for marine biota and humans from discharge of 131I to the marine environment and uptake by algae in Sydney. Australia Journal of Environmental Radioactivity 102:953–963
23. Farhi E, Rivasseau C, Gromova M, Compagnon E et al (2008) Spectroscopic investigation of ionizing-radiation tolerance of a *Chlorophyceae* green micro-alga. J Phys Condens Matter. https://doi.org/10.1088/0953-8984/20/10/104216
24. Idota Y, Harada H, Tomono T, Morimoto K, Kobayashi S, Kakinuma C, Miyajima C, Kasahara F, Ogihara T (2013) Alginate enhances excretion and reduces absorption of strontium and Cesium in rats. Biol Pharm Bull 36:485–491
25. Krejci MR, Finney L, Vogt S, Joester D (2011) Selective sequestration of strontium in desmid green algae by biogenic co-precipitation with barite. Chem Sus Chem 4:470–473

26. Lovett RA (2011) Algae holds promise for nuclear cleanup. Nat News. https://doi.org/10.1038/news.2011.195
27. Rivasseau C, Farhi E, Atteia A, Couté A, Gromova M, de Gouvion Saint Cyr D, Boisson AM, Féret AS, Bligny R (2013a) An extremely radioresistant green eukaryote for radionuclide bio-decontamination in the nuclear industry. Energy Environ Sci 6:1230–1239
28. Rivasseau C, Farhi E, Couté A, Atteia A (2013b) A Novel radioresistant alga of the *Coccomyxa* genus. US patent, US2013/0078707 A1
29. Lisa A Sr (1998) Radon-222 in the Ground Water of Chester County, Pennsylvania. U.S. Department of the Interior, U.S. Geological Survey, Water-Resources Investigations Report 98:4169
30. RSC Periodic Table. http://www.rsc.org/periodic-table/element/86/radon
31. Clever HL (1979) IUPAC solubility data series, vol 2. Xenon and Radon, Pergamon Press, Oxford, England, Krypton
32. Alabdulaaly AI, Maghrawy HB (2010) Radon emanation from radium specific adsorbents. Water Res 44:177–184
33. Lenka P, Sahoo SK, Mohapatra S, Patra AC, Dubey JS, Vidyasagar D, Tripathi RM, Puranik VD (2013) Ingestion dose from ^{238}U, ^{232}Th, ^{226}Ra, ^{40}K and ^{137}Cs in cereals, pulses and drinking water to adult population in a high background radiation area, Odisha, India. Radiat Prot Dosim 153:328–333
34. Hopke PK, Borak TB, Doull J, Cleaver JE, Eckerman KF et al (2000) Health risks due to radon in drinking water. Environ Sci Technol 34:921–926
35. Mir FA, Rather SA (2015) Measurement of radioactive nuclides present in soil samples of district Ganderbal of Kashmir Province for radiation safety purposes. J Rad Res Appl Sci 8:155–159
36. Shivakumara BC, Chandrashekara MS, Kavitha E, Paramesh L (2014) Studies on 226Ra and 222Rn concentration in drinking water of Mandya region, Karnataka state, India. J Rad Res Appl Sci 7:491–498
37. Singh P, Singh P, Singh S, Sahoo BK, Sapra BK, Bajwa BS (2015) A study of indoor radon, thoron and their progeny measurement in Tosham region Haryana, India. J Rad Res Appl Sci 8:226–233
38. Kumar A, Kaur M, Sharma S, Mehra R (2016) A study of radon concentration in drinking water samples of Amritsar city of Punjab (India). Radiat Prot Environ 39:13–19
39. Bonotto DM (2014) 222Rn, 220Rn and other dissolved gases in mineral waters of Southeast Brazil. J Environ Radioact 132:21–30
40. Rangaswamy DR, Srinivasa E, Srilatha MC, Sannappa J (2016) Measurement of radon concentration in drinking water of Shimoga district, Karnataka, India. J Radioanal Nucl Chem 307:907–916
41. Onishchenko A, Zhukovsky M, Veselinovic N, Zunic ZS (2010) Radium-226 concentration in spring water sampled in high radon regions. Appl Radiat Isot 68:825–827
42. Jaseem QKA, Almasoud FI, Ababneh AM, Hobaib ASA (2016) Radiological assessment of water treatment processes in a water treatment plant in Saudi Arabia: water and sludge radium content, radon air concentrations and dose rates. Sci Total Environ 563–564:1030–1036
43. O'Connell JK (1996) Treatment strategies for radon in public water supplies. Water Engineering and Management 143:26–29
44. Alabdulaaly AI, Maghrawy HB (1999) Radon removal from water supplies by diffused bubble aeration system. J Radioanal Nucl Chem 241:3–9
45. Nancy K, Carol L, Gretchen S (1988) Low cost, low technology aeration of radon in drinking water. Pollut Eng 20:52–54
46. Karunakara N, Kumara KS, Yashodhara I, Sahoo BK, Gaware JJ, Sapra BK, Mayya YS (2015) Evaluation of radon adsorption characteristics of a coconut shell-based activated charcoal system for radon and thoron removal applications. J Environ Radioact 142:87–95
47. Lowry JD, Brandow JE (1985) Removal of radon from water supplies. J Environ Eng 111:511–527

48. Tomaszewska B, Bodzek M (2013) The removal of radionuclides during desalination of geothermal waters containing boron using the BWRO system. Desalination 309:284–290
49. Huikuri P, Salonen L, Raff O (1998) Removal of natural radionuclides from drinking water by point of entry reverse osmosis. Desalination 119:235–239
50. Rivasseau C, Farhi E, Compagnon E et al (2016) *Coccomyxa actinabiotis* sp. nov. (Trebouxiophyceae, Chlorophyta), a new green microalga living in the spent fuel cooling pool of a nuclear reactor. J Phycol 52:689–703
51. Jha VN, Tripathi RM, Sethy NK, Sahoo SK, Shukla AK, Puranik VD (2010) Bioaccumulation of 226Ra by plants growing in fresh water ecosystem around the uranium industry at Jaduguda, India. J Environ Radioact 101:717–722
52. Zalewska T, Saniewski M (2011) Bioaccumulation of gamma emitting radionuclides in red algae from the Baltic Sea under laboratory conditions. Oceanologia 53:631–650
53. Masri MSA, Mamish S, Budier Y (2003) Radionuclides and trace metals in eastern Mediterranean Sea algae. J Environ Radioact 67:157–168

Shift in Structural and Functional Diversity of Algal Community: An Ecophysiological Reason

7

Enketeswara Subudhi, Mahendra Gaur, Rajesh Kumar Sahoo, and Mohit Kumar

Abstract

Cyanobacteria are the most ancient lineages of the domain Bacteria and have been playing a crucial role in shaping our planet through their highly proliferating nature in harsh environmental conditions because of their adaptability to grow along with other photosynthetic and heterotrophic microbial community with varied ranges of salinity, pH, temperature, radiation, and water potential. Rise in temperature is reported to be the deciding factor in bringing down the microbial community diversity of hot springs. In the present study, for the first time, we reported the current status of the variability in community structure and predicted metabolic activity among cyanobacteria population of two sulfur hot springs, Atri at 48 °C and Taptapani at 58 °C, from the state of Odisha, Eastern India, using metagenomic approach. We further tried to establish the relationship between the differential occurrences of cyanobacteria clades with those of coexisting non-cyanobacteria clades chloroflexi from our previously published findings of hot spring microbial diversity analysis.

Predominance of thermophilic *Leptolyngbya* (96.25%) in Atri and prevalence of mesophilic *Arthronema* (83.81%) in Taptapani, as discovered through 16S rRNA amplicon sequencing of their community DNA, as a function of temperature, are the interesting features of the present study. Such differential presence of cyanobacteria community in these two hot springs can be correlated with unequal existence of some non-cyanobacteria members' chloroflexi, as well as with possible influence of physiochemical parameters, more specifically temperature. Variation in cyanobacteria diversity and composition of these hot

E. Subudhi (✉) · M. Gaur · R. K. Sahoo
Centre for Biotechnology, Siksha 'O' Anusandhan (Deemed to be University), Bhubaneswar, Odisha, India
e-mail: enketeswarasubudhi@soa.ac.in; rajeshkumarsahoo@soa.ac.in

M. Kumar
Shiv Aastha Hospital & Research Center, Jind, Haryana, India

L. B. Sukla et al. (eds.), *The Role of Microalgae in Wastewater Treatment*,
https://doi.org/10.1007/978-981-13-1586-2_7

springs as revealed through sequence analysis were also evinced by respective differences in richness, evenness, and Shannon's diversity indices. The two tropical sulfur-rich hot springs, Taptapani (48 °C) harboring mesophiles and Atri (58 °C) comprising thermophiles, provide an opportunity to understand the eco-physiological reasons behind the differences in structural and functional profile of cyanobacteria community.

7.1 Introduction

Cyanobacteria, the most ancient domain of Bacteria, have been playing a central role in reshaping our earth because of their exceedingly flourishing nature in severe environmental conditions due to their adaptability to multiply unaffectedly in consort with other microbial community.

Cyanobacteria, widespread on this earth in different environments, could raise many queries regarding (1) their adaptability to newer environment for a long time resulting in huge diversity, (2) the mechanism behind possible linking of ecological diversity with their existing physiological diversity, and (3) the elimination of some progenitors during their adaptation to changing environment. Pertinent replies to above queries may help establishing the diversity and pattern of distribution of cyanobacteria. Whitton and Potts [1] proposed that the cyanobacteria provide an efficient platform to validate above hypotheses as these photo-autotrophic bacteria are known to get proven to thrive in varied terrestrial and aquatic ecological niches with varied ranges of temperature, pH, salinity, water potential, and radiation.

Synechococcus, because of their wide adaptability, are able to invade in several alkaline hot springs of different continents of the world; Africa, Asia, Western North America, as well as Europe [2]. Dvořák et al. [3] established that global elevation in temperature could be linked to the polyphyletic descent of *Synechococcus* from various environmental niches like freshwater, marine water, peat bog, hot springs, etc. Similarly, important role of rise in temperature in reshaping the prolific growth of *Synechococcus* and *Prochlorococcus* was depicted by Flombaum et al. [4]. On the other hand, Callieri et al. [5] established the phylogenetic diversity and global distribution of *Synechococcus* in extreme ecological territories with reference to salinity and oligotrophic conditions from volcanic lakes of Mexico, North Patagonia, and Italy.

Ecophysiological factors such as carbon, sulfur, nitrogen, pyrite, as well as temperature can affect microbial structure building [6, 7]. Reduction in diversity of microbial structure of the community in hot springs due to rise in temperature is reported by Cole et al. [8]. Shift of community from phototropic to chemotrophic was observed with augmentation of temperature by Swingley et al. [9] and De León et al. [10]. From above findings, it can be interpreted that temperature can play an important role in influencing taxonomic diversity and compositions of the community.

With the advent of metagenomics (culture-independent molecular methods), based on the studies of community DNA of the environmental sample, ability to analyze structural and functional complexity and to describe the ecology of the microbial populations is significantly enhanced [11].

The unevenness in cyanobacterial community structure as well as their predicted functional activity of two sulfur hot springs' population – Atri (AT) (longitude, 85°29′ 58.485″E; latitude, 20°13′32.784″N; and elevation, 120 ft. above sea level) at 48 °C and Taptapani (TP) (longitude, 84°24′ 4.6″E; latitude, 19°30′16.8″N; and elevation, 1800 ft. above sea level) at 58 °C of the state of Odisha, Eastern India (Fig. 7.1), having 10 °C difference in temperature using metagenomics approach – is presented for the first time. Metagenomic DNA from the sediments of these hot springs were subjected to analysis through NGS using Illumina platform targeting V3 hyper variable regions of prokaryotic 16S rRNA gene through their amplification and sequencing. We further tried to establish the relationship between the differential occurrences of cyanobacteria clades from two hot springs having variation in temperature with those of coexisting non-cyanobacteria clades chloroflexi from our previous published findings of hot spring microbial diversity analysis [12].

7.2 Materials and Methods

7.2.1 Physicochemical Analysis

Sediment samples were collected from AT and TP hot springs of Odisha (Fig. 7.2a, b). The temperature and pH were measured on-site. Potassium, organic carbon, sulfur, phosphorus, and nitrogen were analyzed by the OUAT, Bhubaneswar [13].

7.2.2 Metagenomic Library Construction and Sequencing

Total genomic DNA from sediment samples of both the hot springs were extracted as described by Sahoo et al. [12]. The primers 341F, 5′-CCTACGGGAGGCAGCAG-3′ and 518R, 5′-ATTACCGCGGCTGCTGG-3′ were used to amplify V3 region of the 16S rRNA using metagenomic DNA [14]. PCR parameters were set as described by Sahoo et al. [12]. The prepared library was sequenced by Illumina GAIIx sequencer at Genotypic Technology Pvt. Ltd (India).

7.2.3 Taxonomic Abundance

All the sequences of both the samples were analyzed using the QIIME v1.9.1-dev [15], and quality control was done by FastQC [16]. QC passed sequences were annotated against Greengenes v13.8 [17] databases. The reference OTUs were picked by UCLUST [18] and taxonomy assignment was done by Mothur [19].

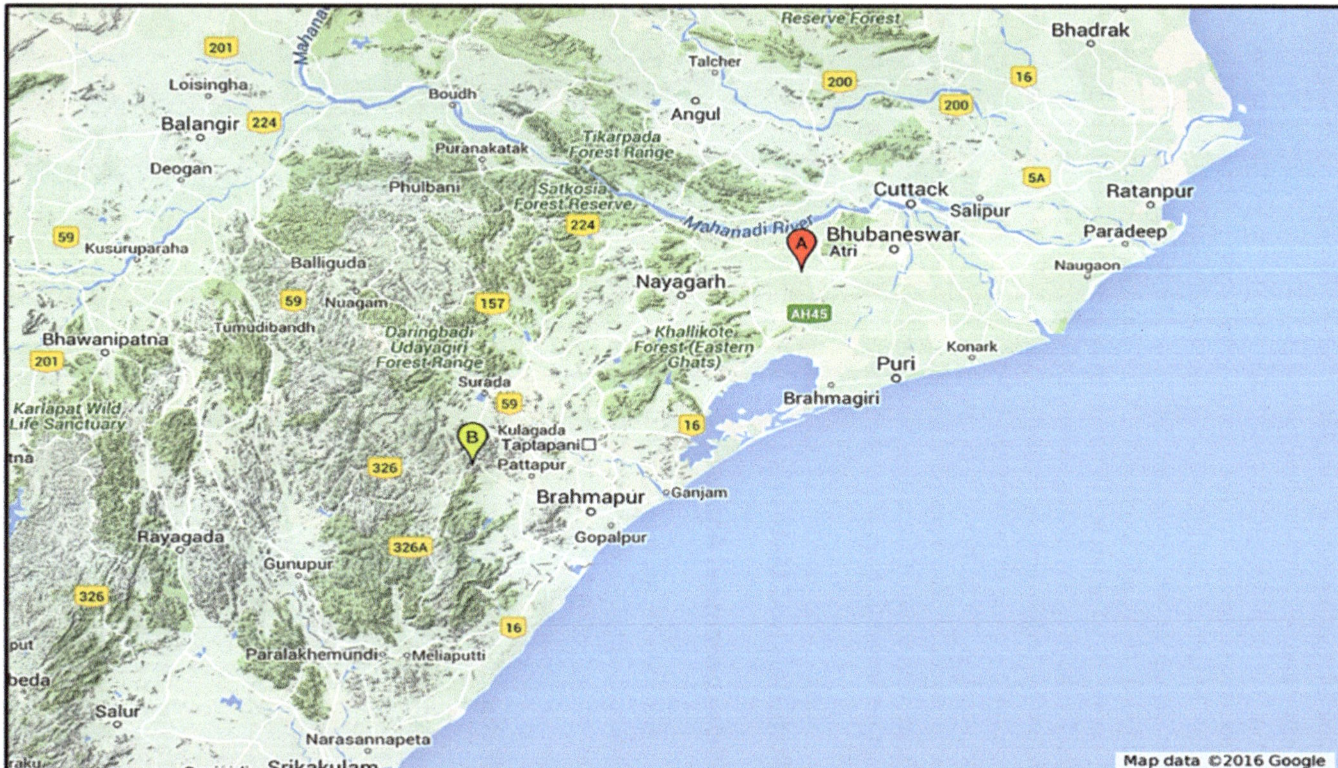

Fig. 7.1 The geographical location map showing AT (red marker) and TP (yellow marker) from which mats were collected. The map was constructed using Google Map Maker (http://www.google.co.in/mapmaker/pulse)

Fig. 7.2 (**a**) Atri hot spring. (**b**) Taptapani hot spring

7.2.4 Predicted Functional Activity

The predicted functional analysis was done by PICRUSt [20] and Tax4Fun [21] to explore the KEGG orthologues (KOs), Clusters of Orthologous Groups of proteins (COGs), and the RNA family (Rfam) in the hot spring metagenome based on the 16S rRNA sequencing data represented in Greengenes database v13.5.

7.2.5 Statistical Analysis

The nonparametric indices, species richness (*S*), Shannon's diversity index (*H′*), Simpson's diversity index (*D′*), Simpson's dominance index (*D*), and Simpson's evenness index (*E*), were calculated from taxonomic abundance for each clone library using the statistical tool PAST v3.02 [22]. With the help of STAMP v2.1.3 [23], predicted functional abundances from PICRUSt and Tax4Fun were compared for statistical significance and biological meaningful differences between each clone library by employing White's nonparametric t-test [24] with a Benjamin-Hochberg FDR multiple test correction [25].

7.3 Results and Discussion

7.3.1 Eco-physiochemical Analysis

The two hot springs are situated at varying altitudes, 120 and 1800 ft. above the Bay of Bengal, respectively (Fig. 7.1), with varying temperatures. A difference of 10 °C in water temperature between two hot springs TP (48 °C) and AT (58 °C) was recorded, but no difference in the pH was found in respective samples ($pH_{TP} = 8.76$ and $pH_{AT} = 8.56$). Physiochemical parameters of sediment samples of the hot springs were different from each other as given in Table 7.1. TP is rich in sulfur and phosphorus as compared to AT. Total organic carbon of AT sample as well as nitrogen and potassium was higher than those of TP (Table 7.1). The dissimilarity in physiological parameters could be due to variation in elevation, phytogeography topography, and environmental conditions surrounding respective hot springs.

7.3.2 Cyanobacteria Community Structure and Composition

Out of 544,887,837 base pairs obtained for hot springs samples targeting the hyper variable V3 region of 16S rRNA gene using the Illumina platform, 35,063,514 bp and 2,056,775 bp of high-quality sequences for AT and TP samples were obtained as analyzed using QIIME. Cyanobacterial diversity and richness were observed to be different in hot spring between TP and AT as observed from their diversity indices (Table 7.2). TP (79322) revealed higher number of sequence reads as compared to AT (4611) and demonstrated highly diverse cyanobacterial populations (1636 OTUs) as compared to AT 133 (OTUs) (Table 7.2) having 10 °C higher temperature. Inverse relationship of hot spring temperature with that of community richness is supported by the findings of Cole et al. [8].

In the analysis, only high-quality sequences were classified using Mothur (cutoff E-value 1e-5) against Greengenes databases. Around 96.42% and 87.35% of the total cyanobacteria of AT and TP hot spring, respectively, belonged to class *Synechococcophycideae* (Fig. 7.3a). At order level, the cyanobacteria members were found to be predominated with *Pseudanabaenales* (87.35%) followed by *Nostocales* (8.19%) and *Chroococcales* (3.58%) in TP (Fig. 7.3b). On the contrary, a highly disproportionate percentage of order *Pseudanabaenales* (96.3%) was observed in AT. Predominance of *Pseudanabaenaceae* (96.3%) at family level was recorded in AT (Fig. 7.3c), while at genus level, TP was found to be prolific in *Arthronema* (83.81%) but AT was with 96.25% *Leptolyngbya* (Fig. 7.3d).

7.3.3 Shift in Cyanobacterial Genera

The cyanobacteria composition from the level of class down to family followed a significant escalation in members of *Pseudanabaenaceae* family. This swing of percentage composition at genus level could be due to 10 °C higher in temperature in

Table 7.1 Different physiochemical parameters[+] of the collected sediment sample

	C (μS/m)	K_2O^a	N^a	OC^a	$P_2O_5{}^a$	$SO_3{}^a$	pH	T (°C)	DOC
Atri	1100	96.6	44.64	4550	5.77	30.42	8.76	58 ± 5	12-June-14
Taptapani	3200	31.79	33.48	2370	21.63	69.8	8.56	48 ± 5	20-June-14

[+]Sahoo et al. (2016)

C conductivity, K_2O dipotassium oxide, *N* nitrogen, P_2O_5 phosphorus pentoxide, SO_3 sulfur trioxide, *T* temperature, *DOC* date of sample collections

[a]All units are in mg/kg

Table 7.2 Diversity indices[a] and richness between AT and TP

	n	*N*	*S*	*H′*	*D′*	*D*	*E*
Atri	3545 (76.83%)	133	28	0.372	0.1068	0.8932	0.0518
Taptapani	74,991 (94.54%)	1636	245	1.957	0.6966	0.3034	0.0289

[a]Subudhi et al. (2017)

n total number of mapped individuals, *N* total number of OUTs, *S* species richness, *H′* Shannon's diversity index, *D′* Shannon's diversity index, *D* Simpson's dominance, *E* Simpson's evenness

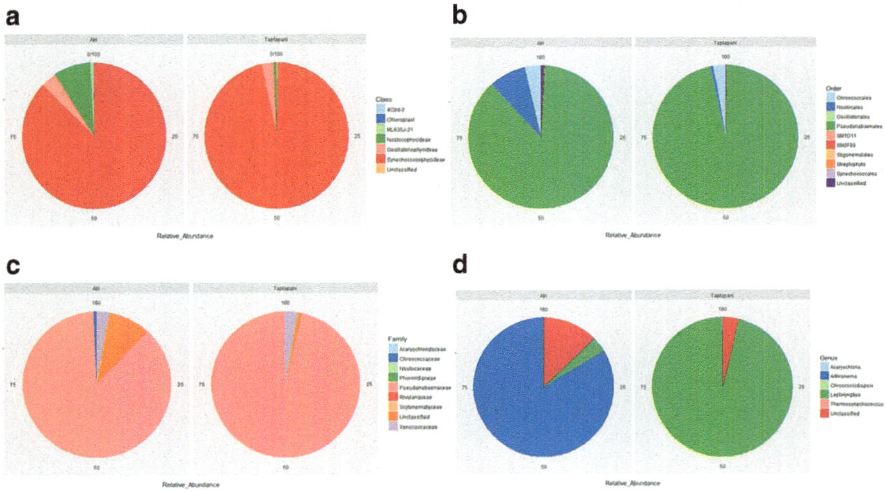

Fig. 7.3 Taxonomic abundance for Atri and Taptapani at (**a**) class, (**b**) order, (**c**) family, and (**d**) genus levels

AT which is elucidated by the mesophilic nature of *Arthronema* sp. [26] and thermophilic nature of *Leptolyngbya* sp. [27]. The hot springs having temperature ranging from 40 °C to 80 °C are chiefly predominated by *Leptolyngbya* as supported by previous sequencing and microscopic study [28–31]. Amarouche-Yala et al. [32] reported worldwide invasion of *Leptolyngbya* sp. in hot springs. They also reported their highest abundance in Algerian hot springs as well as different species such as *L. foveolarum*, *L. laminosa*, and *L. amplivaginata*. *Leptolyngbya* isolated from the Euganean thermal muds exhibited its polyphyletic nature as observed from the clustering of the type species of *Leptolyngbya*, *L. boryana*, and *L. boryana* (ex *L. foveolarum*). It was also observed that only *Leptolyngbya* (ETS-08) and *Spirulina* (ETS-02) were only able to thrive at temperature above 50–55 °C which supports the high proportion of *Leptolyngbya* in AT (58 °C). The phylogenetic relatedness of *Leptolyngbya* strains with those from subaerophytic and geothermal environments has further proven the fact that *Leptolyngbya* are inhabitants of elevated temperature [33].

Amarouche-Yala et al. in 2014 reported the incidence of *Leptolyngbya*, *Synechococcus*, and other cyanobacteria in springs with high temperature varying from 45 to70°C and also suggested *Leptolyngbya* being the most thermophilic. He further observed that the shift in cyanobacteria composition from *Oscillatoriales*, *Stigonematales*, and *Chroococcales* in the springs having temperature 40–55 °C to *Leptolyngbya* dominated hot spring at 70 °C. These findings clearly corroborate our notes on shifting of 83.81% mesophilic *Arthronema* in TP to a thermophilic *Leptolyngbya* dominance in AT (96.25%) as a function of temperature. The shift in community from phototropic dominance to chemotrophic dominance was also documented earlier as an elevation of temperature [9, 10]. Thus, it may be deduced that temperature can influence diversity, structure, richness, and compositions of any community.

7.3.4 Correlation Between Cyanobacteria and Chloroflexi

Degree of difference in cyanobacteria in these two hot springs (AT 96.42%, TP 87.35%) having little difference in other physiological parameters may further be explained by correlating with different amounts of chloroflexi clades (AT 52.39%, TP 7.16%) as also described in our previous report [12]. It was observed that cyanobacteria and chloroflexi were negatively interrelated with 10 °C higher temperature of AT at 58 °C but positively correlated at lower temperature of 48 °C of TP endorsing the existence of complex physiological relationship between these two phyla. However, exactly similar experience was proposed by Wang et al. [34] in their studies on role of temperature in shaping the structure of microbial community in Tibetan plateau hot springs where he found a positive correlation between these two phyla at temperature range 43–55 °C but the reverse in relationship at temperature between 55 and 75 °C and validated the findings from qPCR results. However, role of other physiological parameters cannot be ruled out.

7.3.5 Predicted Functional Analysis

The predicted functional analyses of both the hot springs showed cellular processes, bacterial chemotaxis (3.0817%, 2.7602%), and flagellar assembly (0.33%, 0.21%) as the largest category in the hot springs AT and TP, respectively, as shown in Fig. 7.4. Occurrence of different genes designated with KO accession numbers in different pathways was remarkably high in TP than that of AT. The taxonomic abundance at family level varied in both the hot springs showed absence of phormididaceae completely in AT. Low prevalence of functions (peptidoglycan synthesis, drug metabolism, fatty acid degradation, nitrogen metabolism, amino acid degradation, etc.) in AT and higher activity for cellular processes as compared to TP may be due to10 °C difference in temperature. Differences in the percentage of the functionality of individual subsystem because of difference in the temperature and other physicochemical parameters probably encouraged differential community structure.

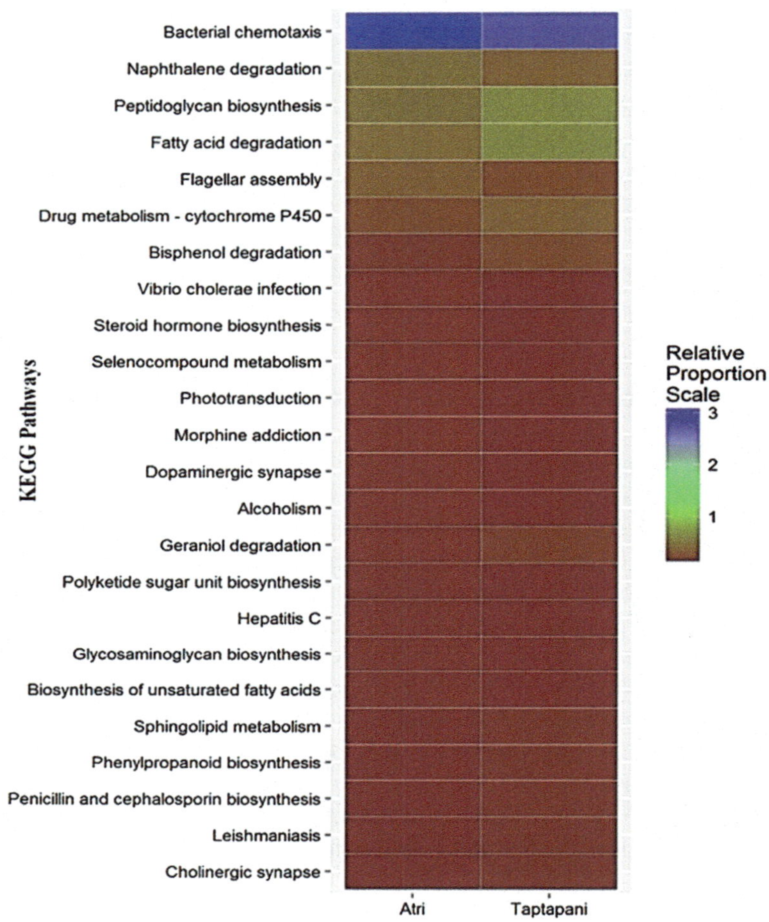

Fig. 7.4 Predicted functional metagenomes at level 3 in both the hot springs

7.4 Conclusion

The present research work discovered the shift in prevalence of *Arthronema* (mesophilic) in TP to *Leptolyngbya* (thermophilic) in AT. Observed community structure variation of both hot springs is further supported by predicted difference in functional profiling in respective environments due to the difference in temperature, altitude, and other physicochemical parameters. Such variation in diversity and composition were also evinced by differences in richness, evenness, and Shannon's diversity indices. However, performance meta-transcriptome profiling probably can throw more light in establishing the correlation between variation in ecophysiology of the hot spring and structural as well as functional cyanobacterial biodiversity.

Acknowledgments We are thankful to our president, Siksha 'O' Anusandhan (Deemed to be University), SPS for their encouragement throughout in carrying out this work.

References

1. Whitton BA, Potts M (2000) The ecology of cyanobacteria: their diversity in time and space, 1st edn. Kluwer Academic Publishers, Dordrecht
2. Castenholz RW (1996) Endemism and biodiversity of thermophilic cyanobacteria. Nova Hedwigia 112:33–47
3. Dvořák P, Casamatta DA, Poulíčková A et al (2014) Synechococcus: 3 billion years of global dominance. Mol Ecol 23:5538–5551. https://doi.org/10.1111/mec.12948
4. Flombaum P, Gallegos JL, Gordillo RA et al (2013) Present and future global distributions of the marine Cyanobacteria Prochlorococcus and Synechococcus. Proc Natl Acad Sci U S A 110:9824–9829. https://doi.org/10.1073/pnas.1307701110
5. Callieri C, Coci M, Corno G et al (2013) Phylogenetic diversity of nonmarine picocyanobacteria. FEMS Microbiol Ecol 85:293–301. https://doi.org/10.1111/1574-6941.12118
6. Hou W, Wang S, Dong H et al (2013) A comprehensive census of microbial diversity in hot springs of Tengchong, Yunnan Province China using 16S rRNA gene pyrosequencing. PLoS One 8:e53350. https://doi.org/10.1371/journal.pone.0053350
7. Huang Q, Jiang H, Briggs BR et al (2013) Archaeal and bacterial diversity in acidic to circumneutral hot springs in the Philippines. FEMS Microbiol Ecol 85:452–464. https://doi.org/10.1111/1574-6941.12134
8. Cole JK, Peacock JP, Dodsworth JA et al (2013) Sediment microbial communities in great boiling spring are controlled by temperature and distinct from water communities. Isme J 7:718–729. https://doi.org/10.1038/ismej.2012.157
9. Swingley WD, Meyer-Dombard DR, Shock EL et al (2012) Coordinating environmental genomics and geochemistry reveals metabolic transitions in a hot spring ecosystem. PLoS One. https://doi.org/10.1371/journal.pone.0038108
10. De León KB, Gerlach R, Peyton BM, Fields MW (2013) Archaeal and bacterial communities in three alkaline hot springs in heart lake geyser basin, yellowstone national park. Front Microbiol. https://doi.org/10.3389/fmicb.2013.00330
11. Bennett AF, Lenski RE (1993) Evolutionary adaptation to temperature II. Thermal niches of experimental lines of Escherichia coli. Evolution (N Y) 47:1–12. https://doi.org/10.2307/2410113
12. Sahoo RK, Gaur M, Das A et al (2016) Comparative analysis of 16S rRNA gene Illumina sequence for microbial community structure in diverse unexplored hot springs of Odisha, India. Geomicrobiol J 0:00–00. https://doi.org/10.1080/01490451.2016.1238980
13. Jackson ML (1962) Soil chemical analysis, 2nd edn. Prentice Hall of Englewood cliffs, Englewood Cliffs
14. Bartram AK, Lynch MDJ, Stearns JC et al (2011) Generation of multimillion-sequence 16S rRNA gene libraries from complex microbial communities by assembling paired-end Illumina reads. Appl Environ Microbiol 77:3846–3852. https://doi.org/10.1128/AEM.02772-10
15. Caporaso J, Kuczynski J, Stombaugh J et al (2012) QIIME allows analysis of high-throughput community sequencing data. Nat Methods 7:335–336. https://doi.org/10.1038/nmeth.f.303
16. Andrews S (2010) FastQC – a quality control tool for high throughput sequence data. http://www.bioinformatics.babraham.ac.uk/projects/fastqc/
17. DeSantis TZ, Hugenholtz P, Larsen N et al (2006) Greengenes, a chimera-checked 16S rRNA gene database and workbench compatible with ARB. Appl Environ Microbiol 72:5069–5072. https://doi.org/10.1128/AEM.03006-05
18. Edgar RC (2010) Search and clustering orders of magnitude faster than BLAST. Bioinformatics 26:2460–2461. https://doi.org/10.1093/bioinformatics/btq461

19. Schloss PD, Westcott SL, Ryabin T et al (2009) Introducing mothur: open-source, platform-independent, community-supported software for describing and comparing microbial communities. Appl Environ Microbiol 75:7537–7541. https://doi.org/10.1128/AEM.01541-09
20. Langille MGI, Zaneveld J, Caporaso JG et al (2013) Predictive functional profiling of microbial communities using 16S rRNA marker gene sequences. Nat Biotechnol 31:814–821. https://doi.org/10.1038/nbt.2676
21. Aβhauer KP, Wemheuer B, Daniel R, Meinicke P (2015) Tax4Fun: predicting functional profiles from metagenomic 16S rRNA data. Bioinformatics 31:2882–2884. https://doi.org/10.1093/bioinformatics/btv287
22. Hammer Ø, Harper DAT, Ryan PD (2001) Paleontological statistics software package for education and data analysis. Palaeontol Electron 4:9–18. https://doi.org/10.1016/j.bcp.2008.05.025
23. Parks DH, Beiko RG (2010) Identifying biologically relevant differences between metagenomic communities. Bioinformatics 26:715–721. https://doi.org/10.1093/bioinformatics/btq041
24. Berger SA, Krompass D, Stamatakis A et al (2012) Statistical methods for detecting differentially abundant features in clinical metagenomic samples. Bioinformatics 27:291–302. https://doi.org/10.1371/journal.pcbi.1000352
25. Benjamini Y, Hochberg Y (1995) Controlling the false discovery rate: a practical and powerful approach to multiple testing. J R Stat Soc 57:289–300
26. Chaneva G, Furnadzhieva S, Minkova K, Lukavsky J (2007) Effect of light and temperature on the cyanobacterium Arthronema africanum – a prospective phycobiliprotein-producing strain. J Appl Phycol 19:537–544. https://doi.org/10.1007/s10811-007-9167-6
27. Verma V, Bhattacharjee MB (2015) Growth and nitrate metabolism in the Thermohalophillic Cyanobacterium Leptolyngbya. J Algal Biomass Util 6:68–75
28. Bohunická M, Johansen JR, Fučíková K (2011) Tapinothrix clintonii sp. nov. (Pseudanabaenaceae, Cyanobacteria), a new species at the nexus of five genera. Fottea 11:127–140. https://doi.org/10.5507/fot.2011.013
29. Dadheech PK, Glöckner G, Casper P et al (2013) Cyanobacterial diversity in the hot spring, pelagic and benthic habitats of a tropical soda lake. FEMS Microbiol Ecol 85:389–401. https://doi.org/10.1111/1574-6941.12128
30. Johansen JR, Olsen CE, Lowe RL et al (2008) Leptolyngbya species from selected seep walls in the Great Smoky Mountains National Park. Arch Hydrobiol Suppl Algol Stud 126:21–36. https://doi.org/10.1127/1864-1318/2008/0126-0021
31. Taton A, Grubisic S, Ertz D et al (2006) Polyphasic study of antarctic cyanobacterial strains. J Phycol 42:1257–1270. https://doi.org/10.1111/j.1529-8817.2006.00278.x
32. Amarouche-Yala S, Benouadah A, El Ouahab BA, López-García P (2014) Morphological and phylogenetic diversity of thermophilic cyanobacteria in Algerian hot springs. Extremophiles 18:1035–1047. https://doi.org/10.1007/s00792-014-0680-7
33. Bruno L, Billi D, Bellezza S, Albertano P (2009) Cytomorphological and genetic characterization of troglobitic leptolyngbya strains isolated from roman hypogea. Appl Environ Microbiol 75:608–617. https://doi.org/10.1128/AEM.01183-08
34. Wang S, Hou W, Dong H et al (2013) Control of temperature on microbial community structure in Hot Springs of the Tibetan Plateau. PLoS One. https://doi.org/10.1371/journal.pone.0062901

Microalgae: An Untapped Resource for Natural Antimicrobials

8

Jayanti Jena and Enketeswara Subudhi

Abstract

Numerous biochemical compounds are synthesized by algae in a wide variety of ecosystems. To date, more than 18,000 new bioactive compounds have been isolated from marine algae; most are still uncharacterized. Therefore, the identification of novel prospective antimicrobials from microalgae presents a unique opportunity. A number of investigations have explored the therapeutic potential of algal extracts and extracellular compounds from a wide range of microalgae; they have confirmed antibacterial, antiprotozoal, antiviral, antifungal, and antiplasmodial activity. Chemical groups such as phenols, fatty acids, indoles, terpenes, acetogenins, and some volatile halogenated hydrocarbons derived from microalgae have shown antimicrobial activity. For example, supercritical extracts of the microalgal *Chaetoceros muelleri* have shown antimicrobial activity due to its lipid composition. Many algal species are also effective against a range of bacteria. For example, *Pithophora oedogonium* targets *Salmonella and Staphylococcus* spp. The algae *Rivularia bullata, Nostoc spongiaeforme, Codium fragile, Colpomenia peregrina Sauvageau, Cystoseira barbata,* and *Zanardiniatypus* are active against many Gram-negative and Gram-positive bacteria.

Multidrug-resistant bacteria pose an increasing challenge to global health, with the future efficacy of antimicrobial drugs being uncertain. Most antimicrobial agents that are successfully used in clinical practice have drawbacks such as toxicity, lack of efficacy, and high costs; furthermore, their frequent use can

J. Jena
Department of Microbiology, IMS & SUM Hospital, Siksha 'O' Anusandhan (Deemed to be University), Bhubaneswar, Odisha, India

E. Subudhi (✉)
Centre for Biotechnology, Siksha 'O' Anusandhan (Deemed to be University), Bhubaneswar, Odisha, India
e-mail: enketeswarasubudhi@soa.ac.in

L. B. Sukla et al. (eds.), *The Role of Microalgae in Wastewater Treatment*,
https://doi.org/10.1007/978-981-13-1586-2_8

result in the emergence of resistant strains of bacteria. Therefore, the development of alternative biodegradable compounds from natural sources with limited side effects is urgently needed. To date, the commercial applications of microalgae-derived compounds has not received as much attention as the fields of antibiotics production, pharmaceuticals, and supplementary biologically active compounds. However, microalgae are destined to become an important raw material for the efficient production of amino acids, vitamins, and other pharmaceuticals. The cultivation of microalgae may provide detailed insights on their practical applications and biotechnological characteristics, which may help researchers develop compounds of interest for their biomedical potential.

8.1 Introduction

The current healthcare system is experiencing a number of clinical problems related to organ transplantations, complicated surgeries, medical device implantation, and chemotherapy. Patients who have undergone these procedures are immunocompromised and thus more susceptible to infections. Furthermore, the global spread of multidrug-resistant bacteria and lack of new antibiotics under development limits the treatment options available to clinicians [81].

The discovery and development of antibiotics are among the most important advances in modern medicine for the life-saving treatment of infectious diseases. However, these “miracle drugs” have lost their efficacy with the appearance of multidrug resistance. Higher rates of morbidity and mortality occur when infectious diseases are caused by multidrug-resistant organisms. In addition, the treatment of these infections is very expensive and requires prolonged hospital stays. This situation is a global epidemiological and public health crisis [13] that is spreading through poor sanitation, person-to-person contact, international travel, and the food chain [91].

The World Health Organization considers multidrug-resistant bacteria to be a major public health concern [103]. Pathogenic bacteria that are resistant to various antimicrobial compounds have been increasing in evolution, prevalence, and distribution. The rapid dissemination of antibiotic-resistant genes through mobile genetic elements, such as plasmids and transposons, has resulted in the emergence of multidrug-resistant strains of many clinically important organisms. Obviously, this situation creates difficulties for clinicians with regard to therapeutic options [37, 62].

8.2 Alternative Sources for Antimicrobial Agents

Bacterial resistance to existing antibiotics, which are mostly derived from bacterial origins, has been increasing rapidly. Thus, there is a need to develop novel efficient compounds using different technologies, including synthetic and semi-synthetic antibiotics [28]. However, the frequently increasing rate of resistance to these antimicrobial compounds, in addition to the paucity of newer drugs, means that continuous investigation is required to find novel molecules and metabolic targets. One promising avenue is the investigation of natural compounds, particularly those from unexploited sources [93]. These alternative antimicrobial agents from natural sources are expected to have minimal side effects, in addition to being environmentally friendly and biodegradable. Researchers are examining bioactive compounds from algae and microalgae as a potential source. A number of functional compounds have been isolated from microalgae. They have the ability to produce a broad range of biologically active compounds, including those with antibacterial, antifungal, enzyme-inhibiting, antiviral, cytotoxic, antiplasmodial, and immunostimulating activities [52].

Microalgae are a rich source of widely distributed bioactive compounds with commercial importance [106]. Microalgal bioactive compounds can be synthesized from secondary metabolism or directly from primary metabolism. These compounds include proteins, vitamins, fatty acids, and pigments with various antimicrobial properties, such as antibiotic, antifungal, antiviral, anticancer, antiprotozoan, antialgal ,and antienzymatic activities [105]. Compounds such as B12, β-carotene, oleic acid, cyanovirin, palmitoleic acid, vitamin E, phycocyanin, linolenic acid, lutein, and zeaxanthin have antimicrobial, antioxidant, and anti-inflammatory properties for the reduction and prevention of diseases [36, 46, 64, 98]. In most microalgae, the bioactive compounds are accumulated in the biomass. In some cases, the metabolites are excreted into the medium; these are known as exometabolites. Bioactive metabolites of microalgal origin are of special interest in the development of new products for the medical, pharmaceutical, cosmetic, and food industries. Further research should be conducted with these bioactive compounds to verify their beneficial effects for humans, their degradability when released into the environment, and their effects when used in animals [106].

8.3 Algae

Algae are simple plants containing chlorophyll for photosynthesis. They may be single- or multi-cellular organisms; they may also exist in colonies, sometimes working together as simple tissues [10]. Algae range from unicellular organisms of 3–10 μm in size to 30-m-long giant kelp [43]. They are found ubiquitously on Earth, including in rivers, lakes, seas, and soils, as well as on walls, plants, and animals. Algae can be divided into two major groups: 1) macroalgae (seaweeds), including green algae, red algae, and brown algae; and 2) microalgae, which are described in the next section [31].

8.4 Microalgae

Microalgae are unicellular organisms consisting of both prokaryotes and eukaryotes. They grow in fresh or salt water and have varied shapes, with a diameter or length of approximately 3–10 μm. Cyanobacteria have very similar structural characteristics to bacteria, but they also contain the chlorophyll *a* required for photosynthesis. Microalgae are distributed all over the biosphere and are responsible for more than 40% of global photosynthesis [20].

Microalgae play a vital role in aquatic ecosystems as the basis of the food chain. They uptake H_2O and CO_2. With the help of solar energy, they synthesize organic compounds, which are then accumulated or secreted as primary or secondary metabolites. Microalgae have the ability to survive under many environmental stress conditions, including salinity, drought, osmotic pressure, photo-oxidation, heat, cold, and ultraviolet exposure [101]. Due to this ability, they can be found in diverse environments, such as fresh water, extreme salinity, blackish water, desert sands, and moist soil. Microalgae have an extra advantage of significant metabolic plasticity, which is dependent on their physiological state (i.e., stressed vs. nonstressed conditions). Therefore, their secondary metabolism can be easily triggered by applying external stress [34].

Until the 1950s, microalgae were not studied for therapeutic purposes. More recently, extensive research efforts have been directed toward microalgae to find novel compounds that might lead to therapeutically useful agents [16, 66, 67]. Microalgae are being investigated as possible antiviral agents [11] to treat infectious diseases caused by previously unexposed viruses that have re-emerged in recent years. A number of algal extracts and extracellular products have proven antifungal, antibacterial, antiprotozoal, antiviral, and antiplasmodial activity [33, 41, 42, 55, 75], as described in the following sections.

8.5 Antimicrobial Activity of Microalgae

The antimicrobial activity of microalgae has been recognized in compounds belonging to several chemical classes, including terpenes, indoles, acetogenins, phenols, volatile halogenated hydrocarbons, and fatty acids [16, 66]. Numerous pressurized extracts from *Dunaliella salina* have shown antimicrobial activity, with the presence of several fatty acids and compounds such as β-cyclocitral, α- and β-ionone, phytol, and neophytadiene [41, 42].

Microalgae are a natural source of highly interesting biologically active compounds. These compounds have received much attention from researchers and manufacturers in recent years due to their potential applications in different life science fields, including as biomass for food/feed and as bioactive compounds for the medical and pharmaceutical industries [36]. Microalgae are promising sources for novel products because of their great biodiversity and recent developments in genetic engineering [46]. The extraction of bioactive compounds has been investigated in a variety of microalgae, including *Botryococcus braunii, Arthrospira (Spirulina),*

Dunaliella salina, Chlorella vulgaris, Haematococcus pluvialis, and *Nostoc* [68, 72, 76], as described in the following sections.

8.5.1 Spirulina

Spirulina (*Arthrospira*) is prokaryotic cyanobacteria that belongs to Cyanophyta. It arose more than 3 million years ago, forming the current oxygen atmosphere, and has been important in the regulation of the terrestrial biosphere [87]. *Spirulina* is the richest source of proteins, containing approximately 60–70% protein [48].

Calcium spirulan (Ca-SP), a novel sulfated polysaccharide extracted with hot water from *Spirulina platensis*, has shown antiviral activity against herpes simplex virus (HSV) type 1, measles virus, human immunodeficiency virus (HIV) 1, and influenza virus [38]. Both extracellular and intracellular spirulan-like molecules from the polysaccharide fractions of *S. platensis* displayed significant antiviral activities against wide range of viruses, including human cytomegalovirus and HIV-1 [1]. Methanolic and aqueous extracts from *S. platensis* reduced HIV-1 viral loads by approximately 50% and 23%, respectively [4]. *Spirulina platensis* and *Spirulina maxima* also demonstrated antiviral activity against HSV-1 and HSV-2, respectively [25, 40].

In an animal study, suspensions of *Escherichia coli* or *Staphylococcus aureus* were injected into 3-week-old chickens; *Spirulina* (0.1%) enhanced the chicken's bacterial clearance abilities by improving the activities of different phagocytotic cells, such as thrombocytes, macrophages, heterophils, and monocytes [85]. In another study, cultures of *S. platensis* displayed antibacterial activity against six *Vibrio* strains: *Vibrio anguillarum, Vibrio parahaemolyticus, Vibrio scophthalmi, Vibrio alginolyticus, Vibrio splendidus* and *Vibrio lentus* [57]. Phycobiliproteins extracted from *Spirulina fusiformis* showed significant antibacterial activity against *Streptococcus pyogenes* and *S. aureus* [70]. Furthermore, the antibacterial activities of purified C-phycocyanin from *S. platensis* clearly inhibited the growth of some multidrug-resistant bacteria, such as *Klebsiella pneumoniae, E. coli, Pseudomonas aeruginosa, and S. aureus* [89].

Spirulina has also exhibited antifungal activity [22]. A butanol extract of *Spirulina* sp. was reported to have activity of 13 mm against *Candida glabrata* [97]. Balb/C mice infected with candidiasis showed a stimulatory effect when *S. platensis* extract was tested [99]. In another study, the antifungal activity of the methanolic extract of *S. platensis* was tested against *Aspergillus flavus*; the reduction of glucosamine production was reported to be nearly 56% [69].

8.5.2 Nostoc

Microalgal biomasses of *Nostoc* have been used in the medical field and as dietary supplements because of their protein, vitamin, and fatty acid content. *Nostoc* contains a spectrum of polyunsaturated fatty acids that include essential fatty acids, such as linoleic, α-linolenic, γ-linolenic, octadecatetraenoic, and eicosapentaenoic acids [108]. Essential fatty acids are precursors of prostaglandins, thus engendering significant interest from the pharmaceutical industry. The medical value of these microalgae has been demonstrated by their use in the treatment of fistulas and some forms of cancer [102].

Nostoc sp. is reported to have a number of secondary metabolites, including antimicrobial compounds. For example, tenuecyclamide a-d was found from *Nostoc spongiaeforme* [111], and noscomin and coniston a-e were found from *Nostoc commune* [50]. The diverse polysaccharides in *N. commune* have been shown to possess antibacterial activity along with antitumor, antiviral, and anti-inflammatory effects [92]. Nostocyclyne A is another antimicrobial compound that has been isolated from *Nostoc* sp. [80]. Cyanovirin, a potential protein molecule produced by a *Nostoc* microalga, showed positive effects in the treatment of HIV and influenza A (H1N1) [98].

8.5.3 Chlorella

Chlorella was discovered by the Japanese, who are the traditional consumers of algae and use it as a food supplement. The microalga *Chlorella* is rich in chlorophyll, vitamins, proteins, minerals, polysaccharides, and essential amino acids. This microalga is 53% (w/w) protein, 23% (w/w) arbohydrate, 9% (w/w) lipids, and 5% (w/w) minerals and oligoelements [49].

Pratt et al. first isolated microalgal active compounds from *Chlorella*; in their study, a mixture of fatty acids (chlorellin) was isolated and demonstrated antibacterial activity against both Gram-negative and Gram-positive bacteria in vitro [82]. Interestingly, the authors also described a practical application during World War II derived from a previous experiment. *Chlorella* spp. were heavily inoculated in open sewage from military installations, rendering it bacteriologically safe for discharge into local streams. There was a reduction in the number of coliforms in the areas where *Chlorella* spp. were present compared with the areas where *Chlorella* spp. were absent [83].

8.5.4 Dunaliella

Dunaliella spp. are green, unicellular, halotolerant microalgae that belong to the Chlorophyceae group. These microalgae are extensively studied because of their diverse nature, including physiological aspects, tolerance of extreme habitats, and many biotechnological applications. *Dunaliella* spp. are a rich source of bioactive compounds, such as carotenoids, glycerol, lipids, enzymes, and vitamins [45, 84]. These microalgae are a major source of natural β-carotene; they are able to produce up to 14% of their dry weight under conditions of high salinity, light, and temperature as well as nutrient limitations [29].

Chang et al. reported that *Dunaliella* cells contained antibiotic substances. The crude extract of this microalga strongly inhibited the growth of *Bacillus cereus*, *S. aureus*, *Enterobacter aerogenes* and *Bacillus subtilis* [17]. In another study, *Dunaliella* microalga also showed antibacterial activity against various microorganisms of importance to the food industry, including *E. coli*, *S. aureus*, *Candida albicans*, and *Aspergillus niger* [41, 42, 45].

Minolenic acid extracted from *Dunaliella primolecta Butcher* (C-525) and *Chlorococcum* sp. (HS-101) [73] showed antibacterial activity against methicillin-resistant *S. aureus* (MRSA). Another study investigated extracts of *Dunaliella* spp. isolated from clean and polluted waters. The authors observed that a heat-labile non-proteinous substance produced by species from the polluted water had the ability to inhibit *E. coli*. It was therefore suggested that microalgae from highly competitive environments are more likely to produce compounds with antimicrobial activity [63] (Tables 8.1, 8.2, 8.3, and 8.4).

Table 8.1 Antibacterial activity of some algae species

Bioactive compound/Microalgae	Targeting bacteria	References
Ambiguine I isonitrile/*Fischrella* sp.	*E. coli* ESS K-12, *Staphyloccocus albus*, *Bacillus subtilis*	[86]
Skeletonema costatum	*Vibrio spp.*	[71]
Carbamidocyclophanes/*Nostoc* sp.	*Staphylococcus aureus*	[12]
γ-lactone malyngolide 14/*Lyngbya majuscule*	*Mycobacterium smegmatis* and *Streptococcus pyogenes*	[15]
Norbietane diterpenoid (20-nor-3a-acetoxyabieta-5,7,9,11,13-pentaene)/*Microcoleus lacustris*	*S. aureus*	[35]
Noscomin/*Nostoc commune*	*Bacillus cereus*, *Staphylococcus* *epidermidis*, *Escherichia coli*	[51]
Phenolic compound/*Nostoc muscorum*	*B. subtilis*, *B. cereus*, *E. coli*, *Salmonella typhi*, *S. aureus*	[24]
Cycloeudesmol/*Chondria oppositiclada*	*S. aureus*, *Candida albicans*	[27]
Hapalindole T/*Fischerella* sp.	*S. aureus*, *Pseudomonas* *P. aeruginosa*, *S. typhi*, *E. coli*	[3]
Euglena viridis	*Pseudomonas*, *Aeromonas*, *E. coli*, *Edwarsiella*	[18]
Padina pavonica	*Enterococcus faecalis*, *S. epidermidis*	[21]
Ulva fasciata, *Chaetomorpha aerea*	*Klebsiella pneumonia*, *P. aeruginosa*, *S. aureus*	[90]
Ulva Lactuca	*B. subtilis*, *B. pumilus*	[77]
Cystoseira sp., *Gelidium latifolium*		

Table 8.2 Antiprotozoan activity of some algae species

Bioactive compounds/Microalgae	Targeting protozoans	References
Ascosalipyrrolidinones/*Ascochyta salicorniae*	*Plasmodium falciarum*	[74]
Viridamide A/*Oscillatoria nigro Virdis*	*Trypanosoma cruzi, Leishmania exicana, Plasmodium falciparum*	[95]
Symplocamide A/*Symploca* sp.	*T. cruzi, Leishmania donovani, P. falciparum*	[60]
Venturamides/*Oscillatoria* sp.	*P. falciparum*	[61]
Snyderol sesquiterpene/*Laurencia obtuse*	*Plasmodium falciarum*	[104]
Ambigol C/*Fischerellambigua*	*T. rhodesiense, P. falciparum*	[112]
Amphidinium sp.	*Trichomonas foetus*	[109]
Dinophysis fortii, Prorocentrum lima	*T. foetus*	[9]
n-hexane, dichloromethane/*Bostrychia tenella*	*T. cruzi* trypomastigotes, *Leishmania amazonensis* promastigotes	[19]

Table 8.3 Antiviral activity of some algae species

Bioactive compounds/ Microalgae	Targeting protozoans	References
Spirulan/*Spirulina* sp.	HIV-1 and HIV-2 (inhibits reverse transcriptase) HSV, influenza	[96]
Nostoflan/*Nostoc flagilliforme*	HSV-1 (HF), HSV-2 (UW-268), human cytomegalovirus (Towne), influenza (NWS), adenovirus (type 2), Coxsackie (Conn-5)	[96]
Cyanovirin-N/*Nostoc ellipsosporum*	HIV-1 (interacts with high mannose groups of envelope glycoproteins, gp120 and blocks its interaction with target cell receptors) HIV-2, HSV-6, Mesles virus	[96]
	Simian immunodeficiency virus, feline immunodeficiency virus	
Tribromo 4,5-dihydroxybenzyl methyl ether/*Symphyocladia latiuscula*	Wild-type HSV-l, APr HSV-I, and TK-HSV-l	[79, 78]
Sulfoquinovosyl diacylglycerol/*Ishige okamurai*	HSV-2	[107]
Dollabelladiene 147, 10,18-diacetoxy – 8-hydroxy	HSV-1and HIV-1	[6, 47]
2,6-dollabeladiene 148/*Dictyota pfaffi*		
8,80-bieckol 151 and 8400-bieckol 152/	HIV-1 reverse transcriptase and protease	[30]
Venustatriol 302, thyrsiferol 303 and thyrsiferyl 23-acetate	Vesicular stomatitis	[88]
	Vesicular stomatitis Indiana virus, HSV-l	
304/*Laurencia venusta*		

Table 8.4 Antifungal activity of some algae species

Bioactive compounds/Microalgae	Targeting fungus	References
Ulva lactuca, Cystoseira sp., Gelidium latifolium	*Candida albicans, Microsporum gypseum, Aspergillus niger*	[77]
Padina pavonica	*Candida spp.*	[21]
Chlamydomonas reinhardtii	*A. niger, Aspergillus fumigatus*	[32]
Trentepohlia umbrina	*A. niger, Trichoderma barsianum*	[94]
Amphidinium sp.	*A. niger*	[109]
Dinophysis fortii, Prorocentrum lima	*A. niger*	[9]

8.6 Natural Compounds

A number of chemical functional groups from algae have been reported to be bacterial inhibitors, including polysaccharides, phlorotannins, peptides, fatty acids, terpenes, and halogenated furanones, as described in the following sections.

8.6.1 Polysaccharides

Fucoidan- and laminarin-like algal polysaccharides have shown antibacterial activity against *E. coli* and *S. aureus* and have been used as oral drugs. They also prevent the adhesion of the biofilm forming *Helicobacter pylori* in gastric mucosa [8, 39, 53, 113]. In Ireland, ultrasound-assisted extraction was used to obtain laminarin from the brown seaweeds *Ascophyllum nodosum* and *Laminaria hyperborean*; the laminarin was shown to be a significant growth inhibitor of *E. coli, Listeria monocytogenes, S. aureus,* and *Salmonella typhimurium* [53]. Hot and cold water extraction was used to obtain polysaccharides from the brown seaweed *Dictyopteris membranacea* and red seaweed *Pterocladia capillacea;* these extracts showed antibacterial activity against Gram-negative *Pseudomonas fluorescens* and *E. coli* and Gram-positive bacteria *B. cereus* and *S. aureus* [2].

Spirulan and Ca-spirulan are the most important anticancer polysaccharides isolated from *Spirulina* spp.; they also showed effective and broad-spectrum activity against HIV-1, HIV-2, and influenza viruses. These sulfated polysaccharides inhibit the reverse transcriptase activity of HIV-1 (like azidothymidine) [26]. Another acidic polysaccharide, nostoflan from *Nostoc flagelliforme*, exhibits potent virucidal activity against HSV-1 [56].

8.6.2 Proteins and Peptides

Lectins are a diverse group of proteins that are found in algae, plants, animals, bacteria, and viruses [5]. They have various biological functions in humans, such as blood-protein regulation, carbohydrate binding, cell adhesion, and immune defense [65].

Lectins extracted from the red algae *Solieria filiformis* have demonstrated inhibitory effects against both Gram-negative and Gram-positive pathogenic bacteria [44]. The inhibition of bacterial growth is thought to occur by the binding of lectin with mannan, which is a linear polymer of the saccharide monomer mannose that arises on the cell surface of Gram-negative bacteria. Mannan acts as a hapten upon binding with a large lectin molecule, producing an immune response. However, it does not seem to inhibit the growth of Gram-positive *S. aureus* or *B. subtilis*, probably due to inappropriate lectin-polysaccharide binding sites on the cell surfaces of these species [100].

In another study, enzymatic hydrolysis was used with trypsin-extracted antibacterial peptides (>10 kDa mass) from *Saccharina longicruris*. Food spoilage from *S. aureus* was inhibited at concentrations of 0.31 to 2.5 mg/mL, indicating that the hydrolysate could be used as a potential agent for food preservation [7].

8.6.3 Fatty Acids

Antibacterial fatty acids, including 13-octadeadienoic acid and cyclopentaneacetic acid, have been obtained by ethanol extraction from *Sargassum vulgare* and by diethyl ether extraction from *Sargassum fusiforme*. Morphological variations were observed in *S. aureus* and *K. pneumonia* cells treated with these seaweed extracts. Transmission electron microscopy showed that the cell walls of both organisms were punctured, resulting in cell wall rupture, protoplasm shrinking, cytoplasmic vacuolation, cytoplasmic seepage, chromatin sprinkling, cell size reduction, and outer cell shape alteration [23]. In another study, long-chain fatty acids extracted from the green microalga *Planktochlorella nurekis* demonstrated antibacterial activity against *Campylobacter jejuni*, *E. coli*, *Salmonella enterica*, and *Lactobacillus johnsonii* [14].

8.6.4 Phlorotannins

The antibacterial activity of phlorotannins is reportedly due to the inhibition of oxidative phosphorylation. Phlorotannins could bind with bacterial proteins, such as cell membranes and enzymes, thus triggering bacterial cell lysis. Phloroglucinol compounds caused bacteriolysis of *Vibrio sp.* when tertiary structures, such as methyl- or acetyl-vinyl, were present [54]. Phlorotannins isolated from *Sargassum thunbergii* algae showed activity against *Vibrio parahaemolyticus* by destroying its cell wall and cell membrane, thus causing membrane permeability destruction and cytoplasm leakage [110].

Lee et al. extracted a wide range of solvents from brown seaweed, *Eisenia bicyclis* (Arame) and investigated them against antibiotic-resistant *Propionibacterium*-related acne. The phlorofucofuroeckol compound (phlorotannin with an alcohol substituent) showed the most potent antibacterial activity, including antimicrobial activity against MRSA [59].

8.6.5 Terpenes

A number of terpene compounds isolated from algae, such as diterpene-benzoate bromophycolides, have the ability to inhibit bacterial growth. Lane et al. extracted bromophycolides (diterpene-benzoate macrolides) from the Fijian red alga *Callophycus serratus* with methanol, dichloromethane, and water. The extracts significantly inhibited MRSA and vancomycin-resistant *Enterococcus faecium* [58].

8.7 Conclusion

Antimicrobial drug resistance is a serious concern, with limited or no treatment options for infections that are emerging globally. Antimicrobial agents may be derived from bacteria and fungi or chemically synthesized. However, resistance to these agents has led to the need for alternative natural sources. Algae are a potential alternative source for antimicrobial agents due to their diversity and ubiquitous nature, along with their ability to produce secondary metabolites that exhibit antimicrobial (i.e., antibacterial, antifungal, antiviral, antimalarial, and antiprotozoan) activities. Algae and their synthesized products have an ability to survive and adapt to a wide range of habitats, even when their environmental conditions are altered or stressed.

To date, there has been quite limited research into these microorganisms, and they predominantly remain an "untapped" resource. Thus, there is obviously a need for further study of the compounds described in this chapter for the treatment and prevention of various diseases, as well as an ongoing search for other undiscovered metabolites. Various technologies are available to assist in the systematic identification and purification of these natural products, which—when combined with in vivo experiments—could lead to novel antimicrobial agents.

References

1. Abdo SM, El-Senousy WM, Ali GH et al (2012) Antiviral activity of freshwater algae. J Appl Pharm Sci 2:21–25
2. Abou Zeid AH, Aboutabl EA, Sleem AA et al (2014) Water soluble polysaccharides extracted from *Pterocladia capillacea* and *Dictyopteris membranacea* and their biological activities. Carbohydr Polym 113:62–66
3. Asthana R, Srivastava A, Singh AP et al (2006) Identification of an antimicrobial entity from the cyanobacterium *Fischerella* sp. isolated from bark of *Azadirachta indica* (Neem) tree. J Appl Phycol 18:33–39
4. Ayehunie S, Belay A, Baba TW, Ruprecht RM (1998) Inhibition of HIV-1 replication by an aqueous extract of *Spirulina* platensis (Arthrospira platensis). J Acquir Immune Defic Syndr Hum Retrovirology 18:7–12
5. Bahar AA, Ren D (2013) Antimicrobial peptides. Pharmaceuticals 6:1543–1575
6. Barbosa JP, Pereira RC, Abrantes JL et al (2004) In vitro antiviral diterpenes from the Brazilian brown alga *Dictyota pfaffii*. Planta Med 70:856–860

7. Beaulieu L, Bondu S, Doiron K et al (2015) Characterization of antibacterial activity from protein hydrolysates of the macroalga *Saccharina longicruris* and identification of peptides implied in bioactivity. J Funct Foods 17:685–697
8. Besednova NN, Zaporozhets TS, Somova LM et al (2015) Review: prospects for the use of extracts and polysaccharides from marine algae to prevent and treat the diseases caused by *Helicobacter pylori*. Helicobacter 20:89–97
9. Bhadury P, Wright PC (2004) Exploitation of marine algae: biogenic compounds for potential antifouling applications. Planta 219:561–578
10. Bold HC, Wynne MJ (1985) Introduction to the algae structure and reproduction, 2nd edn. Prentice-Hall Inc, Englewood Cliffs, pp 1–33
11. Borowitzka MA (1995) Microalgae as sources of pharmaceuticals and other biologically active compounds. J Appl Phycol 7:3–15. https://doi.org/10.1007/BF00003544
12. Bui HTN, Jansen R, Pham HTL, Mundt S (2007) Carbamidocyclophanes A-E, chlorinated paracyclophanes with cytotoxic and antibiotic activity from the Vietnamese cyanobacterium *Nostoc* sp. J Nat Prod 70:499–503. https://doi.org/10.1021/np060324m
13. Bush K, Jacoby GA (2010) Updated functional classification of beta-lactamases. Antimicrob Agents Chemother 54:969–976. https://doi.org/10.1128/AAC.01009-09
14. Cˇ ermák L, Pražáková Š, Marounek M et al (2015) Effect of green alga *Planktochlorella nurekis* on selected bacteria revealed antibacterial activity in vitro. Czech J Anim Sci 60:427–435
15. Cardllina JH, Moore RE, Arnold EV, Clardy J (1979) Structure and absolute configuration of malyngolide, an antibiotic from the marine blue-green alga *Lyngbya majuscula* Gomont. J Organomet Chem 44:4039–4042. https://doi.org/10.1021/jo01337a003
16. Cardozo KHM, Guaratini T, Barros MP et al (2007) Metabolites from algae with economical impact. Comp Biochem Physiol C Toxicol Pharmacol 146:60–78. https://doi.org/10.1016/j.cbpc.2006.05.007
17. Chang T, Ohta S, Ikegami N et al (1993) Antibiotic substances produced by a marine green alga, *Dunaliella primolecta*. Bioresour Technol 44:149–153
18. Das BK, Pradhan J, Pattnaik PK et al (2005) Production of antibacterial from the fresh water alga *Euglena viridis* (Ehren). World J Microbial Biotech 21:45–50
19. De Felicio R, Dealbuquerque S, Young MCM et al (2010) Trypanocidal lei-shmanicidal and antifungal potential from marine red alga Bostrychia tenella J Agardh (Rhodomelaceae, Ceramiales). J Pharm Biomed Anal 52:763–769
20. De Morais MG, Vaz BDS, De Morais EG, Costa JAV (2015) Biologically active metabolites synthesized by microalgae. Biomed Res Int 2015:835761. https://doi.org/10.1155/2015/835761
21. Desbois AP, Lebl T, Yan LM et al (2008) Isolation and structural characterisation of two antibacterial free fatty acids from the marine diatom, Phaeodactylum tricornutum. Appl Microbiol Biotechnol 81:755–764
22. Duda-Chodak A (2013) Impact of water extracts of *spirulina* (WES) on bacteria, yeasts and molds. Acta Sci Pol Technol Aliment 12:33–39
23. El Shafay SM, Ali SS, El-Sheekh MM (2016) Antimicrobial activity of some seaweeds species from Red sea, against multidrug resistant bacteria. Egypt J Aquat Res 42:65–74
24. El-Sheekh MM, Osman MEH, Dyab MA, Amer MS (2006) Production and characterization of antimicrobial active substance from the cyanobacterium *Nostoc muscorum*. Environ Toxicol Pharmacol 21:42–50. https://doi.org/10.1016/j.etap.2005.06.006
25. Emad AS, Sanaa MMS, Vikramjit S (2010) Salt stress enhancement of antioxidant and antiviral efficiency of *Spirulina platensis*. J Med Plant Res 4:2622–2632. https://doi.org/10.5897/JMPR09.300
26. Feldmann SC, Reynaldi S, Stortz CA et al (1999) Antiviral properties of fucoidan fractions from *Leathesia difformis*. Phytomedicine 6:335–340
27. Fenical W, Sims JJ (1974) Cycloeudesmol, an antibiotic cyclopropane containing sesquiterpene from the marine alga, chondria oppositiclada dawson. Tetrahedron Lett 15:1137–1140. https://doi.org/10.1016/S0040-4039(01)82427-8
28. Fernandes P (2006) Antibacterial discovery and development – the failure of success? Nat Biotechnol 24:1497–1503. https://doi.org/10.1038/nbt1206-1497

29. Francavilla M, Trotta P, Luque R (2010) Phytosterols from *Dunaliella tertiolecta* and *Dunaliella salina*: a potentially novel industrial application. Bioresour Technol 101:4144–4150. https://doi.org/10.1016/j.biortech.2009.12.139
30. Fukuyama Y, Kodaama M, Miura I et al (1989) Anti-plasmin inhibitor V. Structures of novel dimeric eckols isolated from the brown alga *Ecklonia* kurome Okamura. Chem Pharm Bull 37:2438–2440
31. Garson J (1989) Marine natural products. Nat Prod Rep 6:143–170
32. Ghasemi Y, Moradian A, Mohagheghzadeh A et al (2007) Antifungal and antibacterial activity of the microalgae collected from paddy fields of Iran: characterization of antimicrobial activity of *Chroococcus dispersus*. J Biol Sci 7:904–910. https://doi.org/10.3923/jbs.2007.904.910
33. Ghasemi Y, Yazdi MT, Shafiee A et al (2004) Parsiguine, a novel antimicrobial substance from *Fischerella ambigua*. Pharm Biol 42:318–322. https://doi.org/10.1080/13880200490511918
34. Guedes AC, Amaro HM, Malcata FX (2011) Microalgae as sources of high added-value compounds-a brief review of recent work. Biotechnol Prog 27:597–613. https://doi.org/10.1002/btpr.575
35. Gutierrez RMP, Flores AM, Solis RV et al (2008) Two new antibacterial norbietane diterpenoids from cyanobacterium *Micrococcus lacustris*. J Nat Med 62:328–331
36. Harun R, Singh M, Forde GM, Danquah MK (2010) Bioprocess engineering of microalgae to produce a variety of consumer products. Renew Sust Energ Rev 14:1037–1047. https://doi.org/10.1016/j.rser.2009.11.004
37. Hawkey PM, Jones AM (2009) The changing epidemiology of resistance. J Antimicrob Chemother https://doi.org/10.1093/jac/dkp256
38. Hayashi T, Hayashi K, Maeda M, Kojima I (1996) Calcium spirulan, an inhibitor of enveloped virus replication, from a blue-green alga *Spirulina platensis*. J Nat Prod 59:83–87. https://doi.org/10.1021/np960017o
39. Hernández AJ, Romero A, Gonzalez-Stegmaier R et al (2016) The effects of supplemented diets with a phytopharmaceutical preparation from herbal and macroalgal origin on disease resistance in rainbow trout against Piscirickettsia salmonis. Aquaculture 454:109–117
40. Hernández-Corona A, Nieves I, Meckes M et al (2002) Antiviral activity of *Spirulina maxima* against herpes simplex virus type 2. Antivir Res 56:279–285. https://doi.org/10.1016/S0166-3542(02)00132-8
41. Herrero M, Ibáñez E, Cifuentes A et al (2006a) Dunaliella salina microalga pressurized liquid extracts as potential antimicrobials. J Food Prot 69:2471–2477
42. Herrero M, Jaime L, Martín-Álvarez PJ et al (2006b) Optimization of the extraction of antioxidants from *Dunaliella salina* microalga by pressurized liquids. J Agric Food Chem 54:5597–5603. https://doi.org/10.1021/jf060546q
43. Hillison CI (1977) Seaweeds, a color-coded, illustrated guide to common marine 1977. Plants of east coast of the United States, Keystone Books. The Pennsylvania State University Press, pp 1–5
44. Holanda ML, Melo VMM, Silva LMCM (2005) Differential activity of a lectin from *Solieria filiformis* against human pathogenic bacteria. Braz J Med Biol Res 38:1769–1773
45. Hosseini Tafreshi A, Shariati M (2009) Dunaliella biotechnology: methods and applications. J Appl Microbiol 107:14–35. https://doi.org/10.1111/j.1365-2672.2009.04153.x
46. Ibañez E, Cifuentes A (2013) Benefits of using algae as natural sources of functional ingredients. J Sci Food Agric 93:703–709. https://doi.org/10.1002/jsfa.6023
47. Ireland C, Faulknar DJ (1977) Diterpenes from *Dolabella californica*. J Organomet Chem 42:3157–3162
48. Ishimi Y, Sugiyama F, Ezaki J et al (2006) Effects of *Spirulina*, a blue-green alga, on bone metabolism in ovariectomized rats and hindlimb-unloaded mice. Biosci Biotechnol Biochem 70:363–368. https://doi.org/10.1271/bbb.70.363
49. Costa JAC, Morais MG (2013) Microalgae for food production. In: Soccol CR, Pandey A, Larroche C (eds) Fermentation process engineering in the food industry. Taylor & Francis, Boca Raton, p 486

50. Jaki B, Orjala J, Heilmann J et al (2000) Novel extracellular diterpenoids with biological activity from the cyanobacterium *Nostoc commune*. J Nat Prod 63:339–343
51. Jaki B, Orjala J, Sticher O (1999) A novel extracellular diterpenoid with antibacterial activity from the cyanobacterium Nostoc commune. J Nat Prod 62:502–503. https://doi.org/10.1021/np980444x
52. Jyotirmayee P, Sachidananda D, Das BK (2014) Antibacterial activity of freshwater microalgae: a review. Afr J Pharm Pharmacol 8:809–818. https://doi.org/10.5897/AJPP2013.0002
53. Kadam SU, O'Donnell CP, Rai DK et al (2015) Laminarin from Irish brown seaweeds *Ascophyllum nodosum* and *Laminaria hyperborea*: ultrasound assisted extraction, characterization and bioactivity. Mar Drugs 13:4270–4280
54. Kamei Y, Isnansetyo A (2003) Lysis of methicillin-resistant *Staphylococcus aureus* by 2, 4-diacetylphloroglucinol produced by *Pseudomonas* sp. AMSN isolated from a marine alga. Int J Antimicrob Agents 21:71–74
55. Kellam SJ, Walker JM (1989) Antibacterial activity from marine microalgae in laboratory culture. Br Phycol J 24:191–194
56. Kenji LK, Lee JB, Hayashi K et al (2005) Isolation of an antiviral polysacharide, nostoflan, from a terrestrial cyanobacteium, *Nostoc flagilliforme*. J Nat Prod 68:1037–1041
57. Kokou F, Makridis P, Kentouri M, Divanach P (2012) Antibacterial activity in microalgae cultures. Aquac Res 43:1520–1527. https://doi.org/10.1111/j.1365-2109.2011.02955.x
58. Lane AL, Stout EP, Lin AS et al (2009) Antimalarial bromophycolides J-Q from the Fijian red alga *Callophycus serratus*. J Organomet Chem 74:2736–2742
59. Lee JH, Eom SH, Lee EH et al (2014) In vitro antibacterial and synergistic effect of phlorotannins isolated from edible brown seaweed *Eisenia bicyclis* against acne-related bacteria. Algae 29:47–55
60. Linnington RG, Edwards DJ, Shuman CF et al (2008) Symplocamide A, a potent cytotototoxin and chymotrypsin inhibitor from marine cyanobacterium. Symploca sp. J Nat Prod 71:22–27
61. Linnington RG, Gonzalez J, Urena L et al (2007) Venturamides A and B: antimalarial constituents of the Panamanian marine cyanobacterium *Oscillatoria* sp. J Nat Prod 70:397–401
62. Livermore DM (2009) Has the era of untreatable infections arrived? J Antimicrob Chemother 64:29–36. https://doi.org/10.1093/jac/dkp255
63. Lustigman B (1988) Comparison of antibiotic production from four ecotypes of the marine alga, *Dunaliella*. Bull Environ Contam Toxicol 40:18–22
64. Markou G, Nerantzis E (2013) Microalgae for high-value compounds and biofuels production: a review with focus on cultivation under stress conditions. Biotechnol Adv 31:1532–1542. https://doi.org/10.1016/j.biotechadv.2013.07.011
65. Maverakis E, Kim K, Shimoda M (2015) Glycans in the immune system and the altered glycan theory of autoimmunity: a critical review. J Autoimmun 57:1–13
66. Mayer AMS, Hamann MT (2005) Marine pharmacology in 2001-2002: marine compounds with anthelmintic, antibacterial, anticoagulant, antidiabetic, antifungal, anti-inflammatory, antimalarial, antiplatelet, antiprotozoal, antituberculosis, and antiviral activities; affecting the cardiova. Comp Biochem Physiol Toxicol Pharmacol CBP 140:265–286. https://doi.org/10.1016/j.cca.2005.04.004
67. Mendes RL, Nobre BP, Cardoso MT et al (2003) Supercritical carbon dioxide extraction of compounds with pharmaceutical importance from microalgae. Inorg Chim Acta 356:328–334. https://doi.org/10.1016/S0020-1693(03)00363-3
68. Mendes RL, Reis AD, Palavra AF (2006) Supercritical CO2 extraction of γ-linolenic acid and other lipids from Arthrospira (Spirulina)maxima: comparison with organic solvent extraction. Food Chem 99:57–63. https://doi.org/10.1016/j.foodchem.2005.07.019
69. Moraes de Souza M, Prietto L, Ribeiro AC et al (2011) Assessment of the antifungal activity of *Spirulina platensis* phenolic extract against *Aspergillus flavus*. Ciencia e Agrotecnologia 35:1050–1058
70. Najdenski HM, Gigova LG, Iliev II et al (2013) Antibacterial and antifungal activities of selected microalgae and cyanobacteria. Int J Food Sci Technol 48:1533–1540

71. Naviner M, Berge J-P, Durand P, Le Bris H et al (1999) Antibacterial activity of the marine diatom Skeletonema costatum against aquacultural pathogens. Aquaculture 174:15–24. https://doi.org/10.1016/S0044-8486(98)00513-4
72. Nobre B, Marcelo F, Passos R et al (2006) Supercritical carbon dioxide extraction of astaxanthin and other carotenoids from the microalga Haematococcus pluvialis. Eur Food Res Technol 223:787–790. https://doi.org/10.1007/s00217-006-0270-8
73. Ohta S, Shiomi Y, Kawashima A et al (1995) Antibiotic effect of linolenic acid from Chlorococcum strain HS-101 and Dunaliella primolecta on methicillin-resistant Staphylococcus aureus. J Appl Phycol 7:121–127. https://doi.org/10.1007/BF00693057
74. Osterhage C, Kaminsky R, Koeing GM et al (2000) Ascosalipyrrolidinone A, an antimicrobial alkaloid, from the obligate marine fungus *Ascochyta salicorniae*. J Org ChemJ Org Chem 65:6412–6417
75. Ozemir G, Karabay NU, Dalay MC, Pazarbasi B (2004) Antibacterial activity of volatile components and various extracts of *Spirulina platensis*. Phytother Res 18:754–757
76. Palavra AMF, Coelho JP, Barroso JG et al (2011) Supercritical carbon dioxide extraction of bioactive compounds from microalgae and volatile oils from aromatic plants. J Supercrit Fluids 60:21–27. https://doi.org/10.1016/j.supflu.2011.04.017
77. Pandian P, Selvamuthukumar S, Manavalan R et al (2011) Screening of antibacterial and antifungal activities of red marine algae *Acanthaphora spicifera* (Rhodophyceae). Biomed Sci Res 3:444–448
78. Park H, Kurokawa M, Shiraki K et al (2005) Antiviral activity of the marine alga *Symphyocladia latiuscula*. Against herpes simplex virus (HSV-1) in vitro and its therapeutic efficacy against HSV-1 infection in mice. Biol Pharm Bull 28:2258–2262
79. Park HJ, Chung HY, Kim I et al (1999) Antioxidative activity of 2, 3, 6-tribromo-4, 5 dihydroxybenzyl methyl ether from *Symphyocladia latiuscula*. J Fish Sci Technol 2:1–7
80. Ploutno A, Carmeli S (2000) Nostocyclyne A, a novel antimicrobial cyclophane from the cyanobacterium *Nostoc* sp. J Nat Prod 63:1524–1526. https://doi.org/10.1021/np0002334
81. Pop-Vicas A, Opal SM (2014) The clinical impact of multidrug-resistant gram-negative bacilli in the management of septic shock. Virulence 5:206–212. https://doi.org/10.4161/viru.26210
82. Pratt R, Daniels TC, Eiler JJ, Gunnison JB et al (1944) Chorellin, an antibacterial substance from chlorella. Science 99:351–352
83. Pratt R, Mautner H, Gardner GM et al (1951) Report on antibiotic activity of seaweed extracts. J Am Pharm Assoc Am Pharm Assoc (Baltim) 40:575–579
84. Preetha K, John L, Subin C, Vijayan K (2012) Phenotypic and genetic characterization of Dunaliella (Chlorophyta) from Indian Salinas and their diversity. Aquat Biosyst 8:27. https://doi.org/10.1186/2046-9063-8-27
85. Quereshi MA, Ali RA, Hunter R (1995) Immuno-modulatory effects of Spirulina platensis supplementation in chickens. In: Proceedings of the 44th Western poultry disease conference. Sacramento, pp 117–121
86. Raveh A, Carmeli S (2007) Antimicrobial ambiguines from the cyanobacterium Fischerella sp. collected in Israel. J Nat Prod 70:196–201. https://doi.org/10.1021/np060495r
87. Romano I, Bellitti MR, Nicolaus B et al (2000) Lipid profile: a useful chemotaxonomic marker for classification of a new cyanobacterium in *Spirulina* genus. Phytochemistry 54:289–294
88. Sakemi S, Higa T, Jefford CW et al (1986) Venustatriol: a new antiviral triterpene tetracyclic ether from Laurencia venusta. Tetrahedron Lett 27:4287–4290
89. Sarada DVL, Kumar CS, Rengasamy R (2011) Purified C-phycocyanin from Spirulina platensis (Nordstedt) Geitler: a novel and potent agent against drug resistant bacteria. World J Microbiol Biotechnol 27:779–783. https://doi.org/10.1007/s11274-010-0516-2
90. Seenivasan R, Indu H, Archana G et al (2010) The antibacterial activity of some marine algae from South East Coast of India. Am-Eur J Agric Environ Sci 9:480–489
91. Sharma A (2011) Antimicrobial resistance: no action today, no cure tomorrow. Indian J Med Microbiol 29:91–92. https://doi.org/10.4103/0255-0857.81774

92. Sheng J, Zeng L, Sun H, Huang A (2001) Biological activities of protein-polysaccharides from *Nostoc commune*. J Guang Acad Sci 17:20–23
93. Sheridan C (2006) Antibiotics au naturel. Nat Biotechnol 24:1494–1496. https://doi.org/10.1038/nbt1206-1494
94. Simic S, Kosanic MM, Rankovic BR (2012) Evaluation of in vitro antioxidant ant antimicrobial activities of green algae *Trentepohloa umbrina*. Not Bot Horti Agro 40:86–91
95. Simmons LT, Engene N, Urena LD et al (2008) Viridamides A and B, lipodepsipeptides with antiprotozoal activity from marine cyanobacterium. Oscillatoria Nigro Viridis. J Nat Prod 71:1544–1550
96. Singh RK, Tiwari SP, Rai AK et al (2011) Cyanobacteria: an emerging source for drug discovery. J Antibio 64:401–412
97. Sivakumar J, Santhanam P (2011) Antipathogenic activity of Spirulina powder. Recent Res Sci Technol 3:158–161
98. Smee DF, Bailey KW, Wong M-H et al (2008) Treatment of influenza A (H1N1) virus infections in mice and ferrets with cyanovirin-N. Antivir Res 80:266–271
99. Soltani M, Khosravi A-R, Asadi F, Shokri H (2012) Evaluation of protective efficacy of Spirulina platensis in Balb/C mice with candidiasis. J Mycol Med 22:329–234. https://doi.org/10.1016/j.mycmed.2012.10.001
100. Strathmann M, Wingender J, Flemming HC (2002) Application of fluorescently labelled lectins for the visualization and biochemical characterization of polysaccharides in biofilms of Pseudomonas aeruginosa. J Microbiol Methods 50:237–248
101. Tandeau de Marsac N, Houmard J (1993) Adaptation of cyanobacteria to environmental stimuli: new steps towards molecular mechanisms. FEMS Microbiol Lett 104:119–189. https://doi.org/10.1016/0378-1097(93)90506-W
102. Temina M, Rezankova H, Rezanka T, Dembitsky VM (2007) Diversity of the fatty acids of the Nostoc species and their statistical analysis. Microbiol Res 162:308–321. https://doi.org/10.1016/j.micres.2006.01.010
103. Lancet T (2009) Urgently needed: new antibiotics. Lancet 374:1868. https://doi.org/10.1016/S0140-6736(09)62076-6
104. Topeu G, Aydogmus Z, Imre S et al (2003) Brominated sesquiterpenes from the red alga Laurencia obtusa. J Nat Prod 66:1505–1508
105. Volk RB (2008) A newly developed assay for the quantitative determination of antimicrobial (anticyanobacterial) activity of both hydrophilic and lipophilic test compounds without any restriction. Microbiol Res 163:161–167. https://doi.org/10.1016/j.micres.2006.03.015
106. Volk RB, Furkert FH (2006) Antialgal, antibacterial and antifungal activity of two metabolites produced and excreted by cyanobacteria during growth. Microbiol Res 161:180–186. https://doi.org/10.1016/j.micres.2005.08.005
107. Wang H, Li YL, Shen WZ et al (2007) Antiviral activity of a sulfoquinovosyldiacylglycerol (SQDG) compound isolated from the green alga *Caulerpa racemosa*. Bot Mar 50:185–190
108. Wang M, Xu YN, Jiang GZ et al (2000) Membrane lipids and their fatty acid composition in Nostoc flagelliforme cells. Acta Bot Sin 42:1263–1266
109. Washida K, Koyama T, Yamada K et al (2006) Karatungoils A and B, two novel antimicrobial polyol compounds, from the symbiotic marine dinoflagellate *Amphidinium* sp. Tetrahedron Lett 47:2521–2525. https://doi.org/10.1016/j.tetlet.2006.02.045
110. Wei Y, Liu Q, Xu C et al (2015) Damage to the membrane permeability and cell death of Vibrio parahaemolyticus caused by phlorotannins with low molecular weight from Sargassum thunbergii. J Aquat Food Prod Technol 25:323–333
111. Whitton BA (2008) Cyanobacterial diversity in relation to the environment. NATO Secur through Sci Ser C Environ Secur 17–43. https://doi.org/10.1007/978-1-4020-8480-5_2
112. Wright JLC, Boyd RK, de Freitas ASW et al (1989) Identification of domic acid, a neuroexcitatory amino acid in toxic mussels from Eastern Pince Edward Island. Can J Chem 67:481–490
113. Yu SH, Wu SJ, Wu JY et al (2015) Preparation of fucoidan-shelled and genipin-crosslinked chitosan beads for antibacterial application. Carbohydr Polym 126:97–107

Algal-Bacterial System: A Novel Low-Cost Biotechnological Initiative in Wastewater Treatment

9

Kalpana Sahoo, Rajesh Kumar Sahoo, Mahendra Gaur, and Enketeswara Subudhi

Abstract

Algal-bacterial process biotechnology is a recent low-cost method toward toxic pollutant removal from wastewater that has received more attention in the present scenario. The pollutants are mainly categorized into inorganic, organic, radioactive, acid/base, etc. The water pollutants mainly include SO_2 from power plants, chemical waste, fertilizers from agricultural use, oil spillage, silt, harmful pesticides, detergents, harmful compounds in cosmetics, pathogenic bacteria from livestock operations and food processing wastes, and chemical compounds found in cosmetics products, effluent outfalls from factories, refineries, waste treatment plants, contaminants from improper disposal of industrial wastes running through rainwater, etc. The proper pretreatment of wastewater needs to be done before disposal to the water bodies unless it would cause serious damage to our entire ecosystem. Algae, mostly behaving as water-purifying agent acting as pollution indicator, can act as a better alternative toward bioremediation through low-cost approach. Due to certain limitations in the algal cell during toxic pollutant accumulation, sometimes it can remediate contaminants up to a certain level. So another emerging concept of algal-bacterial symbiotic system, a less energy consumption technology, was developed which gained much attention toward wastewater treatment in the present scenario. Introduction of cost-effective algal-bacterial consortia treatment technology is reported to treat toxic wastewater effluents from municipal, domestic, industrial, and agricultural activities by using many special unicellular microalgae and pollutant-specific-degrading bacteria. It has been found from recent studies that two important factors such as selection of suitable strain as well as cultivation are responsible for biodegrading toxic chemicals and compounds such as polycyclic aromatic hydrocarbons,

K. Sahoo (✉) · R. K. Sahoo · M. Gaur · E. Subudhi
Biomics & Biodiversity Laboratory, Centre for Biotechnology,
Siksha 'O' Anusandhan (Deemed to be University), Bhubaneswar, Odisha, India
e-mail: kalpanasahoo@soa.ac.in; enketeswarasubudhi@soa.ac.in

L. B. Sukla et al. (eds.), *The Role of Microalgae in Wastewater Treatment*,
https://doi.org/10.1007/978-981-13-1586-2_9

phenolics, and organic solvents. The organic compounds released after algal photosynthesis can ultimately become a food source for a variety of heterotrophic microbes. The growth-promoting substance production by bacteria can promote microalgal growth, whereas few bacteria can promote algal growth by photosynthetic oxygen tension reduction inside the algal cell, e.g., *Pseudomonas diminuta* and *P. vesicularis*. The wastewater can be treated in an open system, i.e., by the construction of artificial ponds, or closed system, i.e., using a bioreactor. From the available reports, it was found that the algal-bacterial consortium may include mainly microalgae and bacteria, e.g., *Flavobacteriia*, *Gammaproteobacteria*, *Bacteroidia*, and *β-proteobacteria*. The harmful dissolved methane in anaerobically treated wastewater can effectively be treated by the methane-oxidizing bacteria and algae. It can be concluded that the conventional algal-bacterial system treatment technology acts as a natural biological treatment method as a viable alternative and is of great importance for achieving good wastewater treatment performance as well as the reduction in energy consumption cost.

9.1 Introduction to the Problem

Pollution is a human intervention that occurs due to the arising concentrations of naturally occurring substances or releasing of nonnatural synthetic compounds (xenobiotics) to the environment. It is also usually caused by the release of organic and inorganic wastes into the environment as a result of industrial, domestic, and agricultural activities, etc. [30, 35]. Clean water is an essential component for humans as well as their environments. The reduction in water quality would also lead to water scarcity that is a major issue around the globe. The fall in quality of water varies from country to country due to a number of factors such as population growth and density, the extent of industrialization quality of nonrenewable water resources, economic situation, and institutional capacity. Water quality maintenance becomes crucial for the protection of the natural habitats of fish, bugs, bird, and plant communities, etc. Operational output is an important aspect in case of treatment methods which lead to innovation of various new techniques in the sector. Nowadays, great advance methods initiate the development of efficient technologies for wastewater treatment, but challenges still remain.

9.2 Causes and Sources of Water Pollution

Water pollution causes death of more than 14,000 people daily and also leads to various risky diseases [43, 55]. In India, water pollution is the main cause of death of an average 580 people due to typhoid, cholera, dysentery, paratyphoid fever, jaundice, amoebiasis, malaria, diarrhea, etc. Developed countries also account for water pollution issues same as water pollution in developing countries. Like in the national water quality report of the United States, 44% of assessed stream miles, 30% of assessed bays and estuarine square, and 64% of assessed lake acres and miles were classified as polluted [18]. Used water from toilets, showers, kitchens, laundries, industries, etc. acts as the major sources of wastewater. Domestic households produce an average of 200–300 L of wastewater per person every day.

The untreated sewage and wastewater cause serious health problems due to the presence of various harmful pathogenic microbes which are the reason for various harmful as well as deadly diseases (http://www.conserve-energy-future.com/sources-and-causes-of-water-pollution.php). Due to the advent of rural development, the rate of waste discharge from supplied extended water channels is increasing year by year. The same also happens in urban areas due to the negligence in sewage channel construction or rehabilitation of the existing ones leading to water pollution problems [1]. The household garbage such as paper, aluminum, rubber, glass, plastic, food, etc. after release into the sea cause pollution and also harm sea animals. A small leakage from the sewer lines causes groundwater contamination making it unfit for drinking water and also becoming the home ground for various insects and mosquitoes. Also, the agricultural drains are contaminated with treated and untreated domestic wastes. The largest amount of water consumption occurs during agricultural activities such as irrigation and chemical fertilizer and pesticide application, discharge containing agricultural residues acting as a major water pollution source for the environment and aquatic life too. The modern urbanization leading to water pollution because of fertilizer application, deforestation, soil erosion, man-made constructional activities, improper sewage management as well as treatment, landfills as more garbage is produced, etc. Mining process also acts as a major source of water contamination by the release of some harmful toxic chemicals, metal waste, and sulfides from the rocks causing serious health issues. A large amount of toxic pollutants pollute the sea and hamper the marine community by oil spillage caused due to the accident of oil-carrying ships. Most of the death occurs in aquatic animals and marine species due to rise in water temperature because of global warming causing water pollution. The nuclear wastes that are produced by radioactive material have serious environmental hazards due to improper disposal. The different sources of water pollution are well described in Fig. 9.1.

9.3 Problems and Processes During Wastewater Treatment

The present scenario of high energy costs and scarcity of natural resources leads to the development of sustainable methods for pollution control with a low energy consumption and the potential for resource recovery. The wastewater treatment was estimated a general consumption of approximately 2–3% of electrical energy per year in treatment plants, which is very costly. The major portion of energy consumption occurs during biological waste treatment in municipal wastewater. Few examples of treatment methods are (a) use of fine screens in primary treatment, (b) membrane technology for the aeration process, (c) direct treatment of high concentration return streams, etc. Due to the high energy consumption in agro-industry

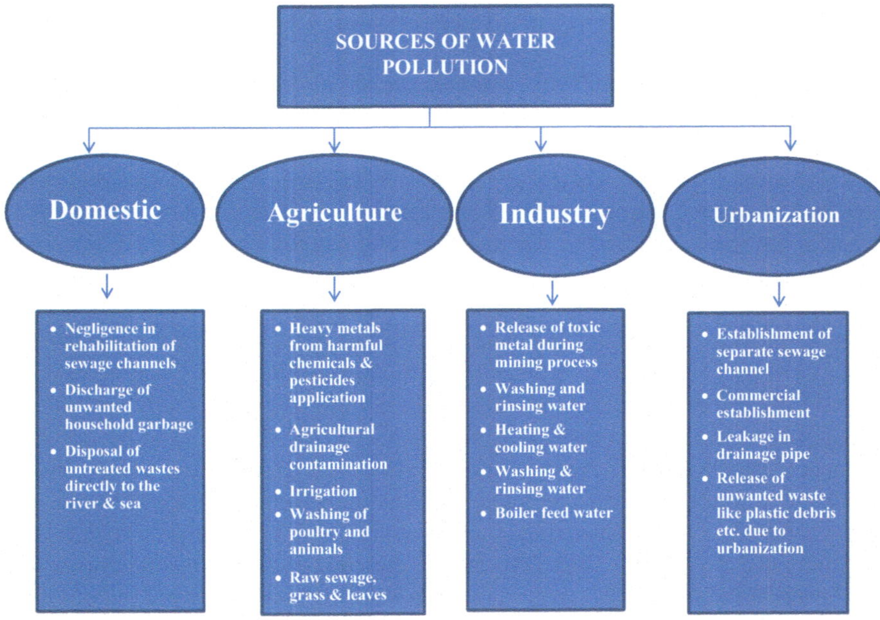

Fig. 9.1 Sources of water pollution

wastewater treatment, implementation of the application in rural areas became difficult. Similarly, the implementation of anaerobic digestion system for biogas production need complex processes such as temperature and waste sludge loading rate became difficult. Therefore, the development of low-cost and eco-friendly wastewater treatment methods became a mandate for achievement in water pollution-related issues.

9.4 Role of Algae in Wastewater Treatment

Algae are the natural food and energy producers, are water refiner or purifier, and are ultimate solutions for the climate challenges facing our world today. Algae have various importance like fuel production, CO_2 recycling, and food and feed for animals as well as humans. Algae act as photosynthetic gas exchangers during space travel [8, 49]. The wastewater quality improvement technology was implemented in the United States for the production of methane from the waste algal biomass [44]. The detail regarding algal pretreatment method for byproduct formation was described in Fig. 9.2. Microalgae act as a major component during the tertiary treatment of domestic wastewater in maturation ponds or the treatment of small- to middle-scale municipal wastewater in facultative or aerobic ponds [2, 6, 33, 41, 42]. They enhance the removal of nutrients, heavy metals, and pathogens and provide O_2 to heterotrophic aerobic bacteria for organic pollutant mineralization, using, in turn, the CO_2 released from bacterial respiration [37].

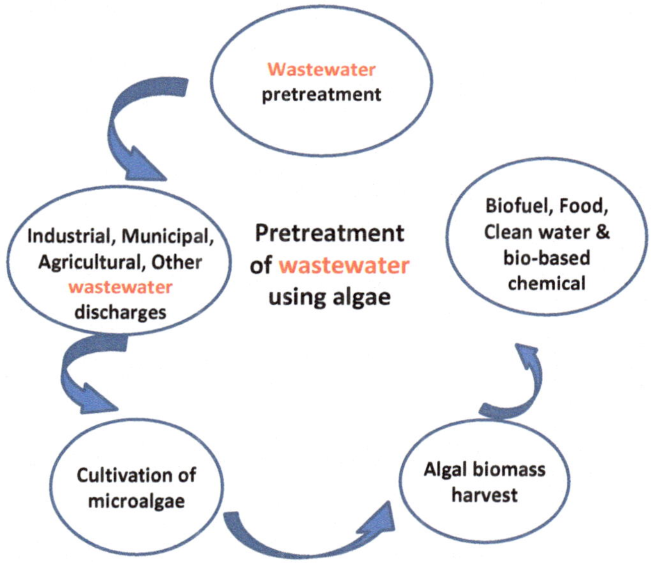

Fig. 9.2 Method for integrated wastewater-based algae cultivation system for by-products

9.4.1 About Algae

Algae, the major primary producers in all kinds of water bodies, and many species flourished in water polluted with organic wastes play an important part in "self-purification of water bodies." They are mainly responsible for tastes and odors in water, e.g., blue-green algae, diatoms, and colored flagellates, e.g., Chrysophyta, Euglenophyta, etc. [48].

9.4.2 Algae in Bioremediation and Water Purification

Wastewater treatment system using algae that has drawn the attention for the last 50–55 years is now widely accepted as an effective conventional treatment system because of accumulation of highly toxic substances like selenium, zinc, and arsenic and higher concentration of radioactive elements from the aquatic environment in their cells and/or bodies, thus acting as an effective tool for bioremediation [48]. It was observed that *Spirogyra* can accumulate radiophosphorus by a factor 850.000 times that of water [32]. The algal cultivation is mainly affected by several environmental factors. Different algal species have different metal removal efficiency. From various studies it was found that individual algae like *Oscillatoria* can remove chromium, *Chlorella vulgaris* can remove cadmium and copper, *Chlamydomonas* can bioremediate zinc, and *Scenedesmus chlorelloides can* successfully remove molybdenum [19, 39, 48, 53]. Microalgae mostly behave as a pollution indicator in wastewater ponds affecting their growth during the treatment. For instance, 10 mg phenanthrene l^{-1} totally inhibited the growth of *C. sorokiniana*, whereas a phenanthrene-degrading *Pseudomonas* strain used to form the consortium easily biodegraded this compound at 25 mg l^{-1}[7].

The quality improvement of wastewater involves chemical, physical, and biological processes in different treatment stages. While the primary treatment of wastewater results in sedimentation of materials, the secondary treatment removes suspended and dissolved organic matter left out in the primary treatment through the biological process. The last process involves purification of nitrates and phosphates, including fine particle removal [16, 48]. Algae can act as a better water-purifying agent by removing all of these contaminants in aquatic as well as marine ecosystem.

9.4.3 Limitations in Algal Bioremediation Technology During Wastewater Treatment

Algae is very sensitive/toxic when exposed to harmful pollutants as well as compounds which include a wide range of toxic and/or persistent substances in the environment [4, 7]. Microalgae are also sensitive to the combined effect of high NH_3 concentrations and high pH leading to decline in algal efficiency [3]. Microalgae are more likely to be inhibited during the treatment of hazardous compounds than their associated target resistant bacteria. For example, the growth of alga *Chlorella sorokiniana* inhibited by the presence of 10 mg phenanthrene l^{-1} and the bacterial strain *Pseudomonas* sp. present in the consortium can biodegrade the pollutant easily [7, 37].

9.5 Algal-Bacterial Symbiotic Relation in Wastewater Treatment

The algae are better source for removal of toxic wastes from the aquatic environment through phytoremediation. But sometimes the heavy contaminants (organic and inorganic pollutants) act as a toxic source in algal metabolism. Heavy metal accumulation in the algal biosystem may lead to lethality leading to the concept of algal-bacterial symbiotic system, acting as a better source for toxic pollutant removal by accumulation of toxic heavy metals into the bacterial biosystems, ultimately causing no harm to the algae and acting as an effective source for wastewater treatment [56]. The algal-bacterial system was first designed by William J. Oswald [41] in wastewater treatment ponds for better utilization of solar energy [31, 40] and further applied for agricultural and industrial wastewater treatment [7, 38]. During this technically and economically viable process, the oxygen is supplied to bacteria by microalgae which oxidize both organic matter and NH_4^+, while carbon dioxide released during bacterial respiration is being utilized by microalgae during photosynthesis (Fig. 9.3). The retention of useful nitrogenous compounds with the avoidance of bacterial nitrification/denitrification [9, 51, 57]. The algal-bacterial biomass

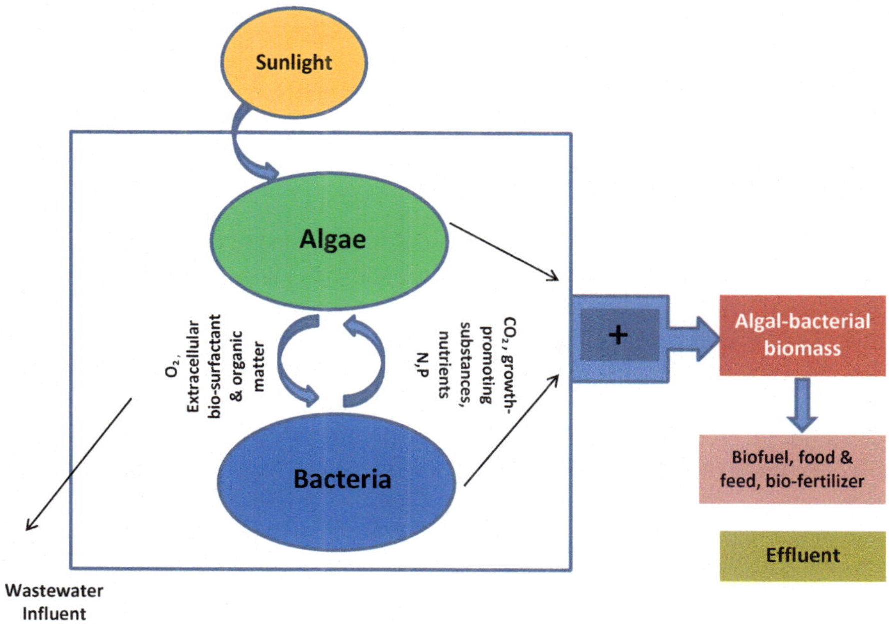

Fig. 9.3 Wastewater treatment plant through algal-bacterial symbiosis

assimilates the nitrogen and phosphorus present in the wastewater causing increase in pH. The higher pH and oxygenation lead to the removal of fecal coliforms from livestock wastewaters. Furthermore, the increase in the number of bacteria can promote algal growth through generating CO_2, releasing growth-promoting substances (e.g., vitamins), and modulating the pH [11, 13, 21].

9.5.1 Mechanism Involved in Algal-Bacterial Symbiosis Behind Wastewater Treatment Process

The biological wastewater treatment in aerobic condition is a new eco-friendly approach. Heterotrophic bacteria convert the organic pollutants into carbon dioxide (CO_2) and water (H_2O) in the presence of oxygen, whereas the process is exactly opposite in case of phototrophic algae. The algae take up carbon dioxide (CO_2) and water (H_2O) and convert this into organic material ($CH_2O \approx$ algae biomass) and oxygen (O_2):

$$CH_2O + O_2 \rightarrow CO_2 + H_2O \quad (9.1)$$

$$CO_2 + H_2O \rightarrow CH_2O + O_2 \quad (9.2)$$

The combination of both the steps (9.1) and (9.2) leads to a net-zero consumption/production of oxygen and carbon dioxide proving the ALBA process as an effective one (Fig. 9.3). It became an effective possible symbiotic approach for purification of wastewater containing various organic and inorganic pollutants such as steroid hormones, polyaromatic hydrocarbons, etc.

9.5.2 Selection of Suitable Strain

Selection of suitable algal as well as bacterial strain is one of the major criteria for wastewater treatment through algal-bacterial approach. The better tolerance of algal species such as *Chlorella* species, *Chlorella sorokiniana*, a higher concentration salicylate tolerant, leads to inhibition of other algal species those would be useful for bioremediation purpose [14, 28, 34, 38]. The complex interaction results in their changing behavior with the phycosphere itself. For example, a polysaccharide gel called sheath produced by photoautotrophic alga *Chlorella sorokiniana* IAM C-212 becomes a suitable habitat for several symbiotic microorganisms as it ensures close proximity [24, 46]. The major raw material used for polycarbonates, epoxy resins, and canned food coatings is bisphenol A (BPA) found in river soils, rivers, industrial wastewater, etc. leading to acute toxicity as well as female hormonal disorders [5, 10, 27, 50]. An algal strain can lower this toxic chemical effect up to a certain level, e.g., *Chlorella sorokiniana* [17]. The waste removal process also depends upon the optimization of proper inoculum concentration for efficient environmental waste removal. The treatment efficiency depends upon the proper inoculum ration of different algae/bacteria indicating that proper inoculum strategy was vital for the treatment performance [52].

9.5.3 Microalgae-Bacteria Consortia Systems

A new method, i.e., microalgae-bacteria consortia system, was developed for assimilation of various contaminants; microalgae can produce various organic substances which are ultimately assimilated by bacteria. In a consortia system, the oxygen released by the microalgae proved useful for bacteria. Factors affecting the performance of this system are light, pH, and selection of microalgal and bacterial species. Since microalgae are suspended and dispersed in the media, harvesting is crucial to achieving a high-quality effluent [25]. The microalgal-bacterial relationship is a complex one. Some species of bacteria can release hormones to promote algal growth, and the presence of a few can improve the algal activity during wastewater treatment [36]. For example, during co-culturing of *Chlorella vulgaris* with *Azospirillum brasilense*, the growth and size of algal cells and colonies increase significantly [21], whereas the algal activity increased in the presence of few heterotrophs such as *Pseudomonas vesicularis* without any plant growth hormone production [36]. In addition to the symbiotic relationship, bactericidal and algicidal activity observed between microalgae and bacteria, respectively, toward certain species. For example, growth inhibition of *Gymnodinium mikimotoi*, a red tide plankton, happens due to the presence of a bacteria, *Flavobacterium* sp. 5N3 [20]. Algae play a dominant role in consortia system which are mostly microalgae-assistant systems as well as algae-dominant systems [25]. During *Chlorella vulgaris* cocultivation with *Bacillus*, the algal nutrient removal efficiency increased significantly [29].

The algal-bacterial symbiotic pretreatment of the effluent sample acts as a better method during salicylate degradation [37]. In algal-bacterial co-culture system, enhancement in algal growth occurred due to the presence of *Pseudomonas diminuta* and *P. vesicularis*, whereas because of anti-algal production, the bacterial strain *Pseudomonas aeruginosa* leads to the growth rate of various green microalgae and cyanobacteria [12]. The pigmentation of algal cell was found to be increased followed by lesser nutrient removal after co-immobilization with nitrogen-fixing bacteria *Phyllobacterium myrsinacearum* [21].

9.5.4 Construction of Algal-Bacterial Artificial Pond: A Case Study

Algal-bacterial pond is an artificial temporary water body constructed for storing and improving wastewater in natural conditions where algal photosynthesis and bacterial decomposition are the principal mechanisms. Waste stabilization ponds were constructed where the dissolved compounds and suspended organic matter were stabilized in various conditions such as aerobic, facultative, matured, etc. The major stabilization pond where algae are used for effective treatment is facultative stabilization pond especially designed for useful purposes like to decrease waste retention time, etc. The major processes in this type of ponds involved which are oxidation, settling, sedimentation, adsorption, and disinfection in the ponds are results of symbiotic relation [47].

A case study for algal-bacterial selenium removal (ABSR) was conducted during agricultural drainage water treatment in the Panoche Drainage District, San Joaquin Valley, since 1997. The effectiveness of this technology was conducted to investigate potential wildlife exposure to selenium at a full-scale facility that will minimize the lifecycle cost for each pound of selenium removed. In the current approach, a series of ponds were designed for promotion of indigenous microbes for nitrate and selenium removal. An affordable reduction of selenium load was observed in San Joaquin River. A 95% and 80% reduction in nitrate and selenium, respectively, was observed by the construction of ABSR plant during 1997 and 1998. The preliminary total cost estimate for a 10-acre-foot per day ABSR facility is less than $200 per acre-foot of treated drainage water [45].

A study was conducted by Tiron et al. [54] by observing the feasibility of treating dairy industry wastewater in a microalgae-bacteria symbiotic system providing the aeration through microalgal photosynthetic without aeration costs which was carried out in a stirred tank bioreactor BIOSTAT®, in batch mode, at a HRT of 96 h with 50 rpm rotation speed, at room temperature (20–31 °C). Around 3 liters of wastewater from dairy industry was used in the study. The physicochemical parameters were maintained inside the bioreactor. Initial microalgal-bacterial biomass was 1.14 g dry weight l^{-1}. A mixed consortium, i.e., *Chlorella* sp.-bacteria, was used for waste treatment in the bioreactor. After 96 h of treatment, the removal efficiencies of organic matter (COD-Cr), total nitrogen (TN), and total phosphorus (TP) observed were 91, 68, and 38%, respectively. The maximum microalgal and microalgal-bacterial system growth rates were 0.13 and 0.10 day^{-1}, respectively. It was concluded from the study that promotion of microalgal-bacterial consortium method for wastewater treatment can be done as a cost-efficient biotechnology.

9.5.5 Challenges and Future Prospects

There are various lacunae/limitations observed in the algal-bacterial wastewater treatment method which need to be focused on the future study. Few examples are (1) the high pH and oxygen produced from algal photosynthesis that lead to the growth reduction of bacteria; (2) antibacterial substances released by algae that may hamper the bacterial growth; (3) weakening of algal cell wall because of the pathogenic bacteria effect ultimately resulting in disruption and cell death; (4) microbial sensitivity to changes in operational and environmental conditions leading to readily occurrence of malfunctioning under adverse conditions [46]; (5) challenges like loss of nitrification, bulking, and foaming due to excessive growth of filamentous bacteria [15, 26]; (6) handling and disposal of sludge generated in a larger quantity huge amount due to conventional treatment methods; and (7) the poor microalgae cells' removal efficiency from the effluent [54].

9.6 Conclusion

Wastewater released from various activities like municipal, increased urbanization, industrial, agricultural, domestic, etc. to the aquatic body leads to water pollution ultimately deteriorating the water quality as well as hampering the food chain by entering the biosystem of aquatic flora and fauna as well as serious damage to the terrestrial life too including human beings. The wastewater containing various harmful chemicals and compounds which are highly toxic gives rise to eutrophication and pollution-related problems. If the wastewater would not be treated properly, they can be hazardous to our ecosystem. So keeping this in our mind, the wastewater needs to be treated properly before their release to the environment as well as water bodies. Separate wastewater treatment plants need to be constructed to mitigate this problem. Algae, considered as water purifier as well as pollution indicator, are accumulating plant nutrients, heavy metals, pesticides, organic and inorganic toxic substances, and radioactive matters in their cells/bodies. Algal utilization in wastewater treatment is considered as an appropriate cost-effective biological method for wastewater treatment. Sometimes the above said method fails due to the bioremediation by algae up to a certain extent because of energy consumption during aeration, harmful heavy metal tolerance, growth hindrance due to other dominating algae, etc. So another most useful technique such as algal-bacterial symbiotic method for wastewater treatment was adopted nowadays which is proved to be most efficient. In this system, none of the living cells got hampered giving ultimately a new insight to cost-effective wastewater treatment technology through biological means.

References

1. Abdel-Raouf N, Al-Homaidan AA, Ibraheem IBM (2012) Microalgae and wastewater treatment. Saudi J Biol Sci 19:257–275. https://doi.org/10.1016/j.sjbs.2012.04.005
2. Abeliovich A (1986) Algae in wastewater oxidation ponds. In: Handbook of microalgal mass culture. CRC Press, Boca Raton, pp 331–338
3. Abeliovich A, Azov Y (1976) Toxicity of ammonia to algae in sewage oxidation ponds. Appl Environ Microbiol 31:801–806
4. Aksmann A, Tukaj Z (2004) The effect of anthracene and phenanthrene on the growth, photosynthesis, and SOD activity of the green alga *Scenedesmus armatus* depends on the PAR irradiance and CO2 level. Arch Environ Contam Toxicol 47:177–184. https://doi.org/10.1007/s00244-004-2297-9
5. Alexander HC, Dill DC, Smith LW et al (1988) Bisphenol a: Acute aquatic toxicity. Environ Toxicol Chem 7:19–26. https://doi.org/10.1002/etc.5620070104
6. Aziz MA, Ng WJ (1993) Industrial wastewater treatment using an activated algae-reactor. Water Sci Technol (United Kingdom) 28:71–76
7. Borde X, Guieysse B, Delgado O et al (2003) Synergistic relationships in algal-bacterial microcosms for the treatment of aromatic pollutants. Bioresour Technol 86:293–300. https // doi.org/10.1016/S0960-8524(02)00074-3
8. Borowitzka MA (1999) Commercial production of microalgae: ponds, tanks, tubes and fermenters. J Biotechnol 70:313–321. https://doi.org/10.1016/S0168-1656(99)00083-8
9. Buys BR, Mosquera-Corral A, Sánchez M, Méndez R (2000) Development and application of a denitrification test based on gas production. Water Sci Technol 41:113–120
10. Cousins IT, Staples CA, Klecka GM, Mackay D (2002) A multimedia assessment of the environmental fate of bisphenol A. Hum Ecol Risk Assess 8:1107–1135. https://doi.org/10.1080/1080-700291905846
11. Croft MT, Lawrence AD, Raux-Deery E et al (2005) Algae acquire vitamin B12 through a symbiotic relationship with bacteria. Nature 438:90–93. https://doi.org/10.1038/nature04056
12. Dakhama A, de la Noüe J, Lavoie MC (1993) Isolation and identification of antialgal substances produced by *Pseudomonas aeruginosa*. J Appl Phycol 5:297–306. https://doi.org/10.1007/BF02186232
13. De-Bashan LE, Moreno M, Hernandez JP, Bashan Y (2002) Removal of ammonium and phosphorus ions from synthetic wastewater by the microalgae *Chlorella vulgaris* coimmobilized in alginate beads with the microalgae growth-promoting bacterium *Azospirillum brasilense*. Water Res 36:2941–2948. https://doi.org/10.1016/S0043-1354(01)00522-X
14. De-Bashan LE, Trejo A, Huss VAR et al (2008) *Chlorella sorokiniana* UTEX 2805, a heat and intense, sunlight-tolerant microalga with potential for removing ammonium from wastewater. Bioresour Technol 99:4980–4989. https://doi.org/10.1016/j.biortech.2007.09.065
15. de-Bashan LE, Bashan Y (2004) Recent advances in removing phosphorus from wastewater and its future use as fertilizer (1997–2003). Water Res 38:4222–4246. https://doi.org/10.1016/j.watres.2003.09.022
16. Droste RL (1997) Relationship of algae to water pollution and waste water treatment. In: Water Treatment. Wiley, New York
17. Eio EJ, Kawai M, Niwa C et al (2015) Biodegradation of bisphenol A by an algal-bacterial system. Environ Sci Pollut Res Int 22(19):15145–15153. https://doi.org/10.1007/s11356-015-4693-2
18. EPA (2009) National water quality inventory: report to congress, 2004 reporting cycle." US Environmental Protection Agency

19. Filip DS, Thomas Peters V, Dean Adams E, Middlebrooks J (1979) Residual heavy metal removal by an algae-intermittent sand filtration system. Water Res 13(3):305–313. https://doi.org/10.1016/0043-1354(79)90211-2
20. Fukami K, Nishijima T, Ishida Y (1997) Stimulative and inhibitory effects of bacteria on the growth of microalgae. Hydrobiologia 358(1-3):185–191. https://doi.org/10.1023/A:1003139402315
21. Gonzalez LE, Bashan Y (2000) Increased growth of the microalga *Chlorella vulgaris* when coimmobilized and cocultured in alginate beads with the plant-growth-promoting bacterium *Azospirillum brasilense*. Appl Environ Microbiol 66(4):1527–1531. https://doi.org/10.1128/AEM.66.4.1527-1531.2000
22. http://Algal-bacterial(ALBA)wastewatertreatment_algae-biogasproject.htm
23. http://www.conserve-energy-future.com/sources-and-causes-of-water-pollution.php
24. Imase M, Watanabe K, Aoyagi H, Tanaka H (2008) Construction of an artificial symbiotic community using a Chlorella-symbiont association as a model. FEMS Microbiol Ecol 63:273–282. https://doi.org/10.1111/j.1574-6941.2007.00434.x
25. Jia H, Yuan Q (2016) Removal of nitrogen from wastewater using microalgae and microalgae-bacteria consortia. Cogent Environ Sci 2:1275089. https://doi.org/10.1080/23311843.2016.1275089
26. Khin T, Annachhatre A (2004) Novel microbial nitrogen removal processes. Biotechnol Adv 22(7):519–532. https://doi.org/10.1016/j.biotechadv.2004.04.003
27. Krishnan AV, Stathis P, Permuth SF et al (1993) Bisphenol-A: an estrogenic substance is released from polycarbonate flasks during autoclaving. Endocrinology 132:2279–2286. https://doi.org/10.1210/endo.132.6.8504731
28. Li T, Zheng Y, Yu L, Chen S (2013) High productivity cultivation of a heat-resistant microalga Chlorella sorokiniana for biofuel production. Bioresour Technol 131:60–67
29. Liang Z, Liu Y, Ge F et al (2013) Efficiency assessment and pH effect in removing nitrogen and phosphorus by algae-bacteria combined system of *Chlorella vulgaris* and *Bacillus licheniformis*. Chemosphere 92:1383–1389. https://doi.org/10.1016/j.chemosphere.2013.05.014
30. Lim SL, Chu WL, Phang SM (2010) Use of *Chlorella vulgaris* for bioremediation of textile wastewater. Bioresour Technol 101:7314–7322. https://doi.org/10.1016/j.biortech.2010.04.092
31. Ludwig HF, Oswald WJ, Gotaas HB, Lynch V (1951) Algae symbiosis in oxidation ponds: I. growth characteristics of "*Euglena Gracilis*" cultured in sewage. Sewage Ind Waste 23:1337–1355. https://doi.org/10.2307/25031742
32. Mackenthum K (1969) The practice of water pollution biology. U. S. Department of the Interior, Federal Water Pollution Control Administration (Now EPA) U.S.D.I., Washington, DC
33. Mara DD, Pearson H (1986) Artificial freshwater environment: waste stabilization ponds. Biotechnology 8:177–206
34. Morita M, Watanabe Y, Saiki H (2000) High photosynthetic productivity of green microalga *Chlorella sorokiniana*. Appl Biochem Biotechnol 87:203–218. https://doi.org/10.1385/ABAB:87:3:203
35. Mouchet P (1986) Algae reactions to mineral and organic micropollutants, ecological consequences and possibilities for industrial-scale application: a review. Water Res 20:399–412
36. Mouget JL, Dakhama A, Lavoie MC, de la Noüe J (1995) Algal growth enhancement by bacteria: is consumption of photosynthetic oxygen involved? FEMS Microbiol Ecol 18:35–43. https://doi.org/10.1016/0168-6496(95)00038-C
37. Munoz R, Guieysse B (2006) Algal-bacterial processes for the treatment of hazardous contaminants: a review. Water Res 40:2799–2815. https://doi.org/10.1016/j.watres.2006.06.011
38. Munoz R, Kollner C, Guieysse B, Mattiasson B (2004) Photosynthetically oxygenated salicylate biodegradation in a continuous stirred tank photobioreactor. Biotechnol Bioeng 87:797–803. https://doi.org/10.1002/bit.20204
39. Nakajima A, Horikoshi T, Sakaguchi T (1981) Studies on the accumulation of heavy metal elements in biological systems XVII. Selective accumulation of heavy metal ions by Chlorella vulgaris. Eur J Appl Microbiol Biotechnol 12:76–83. https://doi.org/10.1007/BF01970040

40. Olguín EJ (2003) Phycoremediation: Key issues for cost-effective nutrient removal processes. Biotechnol Adv 22:81–91. https://doi.org/10.1016/S0734-9750(03)00130-7
41. Oswald WJ (1988) Micro-algae and waste-water treatment. In: Micro-algal biotechnology. Ecotoxicology and environmental safety. Cambridge University Press, Cambridge, pp 305–328
42. Oswald WJ (1995) Ponds in the twenty-first century. Water Sci Technol 31:1–8. https://doi.org/10.1016/0273-1223(95)00487-8
43. Pink DH (2006) Investing in tomorrow's liquid gold. Retrieved from http://finance.yahoo.com/columnist/article/trenddesk/3748
44. Pulz O, Scheibenbogen K (1998) Photobioreactors: design and performance with respect to light energy input. Bioprocess Algae React Technol Apoptosis 59:123–152. https://doi.org/10.1007/BFb0102298
45. Quinn NWT, Leighton T, Lundquist TJ et al (2008) Algal-bacterial treatment facility removes selenium from drainage water. Calif Agric 54:50–56. https://doi.org/10.3733/ca.v054n06p50
46. Rawat I, Gupta SK, Shriwastav A, Singh P, BF KS (2016) Microalgae Applications in Wastewater Treatment. In: Bux F, Chisti Y (eds) Algae biotechnology products and processes. Green energy and technology. Springer International Publishing, Cham
47. Rich GW (2001) Low maintenance—mechanically simple wastewater treatment systems for developing and developed countries. In: Proceedings of the water environment federation
48. Sen B, Tahir M, Sonmez F et al (2013) Relationship of Algae to Water Pollution and Waste Water Treatment. Water Treatment, In. https://doi.org/10.5772/51927
49. Spolaore P, Joannis-Cassan C, Duran E, Isambert A (2006) Commercial applications of microalgae. J Biosci Bioeng 101:87–96. https://doi.org/10.1263/jbb.101.87
50. Staples CA, Dom PB, Klecka GM et al (1998) A review of the environmental fate, effects, and exposures of bisphenol A. Chemosphere 36:2149–2173. https://doi.org/10.1016/S0045-6535(97)10133-3
51. Su Y (2012) Settleable algal-bacterial culture for municipal wastewater treatment.
52. Su Y, Mennerich A, Urban B (2012) Coupled nutrient removal and biomass production with mixed algal culture: impact of biotic and abiotic factors. Bioresour Technol 118:469–476. https://doi.org/10.1016/j.biortech.2012.05.093
53. Ting YP, Lawson F, Prince IG (1989) Uptake of Cadmium and Zinc by the alga *Chlorella vulgaris*: part 1. Individual ion species. Biotechnol Bioeng 34:990–999. https://doi.org/10.1002/bit.260340713
54. Tiron O, Bumbac C, Postolache C (2014) Microalgae bacteria system for biological Wastewater treatment. J Environ Prot Ecol 15:268–276
55. West L (2006) World water day: a billion people worldwide lack safe drinking water. In: Wikipedia
56. Wikfors G, Ukeles R (1982) Growth and adaptation of estuarine unicellular algae in media with excess copper, cadmium or zinc, and effects of metal-contaminated algal food on *Crassostrea virginica* larvae. Mar Ecol Prog Ser 7:191–206. https://doi.org/10.3354/meps007191
57. Zeng RJ, Yuan Z, Keller J (2003) Enrichment of denitrifying glycogen-accumulating organisms in anaerobic/anoxic activated sludge system. Biotechnol Bioeng 81:397–404. https://doi.org/10.1002/bit.10484

Future Prospects of Microalgae in Wastewater Treatment

10

Lala Behari Sukla, Debabrata Pradhan, and Tondepu Subbaiah

Abstract

Wastewaters provide necessary nutrients in aqueous medium for the cultivation of microalgae and a simultaneous removal of pollutants like heavy and toxic metals, TSS, TDS, FOG, BOD, and COD from the wastewater. Another simulated technique of granular activated pellets of microalgae proved promising alternative way for efficient wastewater treatment. Natural lipid, carbohydrate, and protein contents of the microalgae are retained during the enhanced cultivation in the wastewater. These natural contents are suitable for energy production. The high productivity of microalgae coupled with a traditional biofuel production technique would solve the cost- and environmental-related issues with the fossil fuels. Therefore, the designing of suitable high rate algal ponds or photobioreactors for the large cultivation and harvesting of microalgae biomass during the wastewater treatment vis-à-vis biofuels production in an integrated process for the commercial exploration of prospective algal energy.

10.1 Introduction

Microalgae are photosynthetic microplanktons and commonly found in various nutrient-rich aquatic ecosystems like municipal wastewater, industrial wastewater streams, agricultural runoff, shorelines, mine seepage, and concentrated animal feed operations. They are more efficient than terrestrial plants converting solar energy

L. B. Sukla (✉) · D. Pradhan (✉)
Biofuels and Bioprocessing Research Center, Institute of Technical Education and Research, Siksha 'O' Anusandhan (Deemed to be University), Bhubaneswar, Odisha, India
e-mail: lalabeharisukla@soa.ac.in; debabratapradhan@soa.ac.in

T. Subbaiah
K.L. Deemed to be University, Green Fields, Vaddeswaram, Guntur, Andhra Pradesh, India

L. B. Sukla et al. (eds.), *The Role of Microalgae in Wastewater Treatment*,
https://doi.org/10.1007/978-981-13-1586-2_10

and CO_2, results huge biomass production ability. They utilize the dissolved nutrients like N and P in wastewater to convert carbohydrate to energy-rich complex organic lipids and proteins [1]. The complex organic compounds having large active sites both in- and outside of the cells of microalgae transform the inorganic and organic pollutants present in the wastewater by the chemical reactions and adsorption process.

The biomass of microalgae-containing natural organic hydrocarbons is an emerging feedstock biofuel production. It has been reported that the energy value of microalgae biomass is more than 30 times that of the oil plants [2]. The biofuel productivity of microalgae biomass is 100,000 liter per hectare per year, a much higher rate compared to other oil-producing plants. The adaptation power of the microalgae is so high to grow under different physiological stress conditions like marine water or wastewater ponds.

10.2 Wastewater Treatment

Wastewater contains the natural plant nutrients such as N-, P-, and enzyme-producing metals. It is discarded from the anthropogenic activities of industries, municipalities, agricultural runoff, and many more and provides a nutrient-rich media for the high productivity of microalgae biomass. This is evidenced by the high microalgae growth and biomass productivity. The high productivity compensated by the enriched plant nutrient concentration results the high nutrient removal from the wastewater [3]. Therefore, coupling the wastewater treatment with microalgae cultivation may offer a low-cost eco-friendly way for sustainable renewable algae-based biofuel production feedstock. The biomass of microalgae produced in the wastewater retains large amounts of natural hydrocarbons, such as of lipids, carbohydrates, and proteins suitable for biofuel production. However, different physical factors, such as nutrient quantity and quality, light, carbon dioxide, temperature, pH, turbulence, and salinity, are considered for effective microalgae cultivation.

10.3 High Rate Algal Pond

High rate algal pond (HRAP) is an engineered pond that provides the suitable physicochemical condition for the less expensive production of effective algal biomass. Wastewater provides major chemical composition for the growth of microalgae in the HRAP. Microalgal photosynthesis, nutrient uptake, and subsequent growth, coupled with aerobic bacteria degradation of organic compounds, are the broad mechanisms to the wastewater treatment process in the HRAP. Therefore, understanding the role of physicochemical environment in the microalgal growth performance has more impact on the efficiency of the process. Promoting algal production in the HRAP by CO_2 addition enables cost-effective near-tertiary-level wastewater treatment process. Further the enhanced algal biomass, a by-product of wastewater treatment process, can be used for biofuel production. Naturally occurring algae thrive on wastewater providing the oxygen for aerobic bacteria to break down the waste to ammonia, phosphate, and CO_2 which are then assimilated into new algal biomass. Low C:N ratios in wastewater mean that additional CO_2 added to HRAP enable all the wastewater N to be assimilated into algal biomass. CO_2 is easily obtained at the treatment plant as exhaust gas from biogas power generation. CO_2 addition to wastewater treatment HRAP has a further benefit in enhancing bioflocculation of the algal-bacterial biomass to enable low-cost harvest by gravity settling [2].

10.4 Algae Photobioreactor

Algae photobioreactor (APB) assembled with LED lights are designed for the cultivation of microalgae biomass. One of the advantages of the APB over HRAP is that the robustness of the algal strain toward contamination can be eliminated. Pilot-scale APB is used to cultivate an important microalgae species *Chlorella vulgaris* 395 on flue gas from the T.B. Simon Power Plant at Michigan State University [4]. The ABP system is used to facilitate CO_2 sequestration and value-added protein feed production. The effect of flashing light and continuous red LED light on the photosynthetic activity of the microalgae biofilms was investigated in an ABP [5]. In contrast to suspended microalgae cultures, biofilm-based microalgae photobioreactors allowed a uniform exposure of the microalgae to incident radiation throughout the period of microalgae cultivation. The multi-species microalgae biofilms were produced in the photobioreactor.

The PHYCO$_2$'s algae photobioreactor (APB) for enhanced or modified metabolic activity shows great promise for biotechnological exploitation [6]. However, of key concern for many is the safety of genetic modification technology and genetically modified organisms with regard to both the environment and human health, and how these concerns are met will play a key role in ensuring how successful commercialization of genetically modified (GM) algae is achieved. Commercialization opportunities for GM microalgae will inevitably require translation from laboratory to industrial settings, on scales beyond those typically associated with the current biotechnology sector (http://news.algaeworld.org/2017/02/phyco2-msu-make-wastewater-reusable-pure-algae-growth/).

10.5 Hybrid Microalgae Cultivation System

In comparison to HRAP and APB, newly developed hybrid microalgae cultivation system is superior in terms of lipid-rich microalgae production. The hybrid cultivation system enables the separation of two algal activities such as biomass growth and lipid induction phases. This separation facilitates the exponential biomass production and more efficient stress induction techniques simultaneously. The stress induction technique effectively avoids the contamination [7].

10.6 Microalgal Immobilization

In recent years, microalgae biotechnology has been increased for production of consumables like food, cosmetic, aquaculture, and pharmaceuticals. But, during the production process, the small-sized single cellular biomass creates difficulties in the application of biotechnological processes to those organisms. Therefore, cell immobilization technique has been developed to overcome the issues related to the application of biotechnological processes [8]. However, the cell immobilization techniques are more scattered in terms of biotechnological applications. Different passive and active immobilization techniques have been used for the purpose. Effect of immobilization on growth and metabolism of the cells is also reviewed. The immobilization of microalgae has been studied to evaluate different algal activities like metabolite production, culture collection handling, obtaining of energy, and removing of undesired or valuable substances from the media containing nutrients, metals, and different pollutants. Furthermore, the immobilization techniques as well as the living microalgae have been investigated for the applications like biosensors in electronic devices designed to measure the toxicity of effluents.

10.7 Algal Technology for Lake Restoration

Water pollution is a combined result of different anthropogenic activities like industrial effluents, agricultural runoff, sewage discharge, and many more [9]. This leads to eutrophication of water with time. The nutrient-rich eutrophicated water bodies, such as ponds, lakes, and river, develop hazardous algal bloom of blue green algae. The hazardous algal blooms kill the aquatic flora and fauna which is a part of one of the linear food chains of the aquatic ecosystems. Additionally, different floating species of the water do not allow light to enter into the lower water level of the water bodies. This forms an anaerobic condition in the water leading to difficulties for the survival of the aquatic species. There are numerous known disadvantages caused due to eutrophication.

The Phycospectrum Environmental Research Centre (PERC) of Chennai-India has developed a microalgae-based product for restoring the eutrophic water bodies [7]. They have successfully implemented the technique in different water bodies in India and abroad, in terms of sludge reduction, color removal, pH correction, COD and BOD reduction, and odor removal (Table 10.1). PERC has developed the product named as "Phycoplus." The product is designed by addition of customized micronutrients to a few selected robust and natural microalgae. The product breaks the dominance of the polluting algae as well as strips out the essential nutrients for the duckweeds' growth. This process facilitates the disappearance of hazardous algal bloom and floating duckweeds.

The Phycoplus product was administered in a village lake near Aligarh of Northern India. Before starting the restoration, the lake was covered with floating cyanobacterial mats and duckweeds, as shown in Fig. 10.1a. The water of the lake was odorous and black in color. The people of the village abandoned the lake. After

Table 10.1 Parameters of lake water before and after 2 weeks of restoration

	Lake water		
Parameters	Before restoration	After 2 weeks of restoration	% reduction
BOD (mg/L)	220	34	84.5
COD (mg/L)	900	160	82.2
TSS (mg/L)	4720	52	98.9
E. coli	2100	14	99.33
Total coliform	500,000	2500	99.5

Fig. 10.1 (**a**) Appearance of a heavily polluted lake near Aligarh of India before restoration work. (**b**) Appearance of the same lake after 2 weeks of restoration. The boy is spraying Phycoplus

the restoration work by PERC, the lake water became transparent (Fig. 10.1b), COD and BOD reduced to a minimum, and there was a significant reduction of coliform bacteria. The cyanobacterial mats and duckweeds disappeared after 2 weeks of restoration. Periodic addition of the product maintained the lake from further eutrophication and fouling. Different water properties of the lake, before and after 2 weeks of the restoration, are listed in Table 10.1. PERC is planning to evaluate the novel product in more ponds, lakes, and small drains.

Another microalgae technology called "integration of pulsed magnetic field (PMF)" has been jointly developed by PERC and Madras Institute of Magnetobiology of Chennai-India. The technology efficiently enhances the algae-based remediation of wastewater such as reverse osmosis rejects from desalination plants and leather industries' and textile industries' discards. This technology may be a future water remediator, replacing the more expensive and energy-intensive multiple effect evaporators.

10.8 Consortium of Microalgae

The technologies related to microalgal growth, lipid accumulation, and pollutant removal are somehow costly and less efficient. This limitation can be overcome using a microalgae consortium which is better than a monoculture system in terms of biomass and lipid productivity and pollutant removal [10]. The biodiesel production coupled with the use of lipid-extracted algal (LEA) residue provides different advantages such as significant energy value, sustainability, and nutrient recycling and reduces the overall cost of the process integrating tertiary wastewater treatment and microalgal lipid accumulation followed by biodiesel production [11].

10.9 Conclusions

Microalgae convert solar energy and CO_2, resulting in huge biomass production ability. They utilize the dissolved nutrients like N and P in wastewater to convert carbohydrate to energy-rich complex organic lipids and proteins. The complex organic compounds having large active sites transform the inorganic and organic pollutants present in the wastewater by the chemical reactions and adsorption process. This provides the removal of polluting substances from wastewater and huge biomass production, simultaneously. The biomass of microalgae-containing natural organic hydrocarbons is an emerging feedstock biofuel production. Coupling the wastewater treatment with microalgae cultivation may offer a low-cost eco-friendly way for sustainable renewable algae-based biofuel production of feedstock. Different modern techniques, such as high rate algal pond, algae photobioreactor, hybrid microalgae cultivation system, algae-based product, and microalgal immobilization, are used for the simultaneous wastewater treatment and microalgae cultivation. Although several research outputs have been reported by different scientific community, the study is limited to the individual step of the microalgal benefits. Therefore, a comprehensive study is necessary for the pilot plant operation prior to in situ implementation.

Acknowledgments The authors are grateful to Prof. (Dr.) M. R. Nayak, President, Siksha 'O' Anusandhan University, for providing infrastructure and encouragement throughout.

References

1. Aishvarya V, Jena J, Pradhan N, Panda PK, Sukla LB (2015) Microalgae: cultivation and application. In: Sukla LB, Pradhan N, Panda S, Mishra BK (eds) Environmental microbial biotechnology. Springer, Cham, pp 289–312
2. Craggs RJ, Lundquist TJ, Benemann JR (2013) Wastewater treatment and algal biofuel production. In: Borowitzka MA, Moheimani NR (eds) Algae for biofuels and energy developments in applied phycology, vol 5, pp 153–163
3. Narala RR, Garg S, Sharma KK, Thomas-Hall SR, Deme M, Li Y, Schenk PM (2016) Comparison of microalgae cultivation in Photobioreactor, open raceway pond, and a two-stage hybrid system. Frontiers in Energy Research 4:29. https://doi.org/10.3389/fenrg.2016.00029
4. Pavlik D, Zhong Y, Daiek C, Liao W, Morgan R, Clary W, Liu Y (2017) Flashing light effects on CO2 absorption by microalgae grown on a biofilm photobioreactor. Algal Res 25:413–420
5. Girela IM, Curt MD, Fernández J (2017) Flashing light effects on CO2 absorption by microalgae grown on a biofilm photobioreactor. Algal Res 25:421–430 http://news.algaeworld.org/2017/02/phyco2-msu-make-wastewater-reusable-pure-algae-growth/
6. Li XL, Marella TK, Tao L, Peng L, Song CF, Dai L, Tiwari A, Li G (2017) A novel growth method for diatom algae in aquaculture wastewater for natural food development and nutrient removal. Water Sci Technol 75:2777–2783
7. Sivasubramanian V (2017) Lake restoration using microalgal technology. Algae Industry Magazine, February19, 2017
8. Kaparapu J (2017) Micro algal Immobilization Techniques. J Algal Biomass Utln 8:64–70
9. Pradhan D, Sukla LB, Sawyer M, Rahman PKSM (2017) Recent bioreduction of hexavalent chromium in wastewater treatment: a review. J Ind Eng Chem 55:1–20
10. Beacham TA, Sweet JB, Allen MJ (2017) Large scale cultivation of genetically modified microalgae: a new era for environmental risk assessment. Algal Res 25:90–100
11. Chinnasamy S, Bhatnagar A, Hunt RW, Das KC (2010) Microalgae cultivation in a waste water dominate by carpet mill effluents for biodiesel applications. Bioresour Technol 101:3097–3105

Insecticide Toxicity on Indigenous Cyanobacteria from Alluvial Rice Fields

11

Manojit Debnath

Abstract

Pesticides are used for agricultural practices affect the soil's natural beneficial microorganisms like cyanobacteria. Effects of commonly used organochlorine and organophosphate insecticide were studied for growth and remediation potential of rice field indigenous cyanobacterial community under enrichment culture. Inhibition of species richness was noted under insecticide treatment within enriched cyanobacterial community. One unicellular strain of *Aphanothece* sp. and one heterocystous strain of *Nostoc* sp. were subjected to in vitro experiment. Effective concentration 50 (EC 50) of each insecticide was calculated for each strain. When both the strains were studied individually, they showed potentiality to minimize the insecticide concentration in insecticide-spiked growth medium. Organochlorine was found to be the most toxic. To mitigate insecticide pollution, indigenous cyanobacterial population could be used as filter for contaminated agricultural runoff.

11.1 Introduction

Agriculture is an important and integral part of the Indian economy. Agriculture provides food and raw materials to domestic industry of India. On the journey of modernization and industrialization, man has contributed pollution to the life and ecology of plants, animals, and microbes [1]. Increased demand for food and other

M. Debnath (✉)
Integrative Taxonomy and Microbial Ecology Research Group, Department of Biological Sciences, Indian Institute of Science Education and Research Kolkata, Mohanpur, West Bengal, India

Post Graduate Department of Botany, Hooghly Mohsin College, Chinsurah, West Bengal, India

L. B. Sukla et al. (eds.), *The Role of Microalgae in Wastewater Treatment*,
https://doi.org/10.1007/978-981-13-1586-2_11

agro based material has led to the dependency on high-yielding varieties which invite heavy consumption of fertilizers and pesticides [2].

Use of pesticides has seriously environment, agricultural sustainability and human health. The environmental contamination by pesticide residues is of concern due to their persistent nature and bioaccumulation property [1]. The migration of pesticides from agricultural land into surface waters is a major threat to aquatic ecosystems and their biotic communities [3]. Use of pesticide is region specific. Residue of organochlorine and organophosphorus pesticide are high in soils, atmosphere, and in the aquatic system in the tropic and subtropic part of the globe [4]. Bioaccumulation of organochlorine pesticides with in food chain have lead toxicity on immuno-reproductive system [5]. Therefore, the chlorinated pesticides are now largely banned. In India, 100 people died for the parathion-contaminated wheat flour consumption and as a consequences of which the Indian Council of Agricultural Research (ICAR) constituted a committee for possible remedies to combat the pesticide toxicity in the edibles [6]. The 76% of the total pesticides used in India are insecticide compared to herbicides and fungicides [7]. Significant level of pesticides contamination was reported from inland aquatic system and bottled drinking mineral water from mega and major cities as well as rural part of India [8].

Scientific investigation on the effect of insecticides on the nontarget soil microorganisms [9] including diazotrophic cyanobacteria is necessary. Insecticides toxicity on the diazotrophic organisms were related to the insecticide nature and soil type [10]. In the rice field ecosystem, growth and nitrogen fixation of cyanobacteria have received attention in relation to pesticides toxicity [11–14]. Use of cyanobacteria as a biofertilizer for better paddy yield is an old practice [15–17]. These organisms play an important role in soil fertility, plant growth and crop productivity [18]. Not only that, this group of organisms with photoautotrophy and diazotrophy put themselves more adaptable to heavyly contaminated environments [19]. Cyanobacteria have shown ability to mitigate various types of environmental contaminants including pesticides [20], either through their accumulation or degradation.

This chapter summarizes effects of endosulfan, monocrotophos, and phosphamidon (insecticide) on indigenous cyanobacterial community composition and potentiality of individual strain to minimize the insecticide concentration in insecticide-spiked growth medium. In specific, the aim of the present investigation was to reveal whether native cyanobacterial strain (*Aphanothece* sp. and *Nostoc* sp.) isolated from rice field, can be recommended to use as biofertilizer cum biofilter for contaminated agricultural runoff. These findings may contribute for further implementation of photosynthetic microorganisms for effective wastewater management.

11.2 Materials and Methods

11.2.1 Organisms and Culture Conditions: Experimental Organisms Were Selected in a Two-Step Process

Step 1: *Enrichment culture of cyanobacterial community under insecticide stress*. Alluvial rice field soil from Hooghly district, West Bengal, with visible microbial crust was enriched with various concentrations of these three insecticides in liquid BG-11 without nitrogen (-N) medium [21]. Composition of enriched cyanobacterial biomass was examined under light microscope to reveal the community composition compared to control set.

Step 2: *Screening of insecticide-resistant diazotrophic cyanobacterial cultures*. The biomass obtained from insecticide-amended soil enrichment culture was used as inoculums to isolated insecticide-tolerant strain following the spread plate method. Properly homogenized cyanobacterial cultures (1 gm biomass/100ml) were spread on BG-11 (-N) agar plate and incubated until visible colony observed (maximum 15 days). Finally, a non-heterocystous *Aphanothece* sp. and one heterocystous strain of *Nostoc* sp. were selected for the present study. These two strains were selected based on their wider occurrence and faster growth. The organisms were first established in unialgal (using single colony) culture under aseptic condition and maintained in BG-11 (-N).

All isolation and experiments were conducted at a temperature of 28 ± 1 °C with cool white fluorescent light of 25–30 μmol photons m^{-2} s^{-1} in 12/12 h light/dark cycles.

11.2.2 Insecticides Used

Tolerance of cyanobacterial isolates to the insecticides was investigated in BG-11 (-N) medium. The insecticides chosen and concentration used for the study were listed in Table 11.1. Homogenized culture was inoculated and incubated under same conditions as mentioned above. In order to determine the EC 50 (concentration at which 50% growth reduced) growth medium was amended with increasing insecticide concentrations as mentioned in Table 11.1. The flasks were inoculated with 1 gm fresh biomass/100 ml and placed on a rotary shaker-incubator at 120 rpm. After 8 days EC 50 was noted in terms of Chl a in triplicate (n = 3, n is the number of sample taken for the each numerical value). EC 50 of each insecticide was determined using Lagrange's interpolation theory and regression line equation (Table 11.2).

Table 11.1 Detail of pesticides used in the present study

Name of pesticides	Chemical nature	Half-life	Application time in rice plant	Concentration used (ppm)	Field dosage
Phosphamidon (insecticide)	Organophosphate	7 days	Pre-mid-tillering, flowering, after	2, 20, 50, 100	0.1–0.25%
Monocrotophos (insecticide)	Organophosphate	14 days	Nursery, planting to pre-tillering, and panicle initiation to booting	2, 20, 50, 100,	1–1.25%
Thiodan (insecticides)	Endosulfan, organochlorine	10 days	From planting to flowering stage	0.01, 0.05, 5, 25	2 ml/l

Table 11.2 Effective concentrations (EC 50) of studied pesticides in cyanobacteria ($n = 3$)

	Effective concentration (ppm)	
Insecticides	*Aphanothece* sp.	*Nostoc* sp.
Endosulfan	2.7 ± 0.11	4.8 ± 0.19
Phosphamidon	14.6 ± 0.71	18.9 ± 0.83
Monocrotophos	45.30 ± 2.2	76.20 ± 2.8

11.2.3 Assessment of Insecticide Effect

Cyanobacterial biomass was harvested after 8 days for biochemical analysis under insecticide stress and control condition in duplicate ($n = 2$) for evaluating chlorophyll, carotenoid, phycobiliprotein, carbohydrate, and total protein. Emphasis was given to the EC 50 concentration.

11.2.3.1 Chlorophyll a [22]

Cyanobacterial growth was determined in terms of chlorophyll a (Chl a) content from 90% methanol extract. The absorbance of Chl a was measured at 663 nm.

11.2.3.2 Carotenoid [23]

Total carotenoid was extracted using 2 mg fresh biomass in 5 ml 85% acetone. Cyanobacterial cells were disrupted by repeated freezing and thawing. The absorbance was measured at 450 nm.

11.2.3.3 Phycobiliprotein [24]

Cyanobacterial biomass (5gm) in phosphate buffer (0.1 M) was used for extraction process. The absorbance was determined at 562 nm, 615 nm, and 652 nm from supernatant.

11.2.3.4 Carbohydrate [25]

Extraction was done from 1 g fresh biomass in anthrone reagent using a boiling water bath for 10 min. The absorbance of the supernatant was determined at 620 nm. Quantification of carbohydrate was done against the standard curve prepared using glucose.

11.2.3.5 Protein [26]

Cell-free crude extract was used to determine the protein content using bovine serum albumin as the standard.

11.2.3.6 Heterocyst Frequency

Heterocyst frequency (H%) was calculated by a number of heterocyst present per hundred vegetative cells.

$$\mathrm{H\%} = (\text{Number of heterocyst} / \text{Number of cells in filament}) \times 100.$$

11.2.3.7 Inoculation

In all experiments the absorbance of inoculum of all cultures at 760 nm was 0.2. Inoculation was done under aseptic condition in insecticide-containing medium.

11.2.4 Determination of Insecticide Concentration in Growth Medium [27]

Experiments were carried out using 100 ml liquid cultures in 250 ml flasks. Cell-free culture medium was used as control. All flasks were incubated with cotton plugs as mentioned above. The insecticide-amended growth medium of each experiment set was extracted with an equal volume of hexane by vortexing for 30 s twice to estimate remaining/residual insecticide (in the test medium). The organic layer was collected and dried separately. Each dried sample was resuspended in 2 ml hexane and subjected to GC-MS analysis following Lee et al. [27] using standard of each insecticide.

11.3 Results and Discussion

Pesticide (including insecticide) is known to be toxic, only to the targeted pests/insects but not for the nontarget organisms like human. However indiscriminate use of such chemicals has adversely affected the nontarget life forms present to the adjacent ecosystem. These cause serious problem in the soil in many aspects as well as associated water bodies that are receiving the runoff from pesticide-contaminated land mass. In most of the cases, they either kill the soil-inhabiting beneficial microorganisms like diazotrophic cyanobacteria, *Azotobacter*, *Azospirillum*, and many others. The present investigation deals with the effect of three such commonly used insecticides (Table 11.1) on cyanobacterial community composition (species richness), growth, and some biochemical properties of two isolates, which highlight the potentiality to minimize the insecticide concentration within contaminated medium.

11.3.1 Growth Response

Amendment of different concentrations of insecticides in growth medium exhibited varying toxicity to the test cyanobacterial species/strains (Tables 11.2 and 11.3). Only EC 50 concentration of each insecticide was listed to fulfill the aim of this study. The toxicity was proportionate with increasing concentration of insecticides either on community species composition (Table 11.3) or on individual isolated strains (Table 11.2). The degree of toxicity of the studied insecticide was endosulfan greaterthan phosphamidon greaterthan monocrotophos, for *Nostoc* sp. and *Aphanothece* sp. This order is based on EC 50 concentration of tested insecticides (Table 11.3). Beyond the EC 50 level, cyanobacterial growth was inhibited in various degree from day 1 to day 15th (data not shown). Differential permeability of the cyanobacterial cell membrane may cause such variation in the presence of insecticide [28].

The inhibitory effect of monocrotophos could be attributed to the adsorption of this compound on the rich lipid plasma membranes of the algal cells, thus altering the membranes' permeability and diminishing photosynthetic activity as well as increasing reactive oxygen species (ROS) during stress [29]. Table 11.3 indicated that *Nostoc* sp. was more tolerant to different concentrations of studied insecticides than the other strain. This was also reported from previous studies in presence of organophosphorus pesticide monocrotophos up to 150 ppm ([30] and references therein). Kumar et al. [29] study the tolerance of three cyanobacterial strains to endosulfan and record the tolerance in the order of *N. muscorum* greaterthan *A. variabilis* greaterthan *A. fertilissima.* Presence of thick mucillagenous envelop in *Nostoc* spp. may impose higher tolerance of pesticides compared to other heterocystous form with the thin and diffluent mucilage as in *Anabaena variabilis* and weak sheathed *Aulosira fertilissima* [31]. Growth inhibition of cyanobacteria (present study) in presence of three insecticides was more damaging beyond EC 50 (data not shown). In *Anabaena cylindrica* and *Anabaena variabilis* similar observations have been recorded with propanil [32], also for benthiocarb and butachlor in *Anabaena* sp. [33], and with endosulfan in *Nostoc linckia* [34]. In cyanobacteria, inhibition of Chl a driven photosynthesis in presence of insecticides may cause the reduction in growth rate [35]. In *Oscillatoria*, *Hapalosiphon* sp., and *Calothrix braunii* ARM 367 inhibitory effect on growth has been correlated with phototosynthesis, chl a synthesis and diazotrophy in presence of BHC, carbofuran, phorate and malathion [36]. Pesticide-mediated inhibition of electron transport is closely associated with PSII-dependent noncyclic electron acceptor [37]. Pesticides mediated inhibition of electron flow and ATP formation have been shown in cyanobacteria [37]. Under low concentration of studied insecticides, a gradual increase in growth was observed on the 6th/8th day. But in higher concentrations (near to EC 50), growth was usually less than the control set; however, very slow increment was noted between 10–5th

Table 11.3 Comparison of community composition and species diversity of cyanobacterial population under insecticide-amended and non-amended in vitro enrichment

	Insecticides sub EC 50			Insecticides EC 50		
Control (without insecticides)	Endosulfan	Phosphamidon	Monocrotophos	Endosulfan	Phosphamidon	Monocrotophos
Aphanothece sp. [I]						
Gloeocapsa sp. [II]	I, IV, V, VI, IX	I, II, V, VI, VII, IX, X	I, II, IV, V, VI, VIII, IX, X	I, IX, V,	I, II, V, VI, IX, X	I, II, V, VI. VIII, IX, X
O. subbrevis [III]						
Geitlerinema sp. [IV]						
Leptolyngbya sp. 1 [V]						
Phormidium sp. [VI]						
Nostoc sp. [VII]						
Westiellopsis sp. [VIII]						
Nostoc sp. 2 [IX]						
Calothrix sp. [X]						
Species richness						
10	5	7	8	3	6	7

day (data not shown). This may be due to the cyanobacterial degradation/half-life effect of endosulfan, phosphamidon, and monocrotophos or due to the adaptability of cyanobacteria to the insecticides.

11.3.2 Photosynthetic Pigments

In two studied cyanobacteria, growth was maximum for the untreated biomass, with respect to insecticide-amended medium, which might be caused by Chl a synthesis inhibition in insecticide-treated biomass (Tables 11.4 and 11.5). In *Anabaena flos-aquae*, *T. scytonemoides*, and *T. ceylonica* Chl a synthesis has been reported [38–40]. In cyanobacteria inhibition and reduction of Chl a was found to be dose dependent in case of endisulfan [41]. Under low endosulfan doses (2.5 μg/ml), photosynthetic pigments were increased in comparison to respective control [29]. In the present study, maximum phycobiliprotein was detected in *Nostoc* sp. in monocrotophos followed by *Aphanothece* sp. (Tables 11.4 and 11.5). Inhibitory effect of endosulfan at EC 50 was noted in *Aphanothece* sp. which is similar to the *Plectonema boryanum* [41]. Endosulfan toxicity was also reported on *Spirulina platensis* and *Anabaena* sp. by Kumar et al. [42].

11.3.3 Carbohydrate

Little decrease in carbohydrate content compare to control up to EC 50 exposures was noted (Tables 11.4 and 11.5). Significant decrease in the contents of total sugars, reducing sugar, sucrose and polysaccharides have been reported in *Nostoc khilmani* and *Anabaena oscillatoriodes* under thiobencarb [43].

11.3.4 Protein

In *Aphanothece* sp. cellular protein in insecticide-spiked medium was always less compared to control (Table 11.4). In *Nostoc* sp. total protein content was higher under insecticide treatment than control below EC 50 (data not shown). At EC 50, higher protein was only noted in monocrotophos (Table 11.5). Kumar et al. [29] reported protein enhancement in *Aulosira fertilissima*, *N. muscorum*, and *A. variabilis* under pesticides. This suggests that under low concentration of pesticide stimulate the synthesis of stress scavenging proteins. Increase in protein content of *Anabaena sphaerica* due to the effect of 25 μg/ml molinate [44], 2–6 μg/ml benthiocarb [45], 50 μg/ml bavistin, and 1 μg/ml nimbicidin [39] has also been reported. But in the case of *Anabaena* sp. (0.5–2 μg/ml), decrease in protein content was noted [46]. Thus the effect is either insecticide or species specific. In the present investigation decrease in protein content was noted in *Aphanothece* sp. (Table 11.4). Here such activity may be due to insectivcide tolerence limit [29], increased level of ROS or increased protease activity [47]. As a result retardation of growth and carbon-nitrogen assimilation have been noticed [46].

Table 11.4 Effect of insecticides at EC 50 concentration on *Aphanothece* sp. compared to untreated culture ($n = 2$)

Pesticides used	Chl a (μg/ml)	Carotenoid (μg/ml)	Phycobiliprotein (mg/g)	Carbohydrate (mg/g)	Protein (mg/g)	Generation time (h)
Control	3.22	1.7	52	83	1050	108
Endosulfan	0.960	1.2	37	80	525	149
Phosphamidon	1.20	1.4	54	80	609	115
Monocrotophos	1.65	1.9	55	82	651	122

Table 11.5 Effect of insecticides at EC 50 concentration on *Nostoc* sp. compared to untreated culture ($n = 2$)

Pesticides used	Chl a (µg/ml)	Carotenoid (µg/ml)	Phycobiliprotein (mg/g)	Carbohydrate (mg/g)	Protein (mg/g)	Generation time (h)
Control	5.60	2.7	102	95	957	94
Endosulfan	2.35	2.5	124	82	867.5	132
Phosphamidon	2.10	2.9	159	83	926.4	106
Monocrotophos	3.30	2.7	163	87	989	108

11.3.5 Heterocyst Frequency

Heterocyst frequency (H%) of *Nostoc* sp. was calculated in insecticide-treated and insecticide-untreated culture. No observed change in heterocyst frequency was noted under pesticide (at EC 50) treatment compared to their control set. H% observed in *Nostoc* sp. was (6–7%) under both insecticide-treated and insecticide-untreated conditions. In cyanobacteria nitrogen fixation is directly related to heterocyst number. Under insecticide stress destabilization of heterocyst membrane is the main cause for inhibition of dinitrogen fixation in *Anabaena* 7119 [48]. Moreover heterocysts formation was completely inhibited in *N. muscorum* under pesticide treatment [49]. Inhibition of heterocyst differentiation in *N. linckia* in the presence of benthiocarb has also been reported [50].

11.3.6 Minimization of Insecticide by Cyanobacterial Strains

In this present study, assessment of insecticide-spiked medium revealed that the two cyanobacterial strains have the ability to minimize total insecticide content at EC 50 concentrations compared to control set culture medium. In this study, *Nostoc* sp. was more efficient than *Aphanothece* sp. The mean removal values of 35%, 49%, 57% and 28%, 46%, 52% for endosulfan, phosphamidon, and monocrotophos were recorded for *Nostoc* and *Aphanothece* respectively (Fig. 11.1). In the case of all control sets of insecticides, total insecticide content reduced between 6 and 8 percent up to 8 days of incubation. Accumulation and degradation of organochlorine and organophosphorus pesticides were noted in many cyanobacteria including *Synechococcus elongatus*, *Anacystis nidulans*, and *Microcystis aeruginosa* in polluted aquatic ecosystems [51]. Many cyanobacterial genera such as *Synechococcus*, *Microcystis*, *Oscillatoria*, *Nodularia*, *Nostoc* and *Anabaena* have the ability to remove or degrade the lindane residues as reported by El-Bestawy et al. [52].

Degradation of broad-spectrum organophosphorus herbicide glyphosate was reported by *Microcystis* sp., *Lyngbya* sp., *Nostoc*sp and *Anabaena* sp. The utilization of degraded product as phosphate source by these organisms has also been reported [53]. Lipok et al. [54] also demonstrated that *Spirulina* sp. could degrade the glyphosate herbicide. Therefore the cultivation of cyanobacteria in wastewater bodies may have good potential to reduce pollution through degradation of pesticides and further reduction in the BOD and COD through growth support to other microbes [55]. Ibrahim et al. [30] reported that in the absence of inorganic phosphate, cyanobacteria are actively engaged to utilize organophosphate insecticide as a source of phosphate through enzymatic metabolism. They identified the high ability of *N. muscorum* to biodegrade malathion (91%) at different concentrations. Besides this, several studies have shown that pesticides affect the growth of cyanobacteria. In the presence of endosulfan and dimethoate, growth and survivability of *Anabaena doliolum* were decreased [56], whereas hexazinone and atrazine inhibited the growth of *Anabaena flos-aquae* [57]. Therefore, the negative impact of pesticides on the growth of cyanobacterial species can reduce the removal or dissipation of pesticides from ecosystems. In *Anabaena* sp. pathways for metabolic

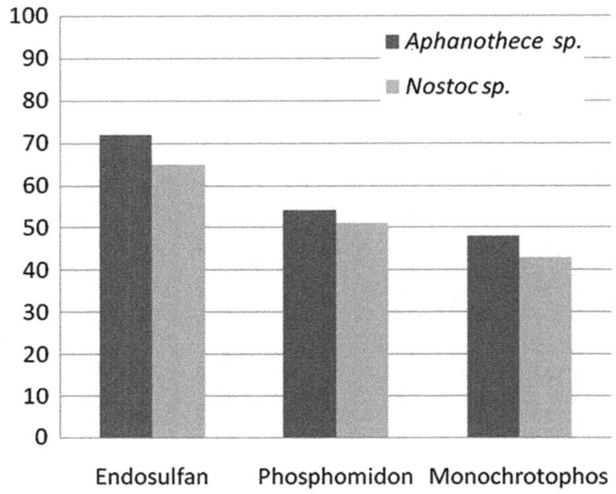

Fig. 11.1 Mean percentage removal of insecticides from culture medium by two cyanobacterial strain

degradation of endosulfan were studied by Lee et al. [27]. This cyanobacterium degrades endosulfan and produced nontoxic endodiol (major product) and trace amount of endosulfan sulfate. Therefore production of nontoxic endodiol may be a detoxification process. However, the question arises "how they produced endodiol?". It might be due to "an increase of pH in the medium, and chemical hydrolysis might influence the rate of endodiol production" [27]. Biological hydrolysis of endosulfan using a strong buffered culture was also investigated [58].

The report in Science Daily [59] stated that the Institute of Natural Resources and Agrobiology of the Spanish National Research Council (CSIC) has suggested a new method of pesticides encapsulation and slow release to prevent the leaching as well as the volatilization of such molecules. Here the pesticide was encapsulated in lecithin liposomes or vesicles leading to the adsorption on clay. This liposomes, pesticide, and clay complex will allow slow release of pesticide when dispersed in water. This entrapment technique may restricts the over spread of pesticides and their residues to other surfaces and aquifers [59]. So, it is also possible to immobilized cyanobacteria for such purpose to control pollution. From the present study and other studies discussed herein, it is clear that cyanobacteria have potential to reduce insecticide load from agriculture field runoff which could be explored for wastewater management.

Acknowledgments The research was supported by Grant DBT-WB (103-Bt/(Estt)/RD/2011-2012) and SERB-DST India (SB/YS/LS-76/2014) to MD. The author is thankful to Dr. P. Bhadury, Department of Biological Sciences, IISER Kolkata, for support and encouragement. The author is thankful to the Head, Department of Botany Hooghly Mohsin College, for necessary support to conduct the study.

References

1. Kumarasamy P, Govindaraj S, Vignesh S, Rajendran RB, James RA (2012) Anthropogenic nexus on organochlorine pesticide pollution: a case study with Tamiraparani river basin, South India. Environ Monit Assess 184(6):3861–3873 2011 July 29
2. Chauhan RS, Singhal L (2006) Harmful effects of pesticides and their control through cowpathy. Int J Cow Sci 2:61–70
3. Bereswill R, Streloke M, Schulz R (2013) Current use pesticides in stream water and suspended particles following runoff: exposure, effects, and mitigation requirements. Environ Toxicol Chem 32:1254–1263
4. Carvalho FP, Fowler SW, Villeneuve JP, Horvat M (1997) Pesticide residues in the marine environment and analytical quality assurance of the results. In: Proceedings of an international FAO/IAEA symposium on the environmental behaviour of crop protection chemicals. IAEA, Vienna, pp 35–57
5. Balakrishnan'G RM, Banerjee BD, Hussain QZ (1985) Effect of dietary protein, dichlorodiphenyltrichloroethane (DDT) and hexachlorocyclohexanes (HCH) on hepatic microsomal enzyme activity in rats. Br J Nutr 54:563–566
6. Wadhwani, Lall IJ (1972) Indian Council of Agricultural Research, New Delhi, pp 44–49
7. Saiyed HN, Bhatnagar VK, Kashyap R (1999) "Impact of pesticide use in India electronic journals: Asia pacific newsletter: 1999–2003. http://www.ttl.fi/Internet/English/Infotion/Electronic+journals/Asian-Pacific+News-let-ter/1999–03/05.htm
8. Agrawal A, Pandey RS, Sharma B (2010) Water pollution with special reference to pesticide contamination in India. J Water Resour Protect 2:432–448
9. Hill IR, Wright SJL (1978) Pesticide Microbiology. Academic Press, London, pp 7–10
10. Nayak DN, Rao RV (1982) Pesticides and nitrogen fixation in paddy soils. Soil Biol Biochem 14:207–210
11. Anand N, Veerappan B (1980) Effect of pesticides and fungicides on blue green algae. Phykos 19:210–212
12. Kolte SO, Goyal SK (1990) Inhibition of growth and nitrogen fixation in *Calothrix marchica* by herbicide *in vitro*. In: Kaushik BD (ed) Proc national symposium on Cyanobacterial nitrogen fixation. Pub Associated Co., New Delhi, pp 507–510
13. Singh PK (1973) Effect of pesticides on blue green algae. Arch Mikrobiol 89:317–320
14. Stratton GW (1987) The effects of pesticides and heavy metals on phototrophic microorganisms. Rev Environ Toxicol 3:71–112
15. De PK (1939) The role of blue-green algae in nitrogen fixation in rice fields. Proc R Soc Lond B 127:121–139
16. Mian MH, Stewart WDP (1985) Fate of nitrogen applied as *Azolla* and blue-green algae (cyanobacteria) in waterlogged rice soils. A ^{15}N tracer study. Plant Soil 83:363–370
17. Relwani LL (1963) Role of blue green algae on paddy field. Curr Sci 32:417–418
18. McCann AE, Cullimore DR (1979) Influence of pesticides on the soil algal flora. Residue Rev 72:1–31
19. Sokhoh NA, Al-Hasan RH, Radwan SS, Hopner T (1992) Self-cleaning of the Gulf. Nature 359:109
20. Megharaj M, Madhavi DR, Sreenivasulu C, Umamaheswari VK (1994) Biodegradation of methylparathion by soil isolates of microalgae and cyanobacteria. Bull Environ Contam Toxicol 53:292–297
21. Stainer RY, Kunisawa R, Mandel M, Cohin-Bazire G (1971) Purification and properties of unicellular blue green algae (order Chrococcales). Bacteriol Rev 35:171–205
22. Mackinney G (1941) Absorption of light by chlorophyll solutions. J Biol Chem 140:315–322
23. Hellebust JA, Craige JS (1978) Handbook of physiological and biochemical methods. Cambridge University Press, Cambridge, pp 64–70
24. Siegelman HW, Kygia JH (1978) Handbook of phycological methods. Cambridge University Press, Cambridge, pp 73–79

25. Spiro RG (1966) Analysis of sugars found in glycoproteins. Methods Enzymol 8:3–26
26. Lowry OH, Rosebrough NJ, Farr AL, Randall RJ (1951) Protein measurement with the folin phenol reagent. J Biol Chem 193:265
27. Lee SE, Kim JS, Kenedy IR, Park JW, Kwon GS, Koh SC et al (2003) Biotransformation of an Organochlorine Insecticide,Endosulfan, by *Anabaena* species. Agric Food Chem 51:1336–1340
28. Lal S, Saxena DM (1980) Cytological and biochemical effects of pesticides on microorganisms. Residue Rev 73:49–86
29. Kumar S, Habib K, Fatma T (2008) Endosulfun induced biochemical changes in nitrogen-fixing cyanobacteria. Sci Total Environ (ELSEVIER) 403:130–138
30. Ibrahim WM, Karam MA, Reda M (2014) Biodegradation and utilization of organophosphorus pesticide malathion by Cyanobacteria. Biomed Res Int 2014:1–6. https://doi.org/10.1155/2014/392682
31. Ahmed MH, Venkataraman GS (1973) Tolerance of Aulosira fertilissima to pesticides. Curr Sci 1973:42–108
32. Wright SJL, Stantharpe AF, Downs JD (1977) Interaction of herbicide propanil and a metabolite, 3, 4-dichloromiline with blue green algae. Acta Pathol Acad Hung 12:51–57
33. Zarger MY, Dar GH (1990) Effect of benthiocarb and butachlor on growth and nitrogen fixation by cyanobacteria. Bull Environ Contam Toxicol 45:232–234
34. Satish N, Tiwari GL (2000) Pesticide tolerance in Nostoc linckia in relation to the growth and nitrogen fixation. Proc Natl Acad Sci India 70:319–323
35. Padhy RN (1985) Cyanobacteria and pesticides. Res Rev:941–944
36. Kaushik BD, Venkataraman GS (1993) Response of cyanobacterial nitrogen fixation to insecticides. Curr Sci 52:321–323
37. Moreland DE (1980) Mechanism of action of herbicides. Annu Rev Plant Physiol 315:97–638
38. Chinnaswamy R, Patel RJ (1983) Effect of pesticide mixtures on the blue green alga *Anabaena flos-aquae*. Microb Lett 24:141–143
39. Rajendran UM, Elango K, Anand N (2007) Effects of a fungicide, an insecticide, and a biopesticdie on Tolypothrix scytonemoides. Pestic Biochem Physiol 87:164–171
40. Roychoudhury P, Kaushik BD (1986) Response of cyanobacterial growth and nitrogen fixation to herbicides. Phykos 25:36–43
41. Prasad SM, Kumar D, Zeeshan M (2005) Growth, photosynthesis, active oxygen species and antioxidants responses of paddy field cynobacterium Plectonema boryanum to endosulfan stress. J Gen Appl Microbiol 51:115–123
42. Kumar S, Jetley UK, Fatma T (2004) Tolerance of Spirulina platensis-S5 and anabaena sp. to Endosulfan an organochlorine pesticide. Ann Plant Physiol 18(2):103–107
43. Mansour FA, Soliman ARI, Shaaban-Desouki SA, Hussein MH (1994) Effect of herbicides on cyanobacteria. I. Changes in carbohydrate content, Pmase and GOT activities in Nostoc kihlmani and Anabaena oscillarioides. Phykos 33:153–162
44. Yan GA, Yan X, Wu W (1997) Effect of herbicide Molinate on mixotrophic growth, photosynthetic pigments and protein content of Anabaena sphaerica under different light conditions. Ecotoxicol Environ Saf 38:144–149
45. Bhunia AK, Basu NK, Roy D, Chakrabarti A, Banerjee SK (1991) Growth, chlorophyll a content, nitrogen fixing ability and certain metabolic activities of N. muscorum: effect of methylparathion and benthiocarb. Bull Environ Contam Toxicol 47:43–50
46. Babu GS, Hans RK, Singh J, Viswanathan PN, Joshi PC (2001) Effect of lindane on the growth and metabolic activities of cyanobacteria. Ecotoxicol Environ Saf 48:219–221
47. Leitao M, das A, Cardozo KHM, Pinto E, Colepicolo P (2003) PCB induced oxidative stress in the unicellular marine dinoflagellates Lingulodinium polyedrum. Arch Environ Comtam Toxicol 45:59–65
48. Orus MI, Marco E (1991) Heterocysts structure alteration and oxygen-mediated inhibition of dinitrogen fixation by trichlorfon in *Anabaena* 7119. J Exp Bot 95:1595–1600
49. Vaishampayan A (1984) Biological effects of rice-field herbicide Monuron on a nitrogen-fixing cyanobacterium *Nostoc muscorum*. Microbios Lett 28:105–111

50. Mishra AK, Pandey AB (1989) Toxicity of three herbicides to some nitrogen- fixing cyanobacetria. Ecotoxicol Environ Saf 17:236–246
51. Vijayakumar S (2012) Potentialapplicationsofcyanobacteriainindustrial effluents-a review. J Bioremed Biodegr 3:1–6
52. El-Bestawy EA, AbdEl-Salam AZ, Mansy AERH (2007) Potential use of environmental cyanobacterial species in bioremediation on lindane-contaminated effluents. Int Biodeterior Biodegrad 59:180–192
53. Forlani G, Pavan M, Gramek M, Kafarski P, Lipok J (2008) Biochemical basis for a widespread tolerance of cyanobacteria to the phosphonate herbicide glyphosate. Plant Cell Physiol 49:443–456
54. Lipok J, Wieczorek D, Jewginski M, Kafarski P (2009) Prospects of in vivo 31PNMR method-ing lyphosate degradation studies in whole cell system. Enzym Microb Technol 44:11–16
55. Singh JS, Kumar A, Rai AN, Singh DP (2016) Cyanobacteria: a precious bio-resource in agriculture, ecosystem, and environmental sustainability. Front Microbiol 7:529. https://doi.org/10.3389/fmicb.2016.00529
56. Mohapatra PK, Mohanty RC (1992) Growth pattern changes of chlorella vulgaris and anabaena dolilum due to toxicity of dimethoate and endosulfan. Bull Environ Contam Toxicol 49:576–581
57. Abou-Waly H, Abou-Setta MM, Nigg HN, Mallory LL (1991) Growth response of freshwas algae, Anabaena flos-aquae and Selenastrum capricornutum to atrazine and hexazinone herbicide. Bull Environ Contam Toxicol 46:223–229
58. Sutherland TD, Horne I, Lacey MJ, Harcourt RL, Russell RJ, Oakeshott JG (2000) Enrichment of an endosulfan- degrading mixed bacterial culture. Appl Environ Microbiol 66:2822–2828
59. Science Daily (2009, March 12) http://www.sciencedaily.com/releases/2009/03/090306084639.htm

Analysis of Growth and Biochemical Contents of Microalgae Grown with Wastewater Effluent of Emami Paper Mill, Balasore

12

P. Dutta, S. Bhakta, E. Sahu, P. Bhuyan, and A. K. Bastia

Abstract

Phyco-remediation of wastewater through microalgae is a promising area; hence, the study focused here is to analyse the growth and biochemical constituents of waste-grown microalgal strain collected from municipality wastewater of Baripada and industrial effluents of Emami Paper Mill, Balasore. Effluent samples were collected and brought to laboratory, and physicochemical parameters of the effluent were characterized. A total of seven microalgal taxa comprising of five cyanobacteria (single species of each genera *Scytonema*, *Leptolyngbya*, *Pseudospongiococcus*, *Hapalosiphon*, *Dolichospermum*) and two green algae (i.e. *Scenedesmus* sp. and *Oocystis* sp.) were isolated and selected for phyco-remediation potential. The strains were inoculated in different dilutions of effluents, viz. 25%, 50%, 75% and 100%, taking the culture media as positive control. The cultures were maintained at controlled condition, i.e. continuous light of 2400 lux with 25 °C temperature. The growth of algae; pigments such as chlorophylls, carotenoids and biliproteins; total carbohydrate and protein content were determined regularly using spectrophotometer up to 20 days. *Scenedesmus* sp. and *Dolichospermum* sp. showed maximum growth in 75% of diluted effluent, but in *Hapalosiphon* sp. it is 50% except *Scytonema* sp. which had a maximum growth in 100% effluent. *Leptolyngbya* showed comparatively maximum response of growth in control to the effluent. The macromolecular content and pigment contents also revealed the same trend as growth for the entire test organism.

P. Dutta · E. Sahu · P. Bhuyan · A. K. Bastia
P. G. Department of Botany, North Orissa University, Baripada, Mayurbhanj, Odisha, India

S. Bhakta (✉)
Botanical Survey of India, Western Regional centre, Pune, Maharashtra, India
e-mail: skbhakta@bsi.gov.in

L. B. Sukla et al. (eds.), *The Role of Microalgae in Wastewater Treatment*,
https://doi.org/10.1007/978-981-13-1586-2_12

12.1 Introduction

The quality of human life on earth is inextricably linked to the overall quality of water resources. Different organic and inorganic contaminants in water bodies from domestic and industrial sources are leading to environmental deterioration [1]. Untreated or partly treated domestic and industrial wastewater is a major concern for the developing nations like India and Bangladesh. It is becoming a challenging task to global society for adopting an effective control measure. The conventional techniques have been used so far for remediation of contaminated water and transport of hazardous material. In this context phyco-remediation employing microalgae is a novel alternative technology of treating industrial effluent. However, the utilization of algal biomass grown through an integrated approach with phyco-remediation is being a stupendous task so far.

Apart from the common wastewater treatment practice, algae are also being used as an important agent of bioremediation in many wastewater facilities. It is becoming more popular to reuse the biomass harvested during wastewater treatment, simultaneously along with CO_2 sequestration, BOD balance and nutrient removal in an integrated manner. Enzymatic breakdown and absorption of macronutrients, micronutrients and trace and heavy metals of wastewater are a part of metabolism of microalgae. However, the strains were always preferred based on their tolerance to wastewater, their nutrient intake capacity and their growth [21]. The dominant form of microalgae in well-aerated cultures has always been the coccoid microalgae that are not readily grazed up to climax of the culture [20]. Microalgae need light, CO_2 and inorganic nutrients like nitrogen and phosphorous (N and P) for their growth and development. There are a few common taxa of microalgae which are preferred most for the phyco-remediation, e.g. *Chlorella*, *Scenedesmus*, *Synechocystis*, *Gloeocapsa*, *Chroococcus*, *Anabaena*, *Lyngbya*, *Oscillatoria*, *Spirulina*, etc., and subsequently utilization of their biomass for food, feed and fuel. Algae got mileage over other phytoremediation entities due to its fast growth in a limited space and its viability in contaminated water using various nutrients.

Unlike general practice of algae cultivation in freshwater nutrient medium, cultivation of algae in wastewater will produce a low-cost biomass as well as treatment of carbon footprint and reduce pollutant load [4]. A successful phyco-remediation of wastewater depends on the extent of contaminants and environmental and ecological conditions. Hence, pre-treatment of wastewater could be fruitful by reducing hardness, TDS through dilution [23, 25] or radiation treatment and so on [5]. Moreover, the selection of native strain is also an essential parameter for phyco-remediation, and wastewater rich in nitrate and phosphate can yield a better biomass which is an additional advantage [3]. Keeping all these in view, the present investigation aims to study the growth, survival, pigment composition and macromolecular contents of two waste-grown algae, isolated from municipality wastewater of Baripada area.

12.2 Materials and Methods

12.2.1 Sample Collection

The algal samples were collected in clean sampling bottles (Tarson, 25 x 50 mm) with the help of forceps and Pasteur pipette from the municipality wastewater of Baripada and Emami Paper Mill, Balasore, and brought to the laboratory for identification, culture and preservation (Fig. 12.1). The industrial effluent of Emami Paper Mill, Balasore, was brought to the laboratory for analysis and further study.

12.2.2 Micrometry, Microphotography and Identification

For identification, two to three slides were prepared from each sample, observed under compound and/or phase contrast microscope, and the characters were enumerated. Photomicrographs were taken using Hund Wetzlar Trinocular Research Microscope with Canon-EOS 550D camera attachment. Micrometry was done using ocular and stage micrometer (Erma, Japan) to determine the cell dimensions. The algal species were identified using the monographs and standard literatures [6, 13–15, 22].

12.2.3 Culture of Algal Isolates

Each algal sample was observed under microscope after 1 or 2 days of collection. The enriched algal species were streaked on agar plates of BG-11 ± N/BB medium and incubated under white fluorescent light. Different culture media were used for culturing different groups of algae. BG-11 medium +N and -N was used for

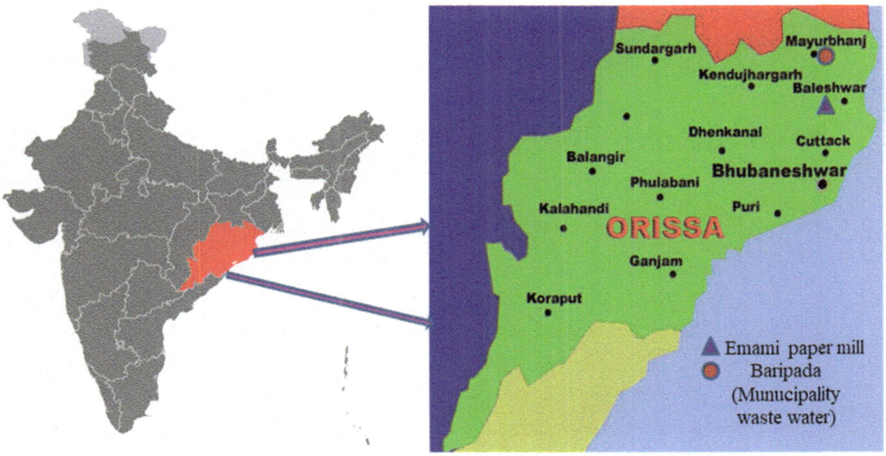

Fig. 12.1 Map of Odisha (Orissa) showing the sampling sites

culturing cyanobacteria [24]. Bold's basal medium [2] was used to culture green algae. For preparation of media, the following chemicals (Merk/Himedia) were used at specific concentrations to make 1 litre with double-distilled water.

Stock solutions of the chemicals (100 ml) were prepared with double-distilled water, sterilized and stored in refrigerator for subsequent use. To prepare 1 litre culture medium, 1 ml from each stock was added dropwise into a clean sterilized volumetric flask of 1 litre capacity containing 600–800 ml distilled water. The contents were stirred continuously during addition using a magnetic stirrer. After addition of nutrient from stock solution, the final volume of the flask was adjusted to 1 litre by adding distilled water. The pH of the culture medium before autoclaving was adjusted by addition of sterile 0.01 N to1 N NaOH/HCl under aseptic condition. The culture medium and the glass wares used were sterilized at 15 lb. pressure for 20 min in an autoclave.

Composition of BG – 11 medium

Ingredients	**g/litre**
1. $MgSO_4 \cdot 7H_2O$	0.075
2. $CaCl_2 \cdot 2H_2O$	0.036
3. a. Citric acid	0.006
b. Ferric ammonium citrate	0.006
c. EDTA (disodium magnesium salt)	0.001
4. Na_2CO_3	0.02
5. K_2HPO_4	0.04
Micronutrients	**mg/litre**
1. H_3BO_3	2.86
2. $MnCl_2 \cdot 4H_2O$	1.81
3. $ZnSO_4 \cdot 7H_2O$	0.222
4. $CuSO_4 \cdot 5H_2O$	0.079
5. $Na_2MoO_4 \cdot 2H_2O$	0.390
6. Co $(NO_3)_2 \cdot 6H_2O$	0.049

N.B. – For culturing non-heterocystous cyanobacteria, KNO_3 at the concentration of 1.5 g/L was added to the above medium

Composition of Bold's Basal Medium

Ingredients	**g/litre**
1. $NaNO_3$	0.25
2. $CaCl_2 \cdot 2H_2O$	0.025
3. $MgSO_4 \cdot 7H_2O$	0.075
4. K_2HPO_4	0.075
5. KH_2PO_4	0.175
6. NaCl	0.025
7. $FeSO_4 \cdot 7H_2O$	0.005
EDTA	0.005

(continued)

Micronutrients	mg/litre
H_3BO_3	2.86
$MnCl_2{\cdot}4H_2O$	1.81
$ZnSO_4{\cdot}7H_2O$	0.222
$CuSO_4{\cdot}5H_2O$	0.079
$Na_2MoO_4{\cdot}2H_2O$	0.39
$Co\ (NO_3)_2{\cdot}6H_2O$	0.04

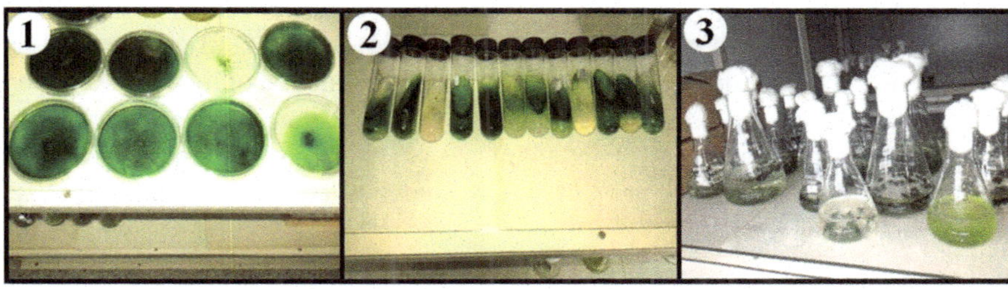

Plate 12.1 Culture and maintenance of waste grown algae in the laboratory 1. Isolation in petri plates, 2. Preservation of isolated taxa in slants, 3. Shake flask culture for mass multiplication

12.2.4 Culture Conditions and Maintenance

The pure cultures of algal strains were isolated from wastewater maintained in agar slants in glass screw-cap tubes containing 1.2% (w/v) agar-agar in the basal inorganic medium and in 100 ml capacity conical flasks containing 50 ml medium before using them in the experiments. They were maintained in culture racks in a temperature-controlled room at 25 ± 1 °C under continuous light intensity of 7.5 W/m^2 fluorescent light of 16:8 light and dark cycles from daylight fluorescent tubes and examined from time to time (Plate 12.1). Cultures at their exponential phase were used as inoculums for the experiments.

12.2.5 Analytical Methods

In the present study, the waste-grown algae were selected as test organism for utilization of factory waste, collected from Emami paper Mill, Balasore. Their growth, pigment composition (chlorophylls, carotenoids and biliproteins) and macromolecular contents (total carbohydrate and protein) were studied at different concentrations of waste (25%, 50%, 75% and 100%) taking growth medium as control. Triplicates were set for each set of experiments, and the cultures were harvested at alternate day up to 20 days of inoculation in alternate light-dark (16:8) for growth and on the last day, i.e. the 20th day, for estimation of pigments and macromolecules.

12.2.5.1 Growth Measurements

Growth experiments were conducted in (18 × 150 mm) hard glass test tubes containing 10 ml of medium for all the experiments conducted during this investigation. The test organisms were cultured in different dilutions of wastewater, taking culture media as a control, i.e. 25% effluent, 50% effluent, 75% effluent and 100% effluent. For growth the cultures were harvested at alternate day up to the 20th day at 25 ± 1 °C under 7.5 W/m^2 cool fluorescent light of 16:8 light and dark cycles. Growth was estimated by measuring the absorbance of the homogenized suspension of the culture in a Systronics 2203 double beam spectrophotometer set at 750 nm [9, 10] with reference to blank containing distilled water. The objective of measuring the growth of pigmented micro-organisms at this wavelength is that absorption by pigments is minimum and the optical density here is a function of light scattering by the organism. Choosing a suitable wavelength for growth measurement is necessary because unless the rate of growth and pigment metabolism is the same, optical density of a microbial suspension at pigment-absorbing wavelength did not truly reflect the growth of the organism [27].

12.2.5.2 Estimation of Pigments

Culture suspension containing green algal samples was centrifuged at 5000 rpm for 10 min. The supernatant was discarded, and the pigments in the pellet were extracted using 10 ml 90% (v/v) methanol for 30 min at 4 °C in refrigerator followed by heating in hot water bath at 60 °C for approximately 1 min. The clear methanol extract was separated from the residue by centrifugation and the process repeated until all the pigments are removed. Then the pellet was discarded, and the entire methanol extracts were combined, and the volume was made 5 ml. The absorbance of the chlorophyll-a pigment was taken in a Systronics 2203 double beam spectrophotometer set at 663 nm and chl-b at 645 nm. The amount of total chlorophyll was determined following [16]. The amounts of total carotenoids were estimated at 475 nm according to [7].

Culture suspension containing blue green algal samples (cyanobacterial) was centrifuged at 5000 rpm for 10 min. The supernatant was discarded, and the pigments in the pellet were extracted using 10 ml 90% (v/v) methanol for 30 min at 4 °C in refrigerator followed by heating in hot water bath at 60 °C for approximately 1 min. The clear methanol extract was separated from the residue by centrifugation and the process repeated until all the lipid soluble pigments are removed. All the methanol extracts are combined, and the volume was made 5 ml. Then absorbance of the chlorophyll-a pigment was taken in a Systronics 2203 double beam spectrophotometer set at 663 nm determined by using the extinction coefficient given by [18]. The amounts of total carotenoids were estimated at 475 nm according to [7]. The water-soluble pigments (phycobilisomes) were obtained as aqueous extract from the methanol extracted filaments. After decanting methanol-soluble pigments, 5 ml of distilled water was noted to the pellet and was kept in the refrigerator at 4 °C overnight. The tubes were brought to room temperature, shaken thoroughly and centrifuged at 4000 rpm for 10 min. Absorbance of water-soluble pigments was taken at 565 nm, 620 nm and 650 nm for phycoerythrin (PE), phycocyanin (PC) and allophycocyanin (APC), respectively. The quality of total phycoerythrin and phycocyanin was calculated using the extinction coefficient given by [11].

12.2.5.3 Estimation of Carbohydrate

The total carbohydrate content of the dried biomass was estimated by anthrone reagent method [12] using glucose as standard taken within the range 10–50 μg/ml. The anthrone in its 'enol' form reacts with furfural or its derivatives produced by the dehydration of sugar to form blue colouration. 5 ml of anthrone reagent (200 mg anthrone powder in 100 ml concentrated H_2SO_4) was added to1ml of algal sample and kept in the ice bucket and allowed to cool for 15 min followed by 10 min in boiling water bath. Samples were cooled and centrifuged at 5000 rpm for 6–8 min. The optical density of the supernatant was measured in a spectrophotometer (Systronics 2203 double beam spectrophotometer) set at 625 nm wavelengths against the anthrone reagent as blank.

12.2.5.4 Estimation of Protein

Protein concentration in samples was determined by following standard method [17] using bovine serum albumin as standard taken within the range 20–100 μg/ml. Folin and Ciocalteu's phenol reagent used is essentially a phosphotungstic phosphomolybdic acid solution, is reduced by phenols to molybdenum blue and is determined spectrophotometrically. Protein reduces the phenol reagent to give a colour complex, and its intensity depends on the amount of tyrosine and tryptophan present in the protein and thus varies for different proteins.

Reagent 'C' was prepared freshly by adding 50 ml of reagent A (2 gm NaOH + 10 gm Na_2CO_3 + 0.1 gm Na-K-tartrate in 500 ml distilled water) and 1 ml of reagent B (0.5% $CuSO_4{\cdot}5H_2O$). Similarly reagent 'D' was prepared by adding 1 ml of Folin-Ciocalteu's phenol (Qualigens, India) to 2 ml of distilled water. 5 ml of reagent C was added to cyanobacterial residue (pellet). The test tubes were incubated at room temperature for 30 min followed by boiling in water bath at 100 °C for 5 min. After cooling, the content was centrifuged at 5000 rpm for 6–8 min, and 0.5 ml of reagent D was added to the supernatant and shaken well to ensure total distribution of the reagent. After keeping the tubes at room temperature for 15 min, the absorbance of the coloured suspension was measured at 750 nm in a Systronics 2203 double beam spectrophotometer against the reagent blank.

12.3 Results and Discussion

12.3.1 Results

A total of seven algal taxa were isolated and identified from two sampling sites, i.e. Baripada municipality wastewater and Emami Paper Mill effluent of Odisha. Cyanobacteria show the occurrence of maximum taxa of five species under five genera followed by green algae having two species under two genera (Plate 12.2, Table 12.1).

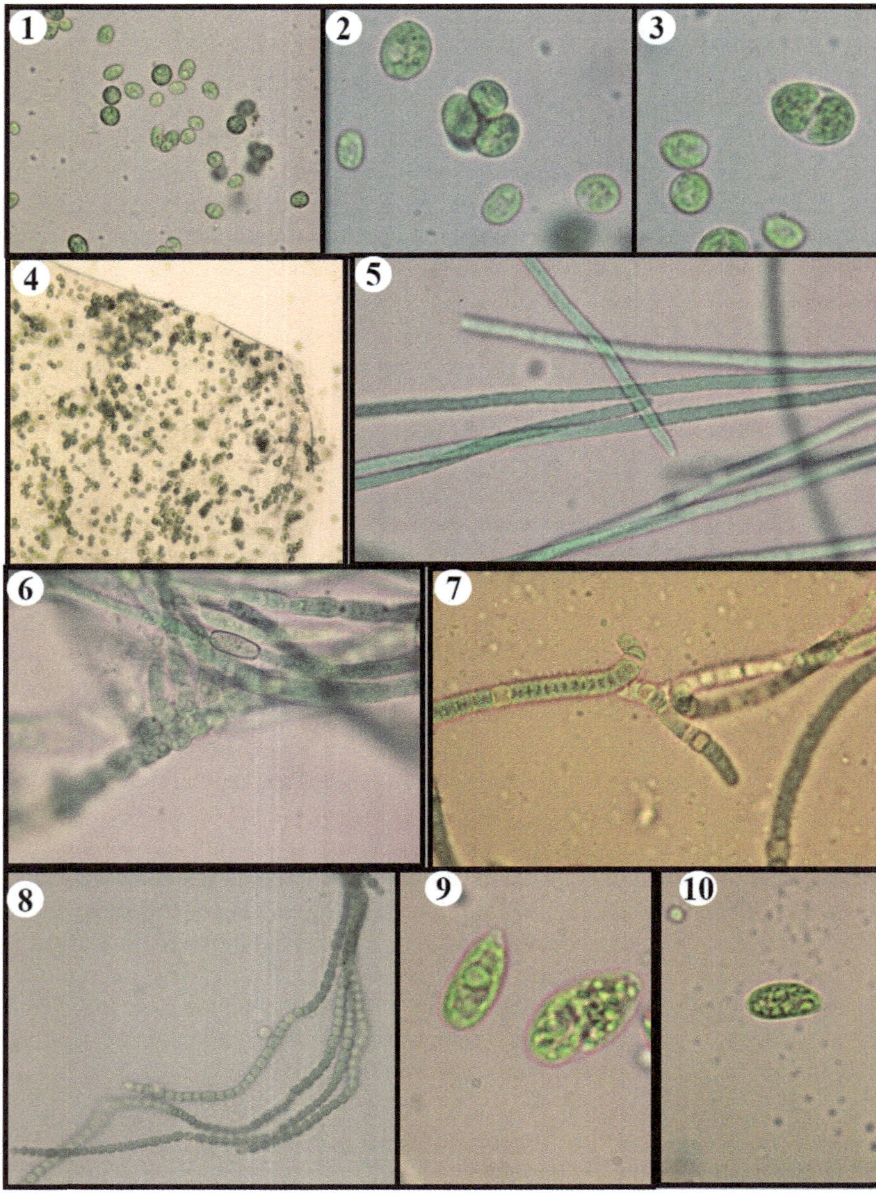

Plate 12.2 1–3. *Oocystis polymorpha*, 4. *Pseudospongiococcus protococcoides*, 5. *Leptolyngbya subuliformis*, 6. *Hapalosiphon hibernicus*, 7. *Scytonema hoffman-bangi*, 8. *Dilichospermum affine*, 9–10. *Scenedesmus sp*

Table 12.1 List of algal taxa isolated from different wastewater sources

Sl. No.	Name of taxa	Municipal waste	Industrial effluent
1	Cyanoprokaryota/cyanobacteria *Pseudospongiococcus protococcoides*	–	+
2	*Leptolyngbya subuliformis*	–	+
3	*Hapalosiphon hibernicus*	–	+
4	*Scytonema Hoffman-bangi*	–	+
5	*Dolichospermum affine*	–	+
6	Chlorophyta *Oocystis polymorpha*	+	–
7	*Scenedesmus* sp.	+	–

Table 12.2 Physicochemical characteristics of wastewater collected from Emami Paper Mill, Balasore

Physical	Temp. °C	pH	Conduct- ivity mS	D.O. mg/L	Chloride mg/l	Total hardness mg/l	Calcium hardness mg/l	Alkalinity mg/l	COD mg/l
Brownish red	29	7.2	1.7	2.03	85.2	290	250	122	48

The physicochemical characteristics of wastewater were also determined in the laboratory following standard procedure as in Table 12.2.

The experiment was conducted to study the growth, pigment composition and macromolecular contents of five algal strains, viz. *Scenedesmus* sp., *Leptolyngbya subuliformis*, *Hapalosiphon hibernicus*, *Scytonema hoffman-bangi* and *Dolichospermum affine*, collected from waste water, in the presence of different concentrations of Emami Paper Mill wastewater. The growth rate was recorded till the 20th day of incubation at alternate day interval, and the pigment composition and macromolecular contents were determined at the end of the 20th day.

12.3.1.1 Effect of Effluent on the Growth

Results on the effect of different concentrations of paper mill effluents on the growth of five algal species indicated that in *Scenedesmus* sp., there were increase in growth with increasing effluent concentration up to 75% and decrease in 100% effluent in all days of harvest. However, the growth is less in control than effluent-treated cultures and increased from the second day to 20th day in all culture conditions. But in *Leptolyngbya* sp., control showed maximum growth compared to effluent-treated media up to the 20th day of incubation. Whereas in *Hapalosiphon* sp., from the second day of incubation, it showed increase in growth in all concentrations of media, more rapid growth was in 100% effluent compared to others. But from the 16th day, it showed maximum growth in 50% dilution compared to others till the 20th day. Like that in *Scytonema* sp., maximum growth was seen in raw effluent from the second day of incubation till the 20th day, which is really effective, but in *Dolichospermum* sp. maximum growth found in 75% effluent compared to other concentrations (Fig. 12.2).

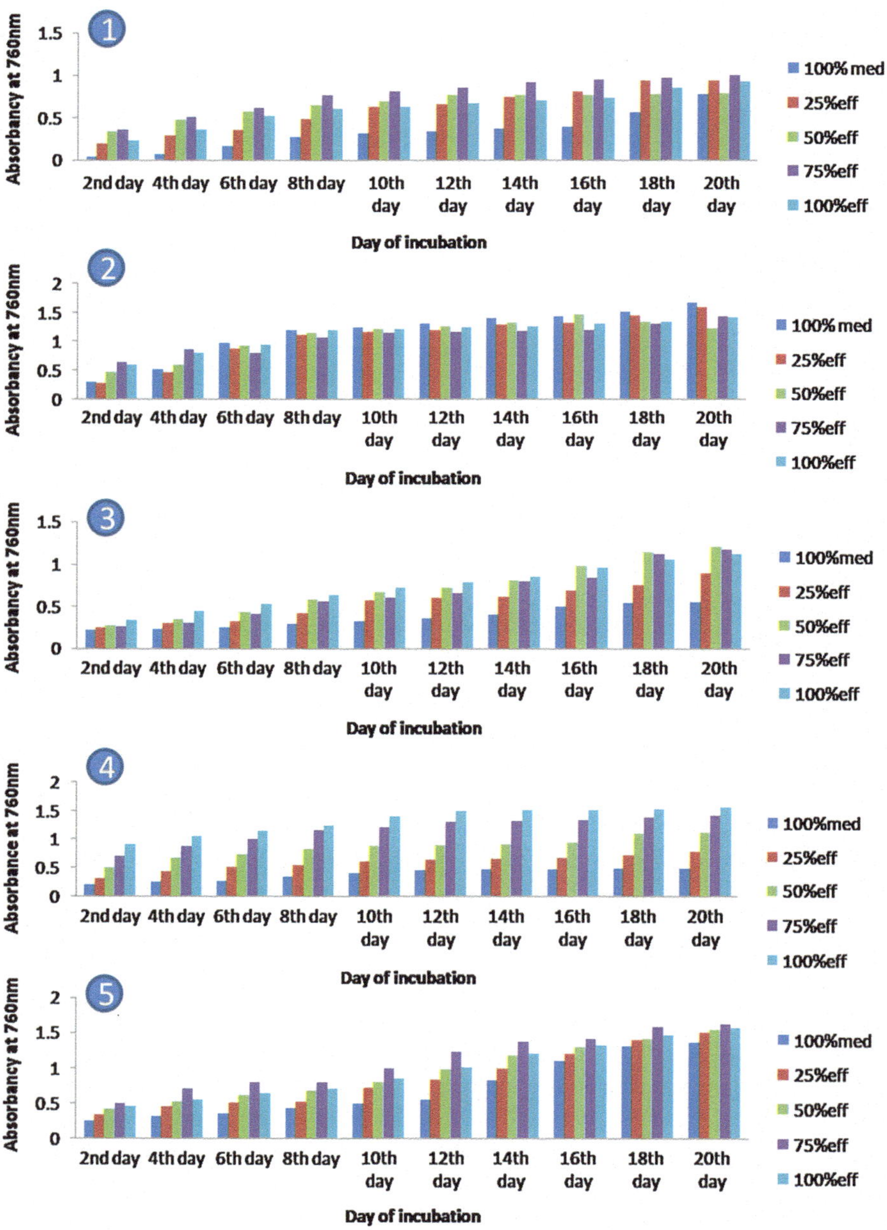

Fig. 12.2 Effect of different concentration of Paper Mill effluent on the growth of algal species incubated up to 20 days. 1. *Scenedesmus sp.*, 2. *Leptolyngbya sp.*, 3. *Hapalosiphon sp.*, 4. *Scytonema sp.*, 5. *Dilichospermum sp*

12.3.1.2 Effect of Effluent on Pigment Composition

The pigment composition of *Scenedesmus* sp. showed that the synthesis and accumulation of methanol-soluble (chl-a, chl-b, carotenoids) pigments increased with increasing effluent concentrations up to 75% and decreased in 100% effluent up to 20 days. However, in *Leptolyngbya* sp. both methanol-soluble (chlorophyll and carotenoids) and water-soluble (PC, PE, APC) pigments decreased with increasing effluent concentration up to 100% and increased in control condition up to 20 days. But in *Hapalosiphon* sp., both methanol-soluble (chlorophyll and carotenoids) and water-soluble (PC, PE, APC) pigments increased with increasing effluent concentration up to 50% and decreased from 75% to 100% effluent. Like that in *Scytonema* sp., both methanol-soluble (chlorophyll and carotenoids) and water-soluble (PC, PE, APC) pigments showed maximum concentration in 100% effluent compared to other effluent-treated media and control. But in *Dolichospermum* sp., both methanol-soluble (chlorophyll and carotenoids) and water-soluble (PC, PE, APC) pigments showed maximum concentration in 75% effluent up to 20 days (Table 12.3, Fig. 12.3).

12.3.1.3 Effect of Effluent on the Macromolecular Contents

The effect of different concentrations of effluent on total carbohydrate and cell protein content of *Scenedesmus* sp. showed that the concentration of macromolecular content increased with increasing effluent concentration up to 75% and decreased in 100% effluent. Moreover both the values are more in effluent-treated culture than the control. But opposite trend was observed in *Leptolyngbya* sp. after 20 days, and the macromolecular contents are found more in control than the effluent. Whereas in *Hapalosiphon* sp., maximum amount of both the total carbohydrate and cell protein content was found in 50% effluent concentration, in

Table 12.3 Pigment content (maximum) of different algal sp. grown under different growth conditions

Sl. No.	Name of the sp.	Chlorophyll content (μg/ml)	Carotenoid content (μg/ml)	Biliproteins content (μg/ml)
1	*Scenedesmus* sp.	75% effluent (chl-a = 0.576 μg/ml and chl-b = 0.32 μg/ml)	75% effluent 0.008 μg/ml	–
2	*Leptolyngbya subuliformis*	Control52μg/ml	Control0.013 μg/ml	Control 269 μg/ml
3	*Dolichospermum affine*	75% effluent 45 μg/ml	75% effluent 0.006 μg/ml	75% effluent 252 μg/ml
4	*Hapalosiphon hibernicus*	50% effluent 14 μg/ml	50% effluent 0.004 μg/ml	50% effluent 165 μg/ml
5	*Scytonema Hoffman-bangi*	100% effluent 10 μg/ml	100% effluent 0.005 μg/ml	100% effluent 115 μg/ml

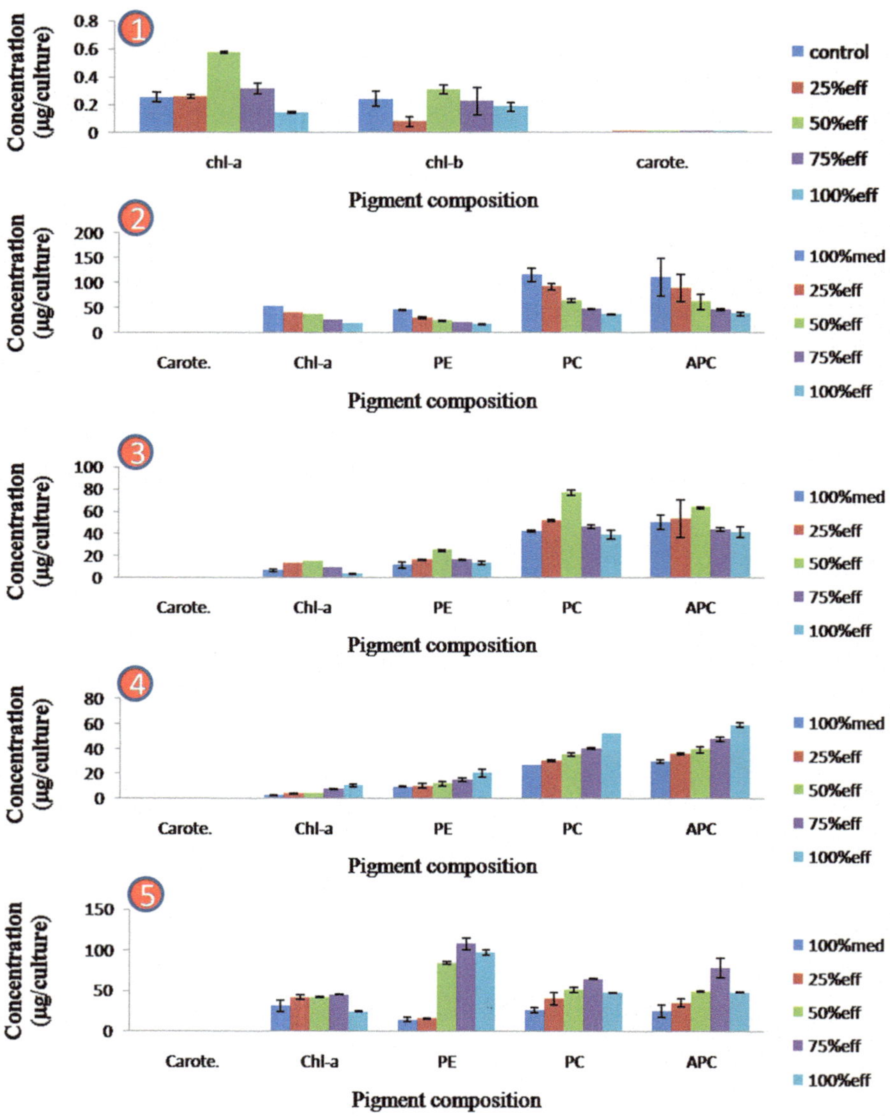

Fig. 12.3 Effect of different concentration of Paper Mill effluent on the pigment composition of algal species incubated up to 20 days. 1. *Scenedesmus sp.*, 2. *Leptolyngbya sp.*, 3. *Hapalosiphon sp.*, 4. *Scytonema sp.*, 5. *Dilichospermum sp*

Scytonema sp. total carbohydrate and cell protein content showed maximum amount in 100% effluent concentration which is a remarkable response. Like that in *Dolichospermum* sp., the macromolecular content (total carbohydrate and cell protein) showed maximum response to 75% effluent than others in 20 days of incubation (Table 12.4, Fig. 12.4).

Table 12.4 Total cell macromolecular content (maximum) of different algal sp. grown under different growth conditions

Sl. No.	Name of the taxon	Carbohydrate (mg/ml)	Protein (mg/ml)
1	*Scenedesmus* sp.	75% effluent 2 mg/ml	75% effluent 1.4 mg/ml
2	*Leptolyngbya subuliformis*	Control 9 mg/ml	Control 3.7 mg/ml
3	*Dolichospermum affine*	75% effluent 6.35 mg/ml	75% effluent 3.17 mg/ml
4	*Hapalosiphon hibernicus*	50% effluent 4.7 mg/ml	50% effluent 1.7 mg/ml
5	*Scytonema Hoffman-bangi*	100% effluent 3.43 mg/ml	100% effluent 1.6 mg/ml

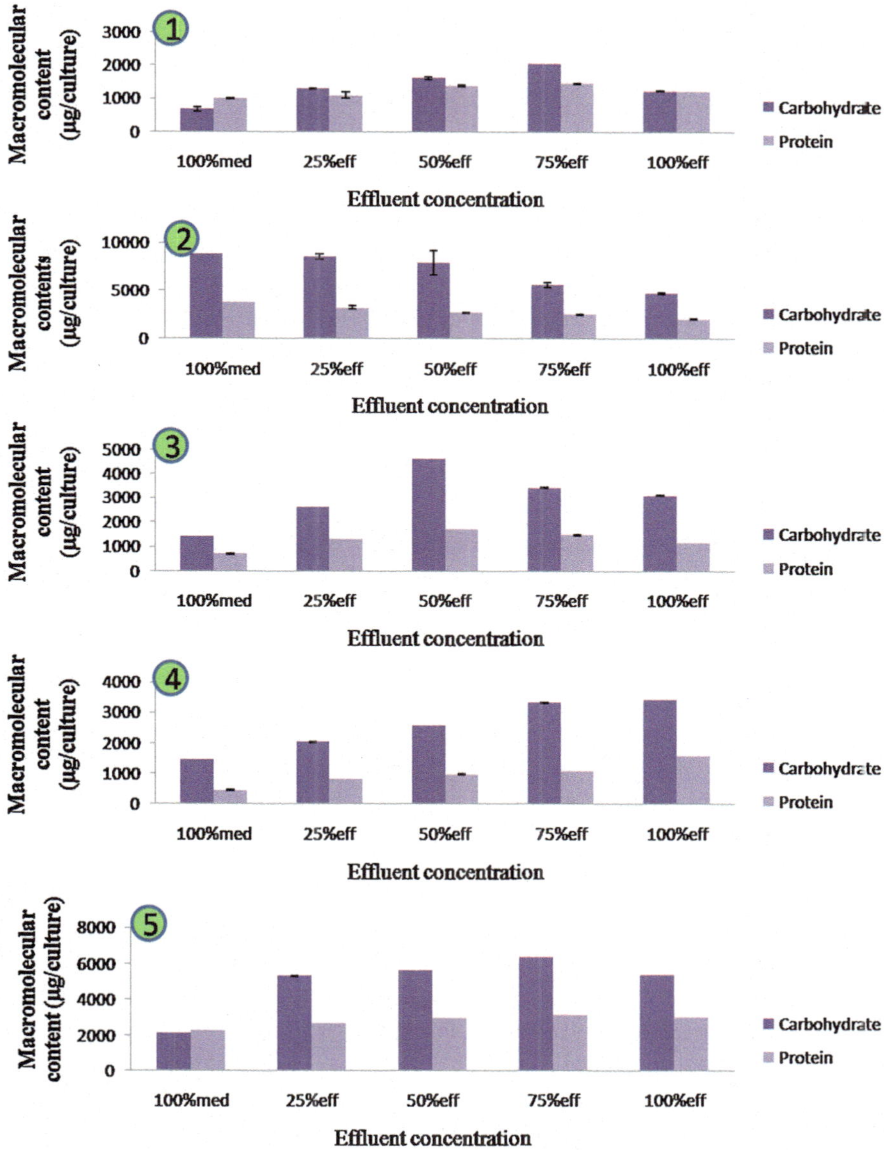

Fig. 12.4 Effect of different concentration of Paper Mill effluent on the macromolecular content of algal species incubated up to 20 days. 1. *Scenedesmus sp.*, 2. *Leptolyngbya sp.*, 3. *Hapalosiphon sp.*, 4. *Scytonema sp.*, 5. *Dilichospermum sp*

12.3.2 Discussion

Microalgae including cyanobacteria and green algae play an important role in the ecology of aquatic systems. The growth of algae in wastewater under controlled conditions aims at maximizing their biomass production. Domestic wastewater is ideal for algal growth since it contains high concentration of all necessary nutrients.

Our study on the growth, pigment composition and macromolecule synthesis of four microalgal strains showed better growth and synthesis of primary metabolites in effluent water than the medium except *Leptolyngbya* sp. In *Leptolyngbya,* maximum growth, pigment and macromolecular content is observed in control condition, whereas in *Scytonema* maximum found to be seen 100% effluent. But in *Scenedesmus* and *Dolichospermum* maximum growth, pigment and macromolecular content were found to be seen in 75% effluent. Like that in *Hapalosiphon* 50% effluent showed maximum growth, pigment and macromolecular content. This may be due to high concentrations of nutrients required for the growth and synthesis and accumulation of photosynthetic pigments. Our results revealed that the increase in growth with the increased effluent concentration is in conformity with the earlier reports [8, 19, 26]. The utilization of effluent varies from microalgal species to species, and to some extent it depends on concentration of effluent used.

12.4 Conclusion

The growth of algae in wastewaters under controlled conditions aims at maximizing their biomass production in order to remove various nutrients which constitute the waste. Identifying oxygen-evolving photosynthetic organisms like algae with high growth rates, high biomass and high utilization potential which could play a dual role of cleansing the wastewater and also serving as a source of feed, fertilizer and fuel is the primary requirement of today's research. The benefit of algae in treatment system is oxygenation and mineralization in addition to their role as producers in the tropic ecosystem.

In the present investigation, it is quite evident that except *Leptolyngbya subuliformis*, four microalgal strains, viz. *Scenedesmus* sp., *Dolichospermum affine*, *Hapalosiphon hibernicus* and *Scytonema hoffman-bangi*, showed better growth in paper mill effluent than the growth medium. In all test species the growth is increased with increasing effluent concentration of up to 50% (*Hapalosiphon*), 75% (*Scenedesmus* and *Dolichospermum*) and 100% (Scytonema), whereas in *Leptolyngbya* the response is more in control condition. The increase in growth may be due to accumulation of nutrients from nutrient-rich wastewater and resulted in the removal of phosphate and nitrogen. In addition to their growth, there is also more synthesis and accumulation of photosynthetic pigments like chlorophylls, carotenoids (minimum synthesis), phycobilisomes and intracellular macromolecular contents like proteins and carbohydrates produced during primary metabolism.

However, the growth, accumulation of pigments and macromolecular contents are occurring the most in *Scytonema* sp. than the other species. From the results it is clear that the extent of utilization of effluents varies from algal species to species and concentrations of the effluent.

Use of oxygen-evolving microalgae and cyanobacteria in particular would be advantageous in many ways. The benefits of algae in treatment of wastewater vis-à-vis their role as producers in the topic ecosystems are noted worthy. The treatment of wastewater and production of sufficient biomass with high-value products using low-cost effluent is no doubt cost-effective. The oxygenic photosynthetic property, high-level tolerance to various pollutants and large surface area are some of the striking features of microalgae which make them highly suitable for effluent treatment and high utilization potential in unproductive ecosystems. More work in this regard is needed for better exploitation and utilization of microalgae.

Acknowledgements We wish to thank the Ministry of Science and Technology, Govt. of India, for financial assistance through DST-INSPIRE fellowship. We also thank the authorities of Botanical Survey of India, MoEFCC, Govt. of India, and head of department, P.G. Department of Botany, North Orissa University, for laboratory work.

References

1. Bernhardt ES, Lawrence EB, Walsh CJ, Berke PE (2008) Understanding, managing, and minimizing urban impacts on surface water nitrogen loading. Ann N Y Acad Sci. The year in ecology and conservation biology 1134:61–96
2. Bischoff HW, Bold HC (1963) Phycological studies.IV. Some soil algae from enchanted rock and related algae species. Univ Texas Pub 6318:1–95
3. Chinnasamy S, Sood A, Renuka N, Prasanna R, Ratha SK, Bhaskar S, Rengasamy R, Lewis DM (2014) Ecobiological aspects of algae cultivation in wastewaters for recycling of nutrients and biofuel applications. Biofuels 5:141–158
4. Chisti Y (2013) Constrains to commercialization of algal fuels. J Biotechnol 167:201–214
5. Cho S, Luong TT, Lee D, Oh Y, Lee T (2011) Reuse of effluent water from municipal wastewater treatment plant in microalgae cultivation for biofuel production. Bioresour Technol 102:8639–8645
6. Desikachary TV (1959) Cyanophyta, I.C.A.R. Monograph on algae. Indian Council of Agricultural Research, New Delhi, p 686
7. Davies BH (1976) Carotenoids. In: Goodwin TW (ed) Chemistry and biochemistry of plant pigments. Academic, New York/London, pp 149–155
8. De la Noue J, Prolux D (1988) Biological tertiary treatment of urban wastewaters with chitosan immobilized *Phormidium*. Appl Microbiol Biotechnol 29:292–297
9. Donald M, Spear GC, Lavin RD, J P, Clesceri NL (1970) Kinetics of algal growh in austere media. In: Jajic JE (ed) Properties and products of algae. Plenum Press, New York/London, pp 97–105
10. Fujita Y, Ohki K, Murakami A (1987) Chromatic regulation of photosystem composition in the cyanobacterial photosynthetic system: kinetic relationship between change of photosystem composition and cell proliferation. Plant Cell Physiol 28:227–234
11. Hattori A, Fujita Y (1959) Photochemically active chromoprotein isolated from blue-green alga *Anabaena cylindrica*. J Biochem 46:633p
12. Herbert D, Phipps PJ, Strange RE (1971) Chemical analysis of microbial cells. Methods Microbiol 5B:209–344

13. Komarek J, Anagnostidis K (1999) Cyanoprokayota,*I.Teil Chroococcales*. Band 19/1, Süßwasserflora von Mitteleuropa. Gaustav Fischer Verlag, Jena, 548 pp
14. Komárek J, Anagnostidis K (2005) *Cyanoprokaryota*, II. Teil Oscillatoriales, band 19/2, Süßwasserflora von Mitteleuropa. Elsevier GmbH, Munchen, 759 pp
15. Komárek J, Fott B (1983) *Das phytoplankton des Süßwassers*, band 7, E. Schweizerbart'sche Verlagsbuchhandlung, Stuttgart, 1001 pp
16. Lichtenthaler HK (1987) Chlorophylls and carotenoids pigment of photosynthetic biomembrane. JMethods Enzymol 148:350–382
17. Lowry OH, Rosebrough NJ, Farr AL, Randall RJ (1951) Protein measurement with the folin phenol reagent. J Biol Chem 193:265–275
18. Mackinney G (1941) Absorption of light by chlorophyll solutions. J Biol Chem 140:315–322
19. Manoharan C, Subramanian G (1992) Sewage-cyanobacterial interaction- a case study. Indian J Env Prof 12:251–258
20. Oswald WJ (1988) Micro-algae and waste-water treatment. In: Micro-algal biotechnology, vol 6. Cambridge University Press, Cambridge, pp 305–328
21. Olguin EJ (2003) Phycoremediation: key issues for cost-effective nutrient removal processes. Biotechnol Adv 22:81–91
22. Philipose MT (1967) Chlorococcales. Indian Council of Agricultural Research, New Delhi, pp 1–365
23. Renuka N, Sood A, Ratha SK, Prasanna R, Ahluwalia AS (2013) Evaluation of microalgal consortia for treatment of primary treated sewage effluent and biomass production. J Appl Phycol 25:1529–1537
24. Rippaka R, Derulles J, Waterbury JB, Herdman M, Stanier RY (1979) Generic assignments, strain histories and properties of pure cultures of cyanobacteria. J Gen Microbiol 111:1–61
25. Singh M, Reynolds DL, Das KC (2011) Microalgal system for treatment of effluent from poultry litter anaerobic digestion. Bioresour Technol 102:10841–10848
26. Subramaninan G, Shanmugasundaram S (1986) Influence of the herbicide 2, 4-d on nitrogen fixation and ammonia excretion by anabaena. Proc Indian Nart sei acad B52:308–312
27. Venkataraman GS (1969) The cultivation of algae. Indian Council of Agriculture Research, New Delhi, p319

Biosurfactants and Bioemulsifiers from Marine Algae

13

Pattanathu K. S. M. Rahman, Aysha Mayat, Joseph Gregory Havelock Harvey, Kamaljeet Sekhon Randhawa, Louise E. Relph, and Michael C. Armstrong

Abstract

The terms biosurfactants and bioemulsifiers have often been used interchangeably to describe surface active biomolecules. However, there are marked differences between the two especially based on their physicochemical properties and physiological roles. Although bioemulsifiers and biosurfactants are both amphiphilic in nature and are produced by a wide range of microorganisms, each exhibits characteristic roles in nature. Biosurfactants continue to receive scientific attention due to their environmentally friendly characteristics relative to chemically derived surfactants. Their unique features of being non-toxic, biodegradable, biocompatible and efficient at low concentrations and their synthesis from natural substrates under mild environmental conditions make them really sought-after compounds. The combination of polysaccharide, fatty acid and protein components in bioemulsifiers confers upon them better emulsifying potential and ability to stabilise emulsions. The aim of writing this chapter is to bring into the fore the biosurfactants and bioemulsifiers from marine microalgae. The chapter presents two case studies and suggests ways to tap into this relatively lesser explored area.

P. K. S. M. Rahman (✉) · A. Mayat · J. G. H. Harvey · K. S. Randhawa
L. E. Relph · M. C. Armstrong
School of Science, Engineering and Design, Teesside University, Middlesbrough, UK
e-mail: P.Rahman@tees.ac.uk

L. B. Sukla et al. (eds.), *The Role of Microalgae in Wastewater Treatment*,
https://doi.org/10.1007/978-981-13-1586-2_13

13.1 Introduction: The Connection Between Biosurfactants and Bioemulsifiers and Their Applications

Since the turn of the twentieth century and the beginning of the industrial age, exploitation of fossil-based resources has led to widespread damage of the Earth's environment [1]. This anthropogenic damage to the environment combined with diminishing reserves of fossil-derived fuels has increased concern around the use of chemically derived surfactant and emulsifying compounds [2]. These factors, coupled with increasingly stringent laws surrounding the release of chemically synthesised products, have led to increased research in the area of bio-derived products [3]. In recent years, due to environmental concerns, research has increased in finding alternative renewable feedstocks to produce substances such as biofuels, surfactants and emulsifiers.

The global market of surfactant production in 2012 was worth approximately 27 billion dollars [4]; however, most of these surfactants are produced through petrochemical processes. To this market, biosurfactants contribute 1.76 billion dollars, and this is expected to rise to 2.8 billion dollars by 2023 [5]. There is an increase in the research and development of environmentally friendly biosurfactants and bioemulsifiers from renewable resources such as bacteria, fungi and yeast.

Biosurfactants and bioemulsifiers are amphipathic molecules; therefore, they have both hydrophobic and hydrophilic regions. The polar portion (hydrophilic) region may be ionic (cationic or anionic), non-ionic or amphoteric (Fig. 13.1). This enables them to possess a variety of beneficial properties including low surface tension, effective hydrocarbon emulsion and antimicrobial activity [7]. These desirable characteristics enable the biosurfactants to have several potential applications within the various industrial sectors such as bioremediation, health care and hygiene. Biosurfactants and emulsifiers are generally found on the surface of the organism producing these compounds as they are secreted extracellularly; they also serve as a protective layer for the microorganisms in adverse environments.

The various types of biosurfactants are classified into five categories: glycolipids (rhamnolipids, sophorolipids, trehalose lipids), fatty acids (mycolic acids, agaricic acids), phospholipids, polymers (glycoprotein, liposan) and lipopeptides [8]. Current research has revolved around the commercialisation of biosurfactants; however, this has reached a bottleneck due to the low productivity and high economic costs [9].

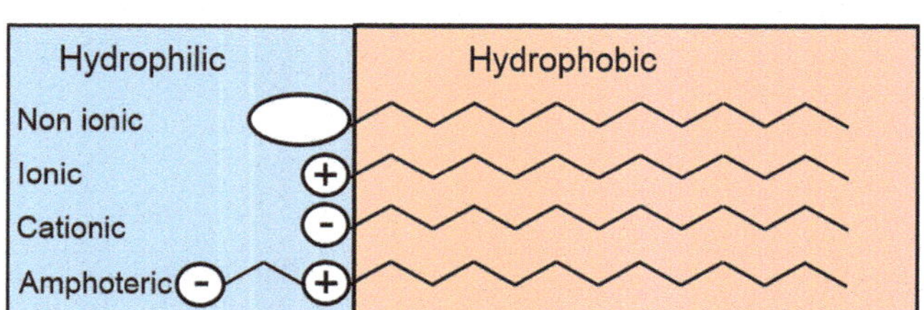

Fig. 13.1 Different structures of amphipathic molecules, with differing hydrophilic moieties [6]. These include non-ionic, ionic, cationic and amphoteric regions

Bioemulsifiers, on the other hand, include high-molecular-weight molecules such as polysaccharides, lipopolysaccharides and lipoproteins [10]. Surfactants reduce surface tension through accumulation of immiscible fluids, which in turn increase the solubility, mobility, bioavailability and contaminant degradation of inorganic compounds [11]. Both bioemulsifiers and biosurfactants contain hydrophobic and hydrophilic moieties, which have varying degrees of polarity. The non-polar (hydrophobic) region is often a hydrocarbon chain.

It has generally been assumed that bioemulsifiers are a class of biosurfactants; however, both have different physiological properties and chemical compositions. Bioemulsifiers have a higher molecular weight than biosurfactants and have more of a complex mixture consisting of lipoproteins and lipopolysaccharides [12]. The main difference between bioemulsifiers and biosurfactants is that the latter reduces surface tension at the gas-solid-liquid level, whereas bioemulsifiers reduce surface tension between immiscible liquids [13]. Biosurfactants often have emulsifying activity; however, emulsifiers do not exhibit all the characteristics of surfactants [14].

Polysaccharides (a type of bioemulsifier) are secreted extracellularly forming a layer around the cells and are involved in the formation of biofilms. They have high-molecular-weight carbohydrate polymers and are generally classified as secondary metabolites [15]. The chemical composition of these compounds has been described as any group of carbohydrates comprising of long chains of simple sugar molecules; examples include alginate, chitin, carrageenan, glucan and agar. There are two types of polysaccharides, the first being exo- or extracellular polysaccharides (EPS) and the second sulphated exopolysaccharides (sEPS). Marine-extracted EPS and sEPS, similarly to biosurfactants and bioemulsifiers, are also used in a multitude of industries including pharmaceutical, food and agriculture [16].

Oceanic biological surface active compounds (or biosurfactants from marine sources) still represent a major untapped and unexplored area of research [17]. Solar energy in the production of polysaccharides has been generally overlooked, despite high product yields and wide variety of polysaccharide production (Table 13.1) [21]. However, due to current market demands for alternatives to synthetic surfactants and emulsifiers, the production of polysaccharides with surface active properties is attracting the attention of researchers [14]. Worldwide production of polysaccharides from marine biomass is between 25,000 and 30,000 tons/year [22]. Dependent on the location within the cells, Roger in 2002 classified polysaccharides into three groups: cytosolic polysaccharides (CPS), which provide carbon and energy sources for the cell; structural polysaccharides (SPS), which make up the cell wall (including peptidoglycans and lipopolysaccharides); and exopolysaccharides (EPS), which are polysaccharides secreted into the extracellular environment in capsules or biofilms.

Algal exopolysaccharides represent a huge range of structures. They are high-molecular-weight structures (10–30 kDa) which encompass homopolymeric and heteropolymeric compositions [23]. EPS structures vary widely between different genera of algae and are generally considered to be related to the environmental conditions of the organism [24]. The majority of EPS formation involves the linking of a nucleotide sugar to a lipid carrier molecule, by glycosyltransferases. Conversion

Table 13.1 Microalgae species and the types of polysaccharides and sugars produced with additional beneficial properties associated with the organism

Algal species	Type of polysaccharides and sugars	Properties	References
Ankistrodesmus angustus	EPS	Antiviral	Raposo et al. [16]
Chaetoceros sp.	EPS	Anticancer	Raposo et al. [16]
Chlorella sp.	sEPS	Anticancer, bioremediation	Raposo et al. [16]
Chlorella stigmatophora	sEPS – glucose, xylose	Anti-inflammatory, immunomodulatory	Raposo et al. [16]
Chlorella vulgaris	sEPS	Antimicrobial, antioxidant	Raposo et al. [16]
Dunaliella primolecta	EPS	Antibiotic, inhibits growth in various *Bacillus*, *Enterobacter* and *Staphylococcus* sp.	Chang et al. [18] and Varshney et al. [19].
Dunaliella salina	EPS	Vitamins, antioxidants	Mishra and Jha [20]
Phaeodactylum tricornutum	sEPS – glucose, mannose	Anti-adhesive, immunomodulating	Raposo et al. [16]
Tetraselmis sp.	sEPS	Antimicrobial, potential probiotics	Raposo et al. [16]

of glucose-6-phosphate to glucose-1-phosphate, a nucleotide sugar precursor, by phosphoglucomutase, is essential to the nucleotide sugar synthesis [25]. The green microalgae *Dunaliella salina* [20] and red algae *Porphyridium cruentum* [26, 27] are receiving attention as robust EPS producers with potential industrial application. The role of photosynthetically synthesised EPS as emulsifiers has been recently explored, with the presence of EPS leading to the metabolising of oil hydrocarbons by cyanobacteria and bacterial consortiums [28].

Bioemulsifiers and biosurfactants from marine environments, in general, are derived from a wide range of organisms including, but not limited to, macroalgae, microalgae, bacteria, diatoms and cyanobacteria. Biosurfactants and bioemulsifiers share many environmental advantages over their chemically synthesised counterparts. They are highly biodegradable and have low toxicity. There is also an abundance of raw materials for the production of these molecules, and they are highly biocompatible [29]. Bioemulsifiers have been associated with a number of potential applications including remediation of oil-polluted water and soil, enhanced oil recovery and clean-up of oil-contaminated vessels and machineries, and heavy metal removal, formation of stable emulsions in food and cosmetic industries and therapeutic activities (antibacterial, antifungal, pesticide and herbicidal agents). Refer to Table 13.2 for more applications. Polysaccharides are widely used as emulsifying, stabilising and gelling agents in food manufacture. At present, two of the most used are alginate, also known as alginic acid and carrageenan. Alginic acid, used in both the food and pharmaceutical industry, is produced by all brown algae,

Table 13.2 A wide range of marine fungi and bacteria have been found to produce biosurfactants and bioemulsifiers with diverse potential applications as surface active molecules

Biosurfactant/ Bioemulsifier	Organisms	Potential application
Exopolysaccharide	*Planococcus maitriensis*	Emulsifier and oil dispersion potential
	Antarctobacter sp.	Emulsification of food oils and metal absorbance
	Halomonas sp.	Emulsification of food oils and stabilisation of emulsions
Glycolipopeptide	*Corynebacterium kutscheri*	Emulsification and degradation of hydrocarbons
	Yarrowia lipolytica	Emulsification of aromatic hydrocarbons
	Halomonas sp. ANT-3B	Remediation of oil spills in cold environments
	Pseudomonas aeruginosa A41	Enhanced oil recovery
Lipopeptide	*Bacillus circulans*	Polyaromatic hydrocarbon solubilisation
Aminolipid	*Myroides* sp.	Emulsification of crude oil

Pepi et al. [30], Amaral et al. [31], Maneerat et al. [32], Thaniyavarn et al. [33], Kumar et al. [34], Thavasi et al. [35], Das et al. [36]

representing between 18 and 40% of the dry matter [37]. Algal EPS also play an integral role in the formation of biofilms. Therefore, much has been made of the potential role of bioemulsifiers and biosurfactants from biofilms. In particular, its applications as a wetting agent, lubricant and solubiliser in the food and environmental biotechnology industry are attracting much attention [14].

The case studies included in this chapter explain the production of biosurfactants and bioemulsifiers from algae isolated from marine or coastal environments. The studies throw light on the optimum conditions in which marine organisms could be harnessed to produce surface active molecules. How the intricacies of sampling, organism screening methods and identification techniques play role in the production of biosurfactants and bioemulsifiers can be understood in these studies.

13.2 Case Studies

13.2.1 Is Marine Microalgae *Chlorella* Capable of Producing Biosurfactants, Bioemulsifiers or Exopolysaccharides?

The genus *Chlorella* belongs to the family of Chlorellaceae which are spherical in shape. This photosynthetic organism generally found in ponds, oceans and lakes, are classified as blue-green unicellular algae. The algae are mainly distributed as a food or nutrition source for both humans and animals as the protein content is approximately 60% which is a common trait amongst microalgal species. There are an array of health benefits associated with this organism, i.e. hepatoprotection, anticancer, antimicrobial, antioxidant and anti-inflammatory properties. They are naturally high in lipid content with approximately 35% fatty acid content [38]. This trait makes the organism desirable for biofuel production. There has been limited research conducted into whether the *Chlorella* sp. produce biosurfactants or bioemulsifier; however, there have been a few species which display similar properties to these compounds. On the other hand, the production of EPS and sEPS from various *Chlorella* as well as other microalgae species has been reported previously (Table 13.1).

In this study, the microalgae *Chlorella* was cultured via shake flask fermentation in BG-11 media; optical density was taken to monitor algal growth at different time intervals (Fig. 13.2). Once the desired optical density (OD) was achieved, the algae were analysed against the chemical surfactant (Tween 20) and a biosurfactant produced by *Bacillus subtilis* to assess the emulsification and surfactant properties.

The emulsification index (E24 percentage) and surface tension were calculated; it was found that microalgae displayed emulsification properties having a higher E24 index in comparison with the blank media; however, it did not seem to lower the surface tension. This suggested that *Chlorella* was more likely to produce a bioemulsifier than a biosurfactant. There was no difference in emulsification activity between the centrifuged and non-centrifuged samples. Surface tension measurements also suggested that the samples are closer to being bioemulsifiers than biosurfactants (Figs. 13.3, 13.4, 13.5, and 13.6).

Next, liquid-liquid extraction was performed on the microalgae, which were then subsequently derivatized using the FAME technique; the samples were ran through a UV spectrophotometer, high-performance liquid chromatography (HPLC) and gas chromatography-mass spectrometry (GC-MS) (Tables 13.3 and 13.4). The UV spectrophotometer found the microalgae can only be identified at a lower wavelength which would make it difficult to analyse under the HPLC. HPLC

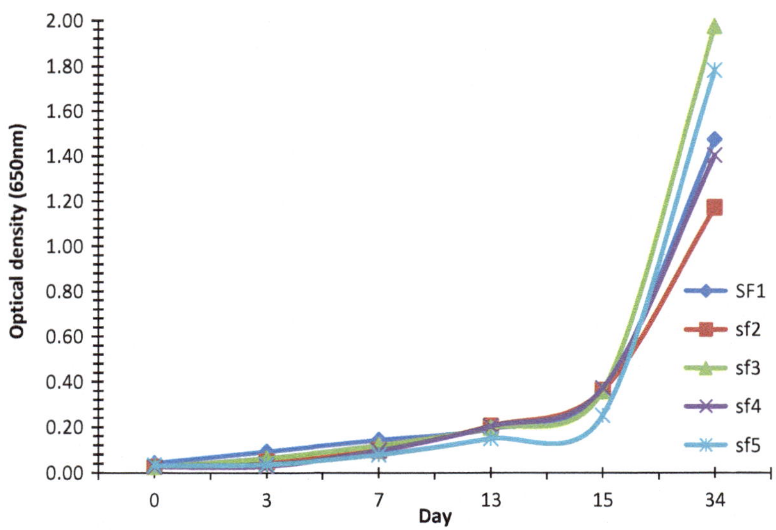

Fig. 13.2 The optical density (650 nm) observed for the algal samples

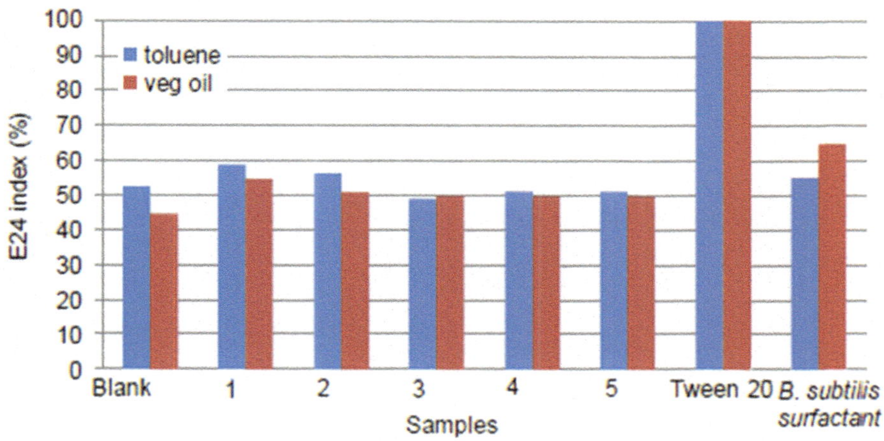

Fig. 13.3 E24 measurements with a blank (BG-11 media), one to five non-centrifuged algal samples, a chemical surfactant (Tween 20) and a biosurfactant from *B. subtilis*

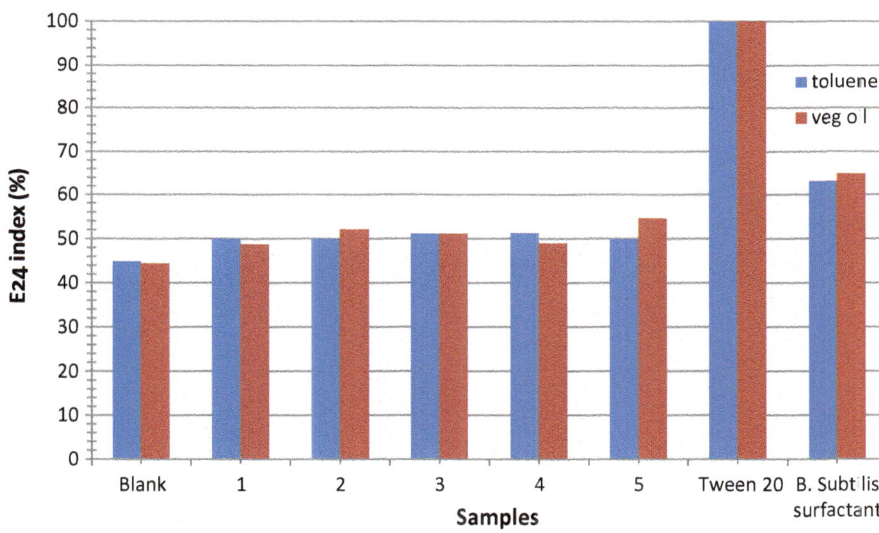

Fig. 13.4 E24 measurements with a blank (BG-11 media), one to five centrifuged algal samples, a chemical surfactant (Tween 20) and a biosurfactant from *B. subtilis*

chromatograms from the blank and algal samples found no indication of biosurfactant, bioemulsifier or exopolysaccharides, whereas GC-MS data suggested the presence of ester- and fluorine-based compounds which seemed to originate from the reagents used in the derivatization of samples; however, the procedure was unable to detect biosurfactant, bioemulsifiers or exopolysaccharides (Figs. 13.7 and 13.8). The limitation of this study has been the usage of only one genus of microalgae, i.e. *Chlorella*, including other algal species such as *Dunaliella* or *Porphyridium* which might have given more definitive results.

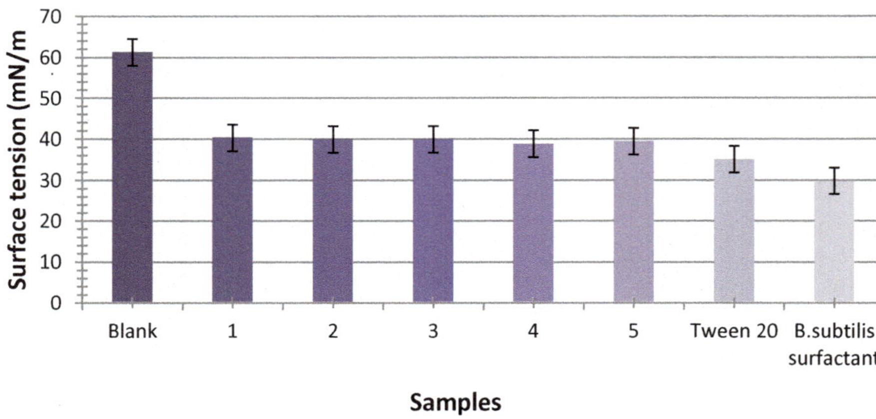

Fig. 13.5 Surface tension measurements obtained from a blank (BG-11 media), one to five non-centrifuged algal samples, a chemical surfactant (Tween 20) and a biosurfactant from *B. subtilis*

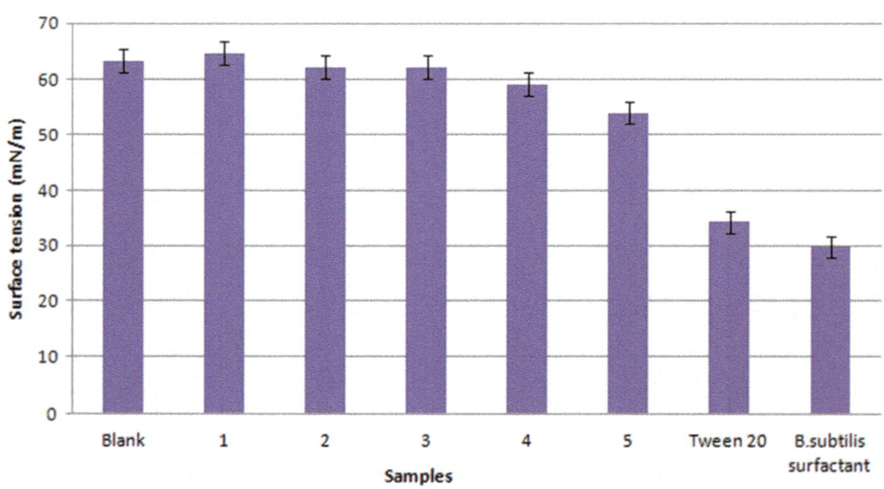

Fig. 13.6 Surface tension measurements obtained from a blank (BG-11 media), one to five centrifuged algal samples, a chemical surfactant (Tween 20) and a biosurfactant from *B. subtilis*

13.2.2 Attempt to Extract Biosurfactants and Bioemulsifiers from Algae Isolated from the Coastal Sites Around North East England

For this case study, the algal samples were taken from the coastal sites in Seaton, Saltburn and Skinningrove of North East England. They were then cultured in F2, 1% F2 and seawater media, before being centrifuged (Tables 13.5, 13.6, and 13.7). Seawater for the media was taken from the rock pool area of Saltburn beach. Each culture consisted of 200 ml of media and 1 g of algal sample with incubation at

Table 13.3 Microalgal, *Chlorella* sp., samples ran on a GC-MS under the surfactant method

Sample no.	Retention time (minutes)	Name of compound	Molecular formula	Molecular weight (g)
BG-11 Media (blank)-A	5.69	1-Fluorododecane	$C_{12}H_{25}F$	188.3253
	28.99	Trimethyl(44–1,1,3,3, tetramethylbutyl)phenoxy)silane	$C_{17}H_{30}OSi$	278.5050
	29.14	Tetrasiloxane	$C_{10}H_{30}O_3Si_4$	310.6854
BG-11 Media (blank-B)	9.75	Tetrasiloxane	$C_{10}H_{30}O_3Si_4$	310.6854
	11.95	Tetrasiloxane	$C_{10}H_{30}O_3Si_4$	310.6854
2A	5.69	2,2,6 trimethyloctane	$C_{11}H_{24}$	156.3083
	11.96	Tetrasiloxane	$C_{10}H_{30}O_3Si_4$	310.6854
2B	14.90	2,2,2-Trifluoromethylamine	$C_2H_4F_3N$	99.0551
	15.13	Cyclohexylmethyl heptyl ester	$C_8H_{16}O_2$	144.2114
	20.96	Tetradecane	$C_{14}H_{30}$	198.3880
3A	5.70	Heptane 4 ethyl 2,2,6,6 tetramethyl	$C_{12}H_{24}$	168.3190
	11.98	Tetrasiloxane	$C_{10}H_{30}O_3Si_4$	310.6854
3B	15.12	Cyclohexane 3,3 dimethylbutyl	$C_{12}H_{24}$	168.3190
	20.09	Tetradecane	$C_{14}H_{30}$	198.3880
4A	5.16	Tris(trimethylsilyl) borate	$C_9H_{27}BO_3Si_3$	278.376
	5.70	1-flurododecane	$C_{12}H_{25}F$	188.3253
	11.96	Tetrasiloxane	$C_{10}H_{30}O_3Si_4$	310.6854
4B	15.13	Cyclohexane 3,3 dimethylbutyl	$C_{12}H_{24}$	168.3190
	20.10	Tetradecane	$C_{14}H_{30}$	198.3880
5A	5.70	1-flurododecane	$C_{12}H_{25}F$	188.3253
	11.96	Tetrasiloxane	$C_{10}H_{30}O_3Si_4$	310.6854
5B	15.13	Cyclohexane 3,3 dimethylbutyl	$C_{12}H_{24}$	168.3190
	20.08	Tetrasiloxane	$C_{10}H_{30}O_3Si_4$	310.6854

The sample number, the retention times of the peaks and the compounds found at those peaks are recorded as well as the molecular weight of those compounds ascertained from the NIST library

20 °C, 150 rpm, for 168 hr. The cell-free supernatants were screened for three parameters as per standard methods: oil displacement tests, surface tension measurements and emulsification activity assays.

Oil displacement tests were carried out using crude oil and were not observed in most samples suggesting the absence of any surface active molecules in these samples.

Surface tension was measured using Du Nouy ring assay on a Kruss K9 tensiometer. A reduction in surface tension was observed in the culture samples in comparison with distilled water; however, it was found to be higher than the controls of pure media (Fig. 13.9). A two-way ANOVA was carried out on the tensiometer measurements, in order to determine if there was a statistically significant difference between the surface tension measurements of samples taken from different sites and cultured using different media. Results show the P-value of the media type was 0.964 which is >0.05, which suggests that there was not a statistically significant difference

Table 13.4 Microalgal, *Chlorella* sp., samples ran on a GC-MS under the FAME method

Sample no.	Retention time (minutes)	Name of compound	Molecular formula	Molecular weight
BG-11 Media (blank)-A	9.79	Tetrasiloxane	$C_{10}H_{30}O_3Si_4$	310.6854
	11.95	Tetrasiloxane	$C_{10}H_{30}O_3Si_4$	310.6854
	13.89	Trimethyl(44–1,1,3,3, tetramethylbutyl)phenoxy) silane	$C_{17}H_{30}OSi$	278.5050
BG-11 Media (blank-B)	9.73	Tetrasiloxane	$C_{10}H_{30}O_3Si_4$	310.6854
	11.95	Tetrasiloxane	$C_{10}H_{30}O_3Si_4$	310.6854
	13.89	Tetrasiloxane	$C_{10}H_{30}O_3Si_4$	310.6854
2A	9.80	Tetrasiloxane	$C_{10}H_{30}O_3Si_4$	310.6854
	11.95	Tetrasiloxane	$C_{10}H_{30}O_3Si_4$	310.6854
2B	11.94	Tetrasiloxane	$C_{10}H_{30}O_3Si_4$	310.6854
	15.12	Cyclohexane 3,3 dimethylbutyl	$C_{12}H_{24}$	168.3190
3A	9.79	Tetrasiloxane	$C_{10}H_{30}O_3Si_4$	310.6854
	11.95	Tetrasiloxane	$C_{10}H_{30}O_3Si_4$	310.6854
	14.90	Trifluoromethylamine	C_3F_9N	221.0244
3B	9.79	Tetrasiloxane	$C_{10}H_{30}O_3Si_4$	310.6854
	15.12	Cyclohexane 3,3 dimethylbutyl	$C_{12}H_{24}$	168.3190
4A	9.78	Tetrasiloxane	$C_{10}H_{30}O_3Si_4$	310.6854
	11.96	Tetrasiloxane	$C_{10}H_{30}O_3Si_4$	310.6854
4B	9.78	Tetrasiloxane	$C_{10}H_{30}O_3Si_4$	310.6854
	15.12	Cyclohexane 3,3 dimethylbutyl	$C_{12}H_{24}$	168.3190
5A	3.25	Perfluorotributylamine	$C_{12}F_{27}N$	671.0920
	5.70	Perfluorotributylamine	$C_{12}F_{27}N$	671.0920
5B	9.81	Tetrasiloxane	$C_{10}H_{30}O_3Si_4$	310.6854
	15.12	Cyclohexane 3,3 dimethylbutyl	$C_{12}H_{24}$	168.3190

The sample number, the retention times of the peaks and the compounds found at those peaks are recorded as well as the molecular weight of those compounds ascertained from the NIST library

caused by the different media type, on the surface tension measurements taken. From the ANOVA, it can be seen that the *P*-value of the sampling location was 0.293, and this suggests that sampling location held no statistical significance in the surface tension of the samples. Data also suggests that there is no statistical significance linking sampling location and media type, as the *P*-value was 0.616 (Table 13.8). A one-way ANOVA was carried out to compare each sample set from each location to the distilled water control, to determine if there was a statistically significant difference. As can be seen in Table 13.9, no statistically significant difference was observed between the culture samples and the distilled water control.

Emulsification activity was measured using three different hydrocarbons: sunflower oil, kerosene and hexane. The emulsification index (E_{24}) was calculated using the following formula:

$$\left(\text{Height of Emulsion / Total height of liquid}\right) \times 100 = \theta E_{24}$$

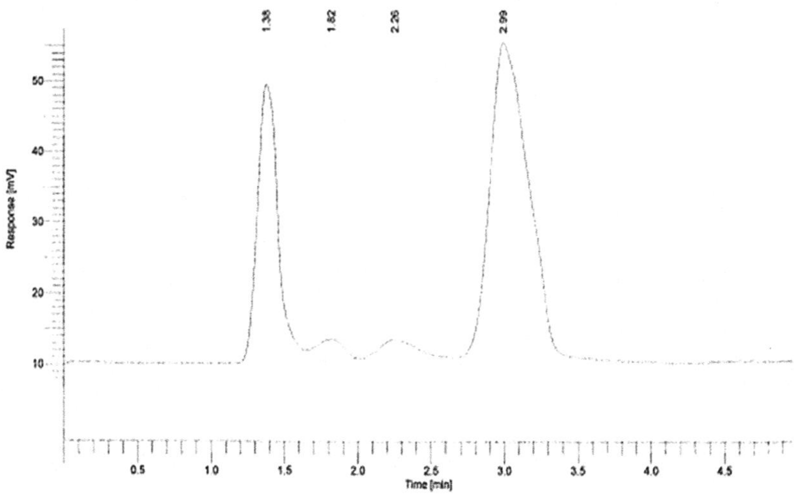

Peak #	Time [min]	Area [mV*sec]	Area [uV*sec]	Height [uV]	Area [%]	Component Name
1	1.38	387.8842	387884	39394	29.9	
2	1.82	42.7337	42734	3093	3.3	
3	2.26	56.8991	56899	2932	4.4	
4	2.99	809.9060	809906	44719	62.4	
		1297.4230	1297423	90137	100.0	

Fig. 13.7 An HPLC chromatogram of the blank (BG-11 media) in a mobile phase of acetonitrile and formic acid (50:50)

Emulsification was observed in sunflower oil in all culture samples and controls (Fig. 13.10), and a small amount of emulsification was observed in some culture samples in kerosene (Fig. 13.11). There was no emulsification observed in hexane. When a two-way ANOVA was carried out (Table 13.10), it was found that there was no statistically significant difference in E_{24} indexes between different locations and media. One-way ANOVA (Table 13.11) also suggests that there was no statistically significant interaction between location and media type. This means that the level of emulsification was not affected by either sampling location or media type. This also means that there was no interaction between the two variants which lead to a change of E_{24} value. Emulsification in the pure media controls suggests that there was a surfactant molecule in the media. Therefore, the emulsification in the culture samples could be due to the presence of media in the cultures or an emulsifying molecule produced by the algal culture.

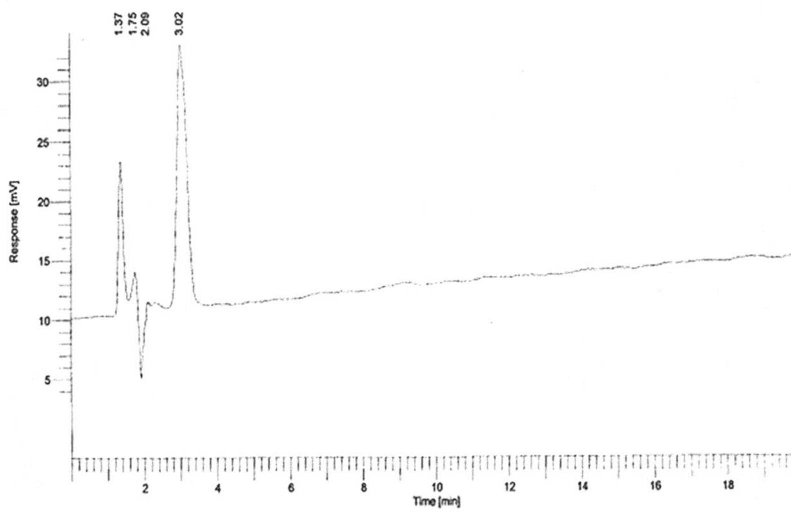

Peak #	Time [min]	Area [mV*sec]	Area [uV*sec]	Height [uV]	Area [%]	Component Name
1	1.37	157.2763	157276	14442	16.5	
2	1.75	105.8694	105869	7815	11.1	
3	2.09	173.4785	173478	5783	18.2	
4	3.02	514.6573	514657	23889	54.1	
		951.2814	951281	51929	100.0	

Fig. 13.8 An HPLC chromatogram of the microalgae, *Chlorella* sp., in a mobile phase of acetonitrile and formic acid (50:50)

Table 13.5 Composition of enriched seawater medium

Component	Amount	Stock solution concentration	Final concentration
Pasteurised seawater	1 L		
Enrichment solution for seawater medium	20 mL/L		

13.3 Present Status and Future Opportunities

The global market for biosurfactants and bioemulsifiers is steadily increasing, and with this the scope for research in this area is also ever increasing. The production of biosurfactants and bioemulsifiers by marine algae is an area with limited literature. The biosurfactants and bioemulsifiers produced by marine or coastal algae are interesting for the researchers due to their structural and functional diversity. There are enormous applications associated with the algal bioactive molecules, and hence

Table 13.6 Composition of 1% F2 medium

Component	Amount	Stock solution concentration	Final concentration
Seawater(non-sterilised)	285 mL		
dH_2O	600 mL		
$NaNO_3$ (Fisher BP360–500)	1 mL/L	7.5 g/100 mL dH20	880 μM
$NaH_2PO_4 \cdot H_2O$(MCIB 742)	1 mL/L	0.5 g/100 mL dH20	36 μM
$Na_2SiO_3 \cdot 9H_2O$ (Sigma 307,815)	1 mL/L	2 g/100 mL dH20	70 μM
Trace metals solution	1 mL/L		
Vitamin B_{12}	1 mL/L		
Biotin vitamin solution	1 mL/L		
Thiamine vitamin solution	1 mL/L		

Table 13.7 Composition of F2 medium

Component	Amount	Stock solution concentration	Final concentration
$NaNO_3$ (Fisher BP360-500)	1 mL	7.5 g/100 mL dH20	880 μM
$NaH_2PO_4 \cdot H_2O$(MCIB 742)	1 mL	0.5 g/100 mL dH20	36 μM
$Na_2SiO_3 \cdot 9H_2O$ (Sigma 307,815)	1 mL	3 g/100 mL dH20	106 μM
Trace metals solution	1 mL/L		
Vitamin B_{12}	1 mL/L		
Biotin vitamin solution	1 mL/L		
Thiamine vitamin solution	1 mL/L		

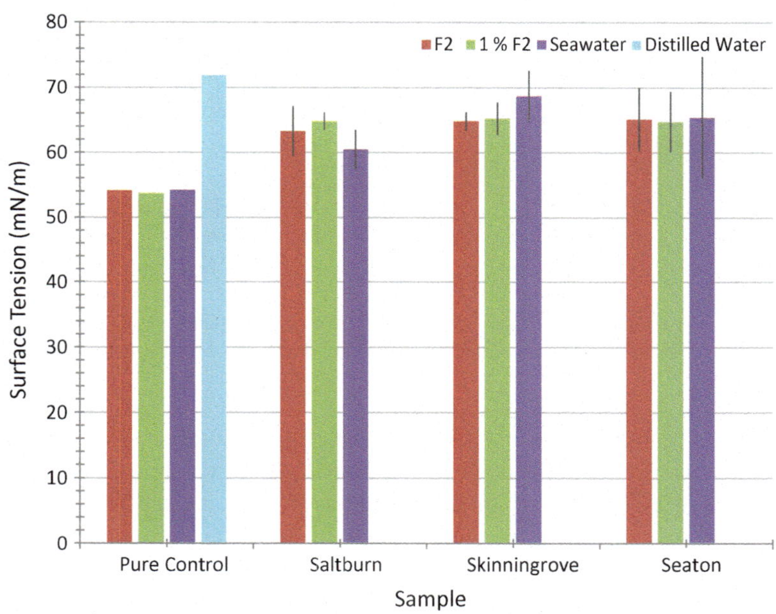

Fig. 13.9 The average surface tension for each culture media type from each site. Both pure media and distilled water were used as controls. Error was calculated using the standard deviation of the original tensiometer readings

Table 13.8 *P*-values obtained from a two-way ANOVA carried out on tensiometer measurements

Variable	*P*-value
Media type	0.964
Sampling location	0.293
Media type and location	0.616

A *P*-value below 0.05 was considered statistically significant

Table 13.9 *P*-values obtained from a one-way ANOVA carried out on surface tension measurements between the culture samples and the distilled water control

Media type	*P*-value
F2	0.261
1% F2	0.257
Enriched seawater	0.993

A *P*-value below 0.05 was considered statistically significant

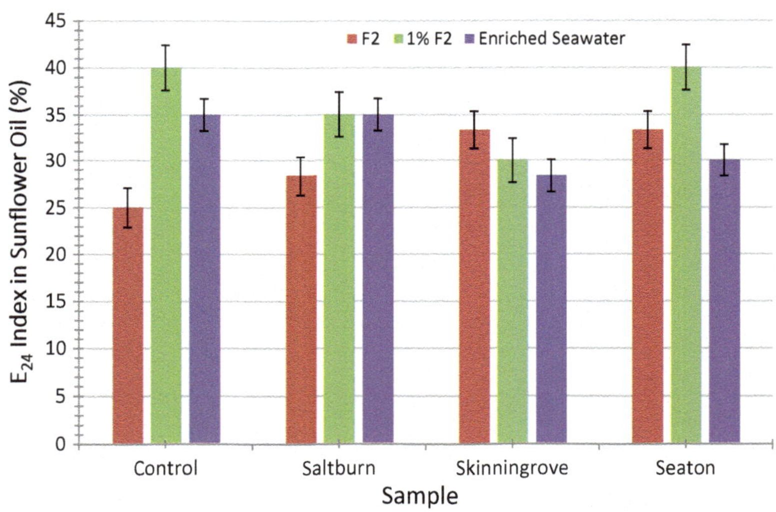

Fig. 13.10 An average E_{24} index (%) in sunflower oil was calculated for each sampling location and media type. Error was calculated using the standard deviation of the E_{24} measurements

focus should be more on the effective algal sampling techniques, effective organism screening, effective identification and characterisation of the bioactive molecules.

The presence of polyaromatic hydrocarbons (PAHs) at sampling sites has previously led to the discovery and isolation of surfactant-producing organisms [28]. This sampling method acts as a sure short way or preliminary screening technique for surface active molecule production, as survival of organisms and utilisation of PAHs as a carbon source are often associated with surfactant or emulsifier production [39].

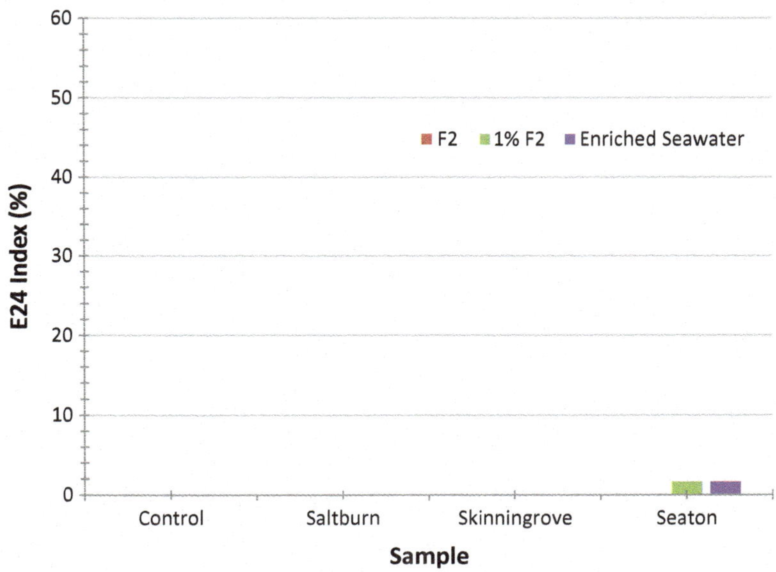

Fig. 13.11 An average E_{24} index (%) in kerosene was calculated for each sampling location and media type. Error was calculated using standard deviation of the E_{24} indexes

Table 13.10 *P*-values obtained from a two-way ANOVA carried out on the sunflower oil E_{24} index measurements

Variable	*P*-value
Media type	0.400
Sampling location	0.365
Media type and location	0.408

A *P*-value below 0.05 was considered statistically significant

Table 13.11 *P*-values obtained from a one-way ANOVA carried out on sunflower oil E_{24} index measurements between culture samples and the control samples of pure media

Media type	*P*-value
F2	0.119
1% F2	0.166
Enriched seawater	0.244

A *P*-value below 0.05 was considered statistically significant

In the study of algal surfactants and emulsifiers, there is little information on the optimisation of culturing conditions. Culture media F2, 1% F2 and enriched seawater media have been used previously. Little information exists on the selectivity of 1% F2 and enriched seawater media for surfactant or emulsifier production; however, Villay et al. [40] found a marked increase in exopolysaccharide production when the microalga *Rhodella violacea* was cultured in F2 media. Alternatively, Bafana [41] found that M1 media was optimised for exopolysaccharide production.

In a study carried out by Rosales-Morales and Paniagua-Michel [28], the addition of hydrocarbons to the algal cultures acted as an early screening technique for the production of bioemulsifiers. Those which exhibited high degradation were screened out, allowing for efficient screening. The addition of hydrocarbons to the algal cultures generally allows for screening of samples to take forward to the next stages of the experiment.

The search for potent biosurfactants and bioemulsifiers is becoming more accessible due to extensive knowledge of properties and the availability of several screening methods [15]. Oil dispersion test is one effective method for the primary screening or detection of biosurfactants [15]. The occurrence of a clear zone in the centre of the oil is an indication of biosurfactant presence [42], and the diameter of the clearing zone is directly correlated to surfactant activity. The addition of pure biosurfactant leads to a linear correlation between quantity of surfactant and clearing zone diameter [7]. The oil dispersion test is rapid and requires no specialised equipment, as well as requiring only a small volume of sample. It can also be applied in cases of low biosurfactant activity and quantity [43]. It is applicable to detecting surfactants in a wide range of microorganisms [43–45]. Additionally, it has been found that there is an inverse linear relationship with surface tension, whereby, increased oil dispersion is correlated with decreased surface tension.

Much work is also being carried out into the specificity of screening methods for biosurfactants and bioemulsifiers. Whilst oil dispersion has been found to be an effective method for the detection of biosurfactants, Satpute et al. [46] also found that this test can be insufficient for the identification of bioemulsifiers. This is due to the fact that bioemulsifiers mainly emulsify liquids, rather than altering interfacial tension between different phases [47]. Early indications in the field of the isolation of surface active molecules from algae are also that only bioemulsifiers are being produced by algal species, rather than biosurfactants. Therefore, it must also be noted that a negative result in oil dispersion tests does not necessarily mean the absence of bioemulsifiers in the culture samples.

Another technique for the measurement of surface tension is using a Du Nouy ring tensiometer method. The method is based on the level of force required to detach the loop from the interface or surface, which is proportional to the interfacial tension [48]. In order for the measurements to be accurate, no contaminants can be present on the platinum ring. This means that the ring must be sterilised before every sample measurement [7]. The Du Nouy ring assay is widely used in the screening of surfactant-producing organisms [49–52].

A culture sample is generally considered to be promising for surfactant presence if it reduces the surface tension of water to 40 mN/m or less [53]. Other studies say that any surface tensiosn reduction >20 mN/m below the surface tension of distilled water (72 mN/m) is counted as 'good', and thereby it is a promising biosurfactant [54]. This technique is highly accurate and easy to carry out however requires specialised equipment and does not allow for simultaneous sample measurement, meaning it is inefficient with large sample numbers [49]. It also requires more sample volume than other screening techniques, using several 'mL' rather than 'μL' used in techniques such as drop-collapse tests. There is also a limit to the surfactant concentration which can be measured using tensiometer measurements, without requiring dilution. It is also not as reproducible as the drop-collapse method [49].

The E_{24} emulsification assays correlate to surfactant concentration and are also widely used as a first-stage screening method for biosurfactant-producing organisms [43, 55]. As mentioned earlier, emulsification capacity and surface activity do not always correlate meaning that E_{24} indexes give only an indication of biosurfactant presence not a confirmation [7]. The use of a known biosurfactant as a positive control allows for easy comparison with the E_{24} measurements of the samples in question.

There is a degree of overlap in the characteristics and functions of biosurfactants and bioemulsifiers. However, it has been found that whilst biosurfactants alter surface tension and emulsify, bioemulsifiers are not known to have an impact on surface tension. Therefore, screening methods based on changes in surface tension may lead to the exclusion of bioemulsifiers [12]. Some alternative confirmatory methods are required in this field which can easily target the detection of bioemulsifiers.

If an algal sample has been found to produce a surface active molecule, which caused a statistically significant change in a screening method, in comparison with the control, the first step might be to use a Tukey test to identify which of the samples has led to the significance. Chemical analysis through a technique such as high-performance liquid chromatography (HPLC) or gas chromatography-mass spectrometry (GC-MS) [15] can be carried out to identify the surface active molecule. Identification of the algal sample in question through DNA sequencing and bioinformatics is also productive.

The identification of algal bioemulsifiers and biosurfactants holds much promise, due to the wide range of molecules produced by algae and the abundance of biomass which can be cultivated. However, much work is needed to identify specific species and molecules which have surface active properties. This includes optimisation of sampling and culturing conditions as well as appropriate screening methods. Whilst the oil dispersion tests, surface tension measurements and E_{24} emulsification assays are widely used in the detection of biosurfactants, future studies should consider the use of techniques which are specifically targeted at bioemulsifiers, as the change in interfacial tension which these techniques are reliant upon is not always exhibited by bioemulsifiers.

Acknowledgement Authors wish to thank the BBSRC HVCfP Network for the Business Interaction Voucher. PKSMR wishes to thank BBSRC International Workshop Fund for establishing a Joint UK-NZ Partnership in Industrial Biotechnology and Bioenergy and Dr. Steve Thompson, British High Commissioner (Wellington, New Zealand), for helping to establish UK/NZ Algal network.

References

1. Wuebbles D, Jain A (2001) Concerns about climate change and the role of fossil fuel use. Fuel Process Technol 71(1–3):99–119
2. Muller MM, Kugler JH, Henkel M, Gerlitzki M, Hormann B, Pohnlein M, Syldatk C, Hausmann R (2012) Rhamnolipids-next generation surfactants? J Biotechnol 162(4):366–380
3. Reis RS, Pacheco GJ, Pereira AG, Freire DMG (2014) Biosurfactants: production and applications. In: Chamy R, Rosenkranz F (eds) Biodegradation – life of science. CC BY, Sydney
4. Geys R, Soetaert W, Van Bogaert I (2014) Biotechnological opportunities in biosurfactant production. Curr Opin Biotechnol 30(0):66–72
5. Sekhon KK, Rahman PK (2014) Rhamnolipid biosurfactants – past, present and future scenario of global market. Front Microbiol 5:454
6. Campos JM, Stamford TLM, Sarubbo LA, Luna JM, Rufino RD, Banat IM (2013) Microbial biosurfactants as additives for food industries. Biotechnol Prog 29(5):1097–1108
7. Walter V, Syldatk C, Hausmann R (2013) Screening concepts for the isolation of biosurfactant producing microorganisms. In: Madame curie bioscience. Landes Bioscience Austin, Texas
8. Karsa DR (1998) New products and applications in surfactant technology. Academic, Sheffield, pp 88–108
9. Das P, Mukherjee S, Sen R (2009) Substrate dependent production of extracellular biosurfactant by a marine bacterium. Bioresour Technol 100(2):1015–1019
10. Smyth TJP, Perfumo A, Marchant R, Banat IM (2010) Isolation and analysis of lipopeptides and high molecular weight biosurfactants. In: Timmis KN (ed) Handbook of hydrocarbon and lipid microbiology. Springer, Berlin
11. Singh SN, Kumari B, Mishra S (2012) Microbial degradation of xenobiotics environmental science and engineering. In: Singh SN (ed) Microbial degradation of alkanes. Springer, Berlin
12. Uzoigwe C, Burgess JG, Ennis CJ, Rahman PKSM (2015) Bioemulsifiers are not biosurfactants and require different screening approaches. Front Microbiol 6(1):1–6
13. Fracchia L, Cavallo M, Martinotti MG, Banat IM (2012) Biosurfactants and bioemulsifiers biomedical and related applications – present status and future potentials. Biomed Sci Eng Technol 14(1):325–370
14. Paniagua-Michel JDJ, Olmos-Soto J, Morales-Guerrero ER (2014) Algal and microbial exopolysaccharides: new insights as biosurfactants and bioemulsifiers. Adv Food Nutr Res 73(1):221–257
15. Satpute SK, Banpurkar AG, Dhakephalkar PK, Banat IM, Chopade AB (2010) Methods for investigating biosurfactants and bioemulsifiers: a review. Crit Rev Biotechnol 1(1):1–18
16. Raposo MFDJ, Morais RMSCD, Morais AMMBD (2013) Bioactivity and applications of Sulphated polysaccharides from marine microalgae. Mar Drugs 11(1):233–252
17. Susilaningsih D, Okazaki F, Yopi Y, Widyastuti Y, Harayama S (2013) Isolation and screening of surfactant-producing bacteria from Indonesian marine environments and its application on bioremediation. Annales Bogorienses 17(2):43–53
18. Chang T, Ohta S, Ikegami N, Miyata H, Kashimoto T, Kondo M (1993) Antibiotic substances produced by a marine green alga, *Dunaliella primolecta*. Bio/Technology 44(2):149–153
19. Varshney P, Mikulic P, Vonshak A, Beardall J, Wangikar PP (2015) Extremophilic micro-algae and their potential contribution in biotechnology. Bioresour Technol 184:363–372
20. Mishra A, Jha B (2009) Isolation and characterization of extracellular polymeric substances from micro-algae *Dunaliella salina* under salt stress. Bioresour Technol 100(1):3382–3386
21. Laurienzo P (2010) Marine polysaccharides in pharmaceutical applications: an overview. Mar Drugs 8(9):2435–2465
22. Pichavant L (2009) Synthese et reactivite de monomere issus de resources renouvelables pour la polymerisation radicalaire, PhD Thesis. Technologie et sante Universite de Reims Champagne-Ardenne 1(1):259
23. Singh A, Van Hamme JD, Ward OP (2007) Surfactants in microbiology and biotechnology: part 2. Application aspects. Biotechnol Adv 25(1):99–121

24. Pawar ST, Bhosale AA, Gawade TB, Nale TR (2013) Isolation, screening and optimization of exopolysaccharide producing bacterium from saline soil. J Microbiol Biotechnol Res 3(3):24–31
25. Patel AK, Michaud P, Singhania RR, Soccol CR, Pandey A (2010) Polysaccharides from probiotics: new developments as food additives. Food Technol Biotechnol 48(1):451–463
26. Fuentes MMR, Sanchez JLG, Sevilla JMF, Fernandez FGA, Perez JAS, Grima EM (1999) Outdoor continuous culture of *Porphyridium cruentum* in a tubular photobioreactor: quantitative analysis of the daily cyclic variation of culture parameters. J Biotechnol 70(1):271–288
27. Patel AK, Laroche C, Marcati A, Ursu AV, Jubeau S, Marchal L, Petit E, Djelveh G, Michaud P (2013) Separation and fractionation of exopolysaccharides from *Porphyridium cruentum*. Bioresour Technol 145(1):345–350
28. Rosales-Morales A, Paniagua-Michel J (2014) Bioremediation of hexadecane and diesel oil is enhanced by photosynthetically produced marine biosurfactants. J Bioremed Biodegr 4(1):1–5
29. Calvo C, Toledo FL, Pozo C, Martinez-Toledo MV, Gonzalez-Lopez J (2004) Biotechnology of bioemulsifiers produced by microorganisms. J Food Agric Environ 2(3–4):238–243
30. Pepi M, Cesaro A, Luit G, Baldi F (2005) An Antarctic psychrotrophic bacterium *Halomonas* sp. ANT 3-b, growing on n-hexadane, produces a new emulsifying glycolipid. FEMS Microbiol Ecol 53(1):157
31. Amaral PFF, Da Silva JM, Lehocky M, Barros-Timmons AMV, Coelho MAZ, Marrucho IM (2006) Production and characterisation of a bioemulsifier from *Yarrowia lipolytica*. Process Biochem 41(8):1894
32. Maneerat S, Bamba T, Harada K, Kobayashi A, Yamada H, Kawai F (2006) A novel crude oil emulsifier excreted in the culture supernatant of a marine bacterium, *Myroides* sp. strain SM1. Appl Microbiol Biotechnol 70(1):254
33. Thaniyavarn J, Chongchin A, Wanitsuksombut N, Thaniyavarn S, Pinphanichakarn P, Leepipatpiboon N (2006) Biosurfactant production by *Pseudomonas aeruginosa* A41 using palm oil as carbon source. J Gen Appl Microbiol 52(4):215
34. Kumar SA, Mody K, Jha B (2007) Evaluation of biosurfactant/bioemulsifier production from marine bacterium. Bull Environ Contam Toxicol 79(6):617
35. Thavasi R, Jayalakshmi S, Balasubramanian T, Banat IM (2007) Biosurfactant production by *Corynebacterium kutscheri* from waste motor lubricant oil and peanut oil cake. Lett Appl Microbiol 45(6):686
36. Das P, Mukherjee S, Sivapathasekaran C, Sen R (2010) Microbial surfactants of marine origin: potentials and prospects. In: Sen R (ed) Biosurfactants. Landes Bioscience and Springer Science + Business Media, New York
37. Reyes-Tisnado R, Hernandez-Carmona G, Montesinos ER, Higuera DL, Lopez Guttierrez F (2005) Food grade alginates extracted from the giant kelp *Macrocystis pyrifera* at pilot-plant scale. Rev Investig Mar 26(3):185–192
38. Kantas D, Papadopoulos S, Papapolymerou G, Gougoulias N, Karayannis V, Spiliotis X (2015) The macronutrient content of *Chlorella vulgaris* microalgae grown in laboratory and small pilot scale bioreactors. In: Proceedings of the fifth international conference on environmental management, engineering, planning and economics, pp 467–474
39. Marcade M, Monleon L (1996) Screening and selection of surfactant producing bacteria from waste lubricating oil. J Appl Bacteriol 81(2):161–166
40. Villay A, Laroche C, Roriz D, El Alaoui H, Delbac F, Michaud P (2013) Optimisation of culture parameters for exopolysaccharides production by the microalga *Rhodella violacea*. Bioresour Technol x(x):xx–xx
41. Bafana A (2013) Characterization and optimisation of production of exopolysaccharide from *Chlamydomonas reinhardtii*. Carbohydr Polym 95(2):746–752
42. Morikawa M, Hirata Y, Imanaka T (2000) A study on the structure-function relationship of lipopeptide biosurfactants. Biochem Biophy Acta 1488(1):211–218
43. Plaza G, Zjawiony I, Banat I (2006) Use of different methods for detection of thermophilic biosurfactant-producing bacteria from hydrocarbon-contaminated bioremediated soils. J Pet Eng 50(1):71–77

44. Huy N, Jin S, Amada K (1999) Characterisation of petroleum-degrading bacteria from oil-contaminated sites in Vietnam. J Biosci Bioeng 88(1):100–102
45. Youssef N, Duncan K, Nagle D et al (2004) Comparison of methods to detect biosurfactant production by diverse microorganisms. J Microbiol Methods 56(3):339–347
46. Satpute SK, Bhawsar BD, Dhakephalkar PK, Chopade BA (2008) Assessment of different screening methods for selecting biosurfactant producing marine bacteria' Indian. J Mar Sci 37(3):243–250
47. Ellaiah P, Prabhakar T, Sreekanth M, Taleb AT, Raju PB, Saisha V (2002) Production of glycolipids containing biosurfactants by Pseudomonas species. Ind J Exp Biol 40(1):1083–1086
48. Tadros T (2005) Adsorption of surfactants at the air/liquid and liquid/liquid interfaces. In: Applied surfactants: principles and applications. Wiley VCH, Weinheim, pp 81–82
49. Bodour A, Miller-Maier RM (1998) Application of a modified drop-collapse technique for surfactant quantitation and screening of biosurfactant-producing microorganisms. J Microbiol Methods 32(3):273–280
50. Das M, Das S, Mukherjee R (1998) Surface active properties of the culture filtrates of a *Micrococcus* species grown on n-alkanes and sugars. Bioresour Technol 63(1):231–235
51. Makkar R, Cameotra S (1997) Biosurfactant production by a thermophilic *Bacillus subtilis* strain. J Ind Microbiol Biotechnol 18(1):37–42
52. Persson A, Molin G (1987) Capacity for biosurfactant production of environmental *Pseudomonas* and *Vibrionaceae* growing on carbohydrates. Appl Microbiol Biotechnol 26(5):439–442
53. Cooper D (1986) Biosurfactants. Microbiol Sci 3(5):145–149
54. Willumsen P, Karlson U (1997) Screening of bacteria, isolated from PAH-contaminated soils, for production of biosurfactants and bioemulsifiers. Biodegradation 7(5):415–423
55. Christova N, Tuvela B, Lalchev Z (2004) Rhamnolipid biosurfactants produced by *Renibacterium salmoninarum* 27BN during growth on n-hexadecane. Z Naturforcsh 59(1–2):70–74

Production of High-Quality Biodiesel by *Scenedesmus abundans*

14

S. K. Mandotra, A. S. Ahluwalia, and P. W. Ramteke

Abstract

The ever-increasing need of energy both in the domestic and industrial front has augmented the consumption of fossil fuel; consequently, complexity arises owing to exhausting fuel supplies and due to their contribution to climate change by the emission of large quantities of greenhouse gases. The renewable, economic, and carbon-neutral biofuel from algae has made it a promising feedstock that can curtail global dependency on rapidly depleting fossil fuel-based petro diesel. Moreover, higher biomass and cellular lipid accumulation competence and economic sustainability even in large-scale production make algae a better choice than other existing oil crops. There are quite a few studies reporting number of green microalgae as a potential feedstock for biofuel production. Accumulation of lipid in microalgae is species dependent, and in potential strains it ranges from 25% to 60% of dry cell weight, in modified growth conditions; however, some microalgae are reported to accumulate more than 60% of cellular lipid content. The present chapter is specifically aimed to review freshwater green microalga *Scenedesmus abundans* as a prospective feedstock for high-quality biofuel production.

S. K. Mandotra · A. S. Ahluwalia (✉)
Department of Botany, Panjab University, Chandigarh, India
e-mail: phykos@pu.ac.in

P. W. Ramteke (✉)
Department of Biological Sciences, Sam Higginbottom University of Agriculture, Technology and Sciences, Allahabad, Uttar Pradesh, India

L. B. Sukla et al. (eds.), *The Role of Microalgae in Wastewater Treatment*,
https://doi.org/10.1007/978-981-13-1586-2_14

14.1 Introduction

Extensive consumption of fossil fuels in industries and in transport sector has elevated the air pollutants to their critical level; they are disturbing the ecological equilibrium that not only causing climate change but also posing serious threats to human as well as animal health. At present, 90% of global energy demand is fulfilled by the use of fossil fuel-based petro diesel. Their use is increasing year by year with the increase in human population and industrialization; with this ever-increasing pace, it has been anticipated that the conventional fossil fuel reserves will be completely exhausted by the year 2050 [17, 24]. To fulfill global energy demand and to protect environment as well as human health, researchers are facing immediate challenge to explore nonconventional sources of energy that are renewable, environment friendly, and cost-effective in nature.

To combat energy crises, first-generation biofuel came into existence; they are basically plant-based high-energy organic matter that can be processed into biofuel. But the major issue associated with first-generation biofuel is the nature of feedstock which is also consumed by humans as food products. Food crops such as soybean, sunflower, corn, sugarcane, sugar beet, etc. are good source of oil, sugar, and starch which can be converted into biodiesel, ethanol, and biogas. Second-generation biofuel is derived from agricultural waste or residue; unlike first-generation biofuel, second-generation biofuel are nonfood feedstock. The major disadvantage with second-generation biofuel is high processing cost associated with it; apart from this, technology barriers and feedstock collection are also drawbacks [20, 21, 31, 45, 46]. Most recent and advanced biofuel are third generation. These fuels are specifically derived from algae. By using algae as a feedstock, a number of fuels such as biodiesel, biohydrogen, ethanol, butanol, and jet fuel can be obtained (Fig. 14.1). Unlike first- and second-generation biofuels, third-generation biofuels have several advantages due to higher growth rate and lipid content of the algae; more importantly algae do not require arable land for its growth and do not compete with other food crops [11, 25, 26].

About 200,000–800,000 algal species are in existence, out of which about 50,000 species have been identified and documented [24, 40]. Several potential algal (microalgae, macroalgae, and cyanobacteria) species have been reported for biofuel production. With slight modification in growth conditions, the lipid content of algae can be enhanced as much as 80% of their dry cell weight [1, 18]. Apart from biofuel, algae are capable in producing high-value coproducts that have nutritional and pharmaceutical importance; with this advantage, the overall downstream processing becomes more economic and sustainable at industrial scale. Therefore, the present chapter deals with one of the most widely distributed potential biofuel-producing microalga *Scenedesmus abundans*. Its biofuel potential in light of fatty acid-derived biodiesel properties and various growth conditions pertaining to higher growth and lipid content will be discussed.

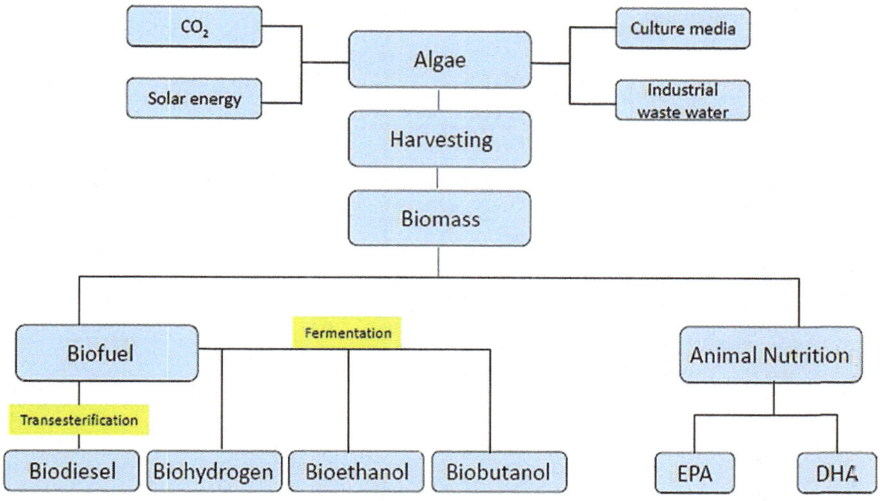

Fig. 14.1 Overview of biofuel production form microalgae

14.2 Microalgal Lipid Content

Different algal species have different growth rate and biomass-accumulating potential; moreover, cellular lipid content of different algal species varied significantly. The lipid content of green microalgae is generally higher than that of blue-green algae (cyanobacteria) [20, 21, 50]. Therefore, the selection of algae is of prime importance while considering it for biofuel production. The biofuel potential of algae does not only depend on the cellular lipid content; instead, it is also dependent on the biomass content of the algae; the overall lipid content (g/L) significantly affected the biomass content [25].

Few microalgal species such as *Botryococcus braunii*, *Neochloris oleoabundans*, *Dunaliella tertiolecta*, and *Chlorella emersonii* have been reported to have very high amount of lipid content (>60%) of their dry weight biomass [24]. The natural oil percentage (% dry cell weight) of various microalgal species studies by different researchers are given in Table 14.1.

14.3 Lipid Accumulation Potential of *Scenedesmus abundans*

S. abundans is a freshwater green microalgae belonging to the genus *Scenedesmus* in the class Chlorophyceae. Cells are nonmotile in nature, and their colonies are usually two to four celled arranged in a linear series; rarely, they are found in eight-celled colonies (Fig. 14.2). In addition to the spines present in all the four corners, cells also contain one or more median lateral spines which form the outer face [34].

A study carried out by Mandotra et al. [25] explored the biofuel potential of *S. abundans* isolate from the culture collections of Dal Lake, Kashmir, India. On the

Table 14.1 Lipid content of selected algal species

Algae species	Alga type	Lipid/oil percentage (% dry cell weight)	References
Botryococcus braunii	Green alga	25–75	[8]
Neochloris oleoabundans	Green alga	35–65	[15]
Crypthecodinium cohnii	Red alga	20–51	[8]
Scenedesmus quadricauda	Green alga	2–18	[41]
Dunaliella tertiolecta	Green alga	17–71	[28]
Chlorella emersonii	Green alga	25–63	[15]
Haematococcus pluvialis	Red alga	15–32	[10]
Chlorella pyrenoidosa	Green alga	11–26	[32]
Spirulina maxima	Green alga	4–9	[27]
Dunaliella salina	Green alga	6–25	[1]
Euglena gracilis	Green alga	14–20	[5]
Phaeodactylum tricornutum	Diatom	18–57	[8]
Skeletonema costatum	Diatom	13–51	[41]
Dunaliella primolecta	Green alga	23	[8]
Scenedesmus obliquus	Green alga	11–55	[1]
Oocystis pusilla	Green alga	10	[27]
Arthrospira maxima	Blue-green alga	20	[4]
Scenedesmus abundans	Green alga	37	[25]
Nannochloropsis oculata NCTU-3	Green alga	31–50	[9]
Isochrysis galbana	Prymnesiophytes	7–40	[30]

Fig. 14.2 Microscopic image of microalga *S. abundans*

basis of preliminary analysis, the alga was found to have the biomass content of 0.98 g/L along with 37% lipid content of their dry cell weight. Nutrient limitation in the culture medium is one of the strategies to enhance cellular lipid content of the algae. Nitrogen is one of the major intracellular components that plays vital role in protein synthesis along with cell division of algae [3, 43]. It is well known that nitrogen deprivation in the growth medium leading to higher lipid accumulation compromises biomass content of the alga. However, in a few studies, it has been reported that there is optimum nitrogen-limited condition in which algae suffer with minimal stress condition which is sufficient to induce lipid accumulation without compromising biomass content. A study carried out by Mandotra et al. [25] with *S. abundans* demonstrated the effect of nitrogen limitation. Six different concentrations (0.0 g/L, 0.08 g/L, 0.16 g/L, 0.24 g/L, 0.32 g/L, and 0.4 g/L) of nitrogen (KNO_3) were taken in the growth medium. Nitrogen concentration of 0.4 g/L was taken as positive control with standard concentration of media, whereas 0.0 g/L was taken as negative control with the absence of nitrogen in the growth medium. The result demonstrated that the higher biomass and lipid content of 1.11 g/L and 489 mg/L, respectively, was found in the culture medium supplemented with the nitrogen concentration of 0.32 g/L. Although the growth condition (0.32 g/L) was slight nitrogen limiting, still the biomass content was higher in comparison with standard nitrogen concentration (0.4 g/L) of the growth medium. The study confirmed that, even though nitrogen content in the growth medium favors higher growth, but beyond certain limit, it shows inhibitive effect on biomass content of the algae [22, 25]. Yet in another study, different sources of nitrogen, viz., ammonium nitrate, ammonium sulfate, and sodium nitrate, were examined to grow *S. abundans* in autotrophic culture condition; as a result, maximum lipid yield of 3.55 mg/L/day was achieved with the culture medium having ammonium nitrate as a nitrogen source [14].

Carbon dioxide (CO_2) is one of the important factors that significantly enhance the yield of algal biomass. In open pond cultivation, about 1 kilogram of dry algal biomass utilizes 1.83 kilogram of CO_2 [6, 8, 35]. Chellamboli and Perumalsamy [6] performed a study using *S. abundans* by central composite design-response surface methodology (CCD-RSM); in their study they have considered different culture time, inoculum concentration, and sodium bicarbonate content. As a result, maximum biomass yield of 39.1 mg/L/day with highest lipid content of 26.2% dry cell weight was recorded with 8 g/L sodium bicarbonate and 10% inoculum on 30th day of the culture period.

Phosphate is another macronutrient that plays vital role in the growth of the algae. It is one of the important constituents of nucleic acid metabolism, ATP synthesis, signal transduction, and phospholipid metabolism [3]. Phosphate-limiting condition significantly affects the cell division process of the algae; synthesis of most of the intracellular molecules such as carbohydrate, protein, and chlorophyll pigment ceases and carbon flux is directed toward the lipid synthesis [23, 29, 38, 48]. Therefore, like nitrogen limitation, phosphate limitation is also one of the widely studied methods of lipid accumulation. A study carried out with microalga *S. abundans* with different phosphate (K_2HPO_4) concentration (20 mg/L, 40 mg/L,

60 mg/L, and 80 mg/L) in the culture medium showed significant effect on the growth and lipid content of the alga. The study demonstrated the growth of alga on all phosphate concentrations; however, the highest biomass content (770 mg/L) was recorded on the culture supplemented with 60 mg/L K_2HPO_4. On the other hand, highest lipid percentage was observed in culture with least amount (20 mg/L) of K_2HPO_4 content. Lipid concentration (mg/L) of the culture is biomass dependent; it is the total amount of lipid content extracted from the biomass present in 1 liter of the culture medium. Therefore, the highest lipid concentration of 176 mg/L was recorded in the culture with 60 mg/L of K_2HPO_4 [26].

Like different nitrogen and phosphate concentration, the effect of other growth conditions such as pH and light intensity was also studied for *S. abundans*. By keeping all the culture conditions constant, different pH (pH 5, pH 6, pH 7, pH 8, and pH 9) and light intensities (3000, 4000, 5000, and 6000 lux) were studied to assess the effect on biomass accumulation and lipid production [26]. Among different light intensities, culture with the light intensity of 6000 lux has shown highest biomass concentration of 742 mg/L. Light intensity is one of the crucial parameters; at low light intensity, biomass content compromises, whereas, extremely high light intensities damage photosynthetic apparatus (PSI and PSII) [42, 44]. Higher light intensity required by *S. abundans* can be explained by its higher growth rate; during log phase of the growth, the alga culture becomes dense; to overcome self-shading effect, the alga requires higher light intensities, so that every cell gets enough amount of light to carry out normal physiological processes. Lipid content of *S. abundans* increased with increased light intensities. Highest lipid percentage (32%) with lipid concentration of 243 mg/L was recorded in the culture illuminated with 6000 lux light intensity [26]. At higher light intensities, light energy is converted and stored into chemical energy in the form of photo assimilates to overcome photooxidation [37, 47].

Biomass content of *S. abundans* increases with increase in pH from 5 to 8; the growth, however, decreased at pH 9. Highest biomass concentration of 769 mg/L was observed at pH 8. As far as lipid percentage was concern, highest lipid percentage (26%) with the lipid concentration of 179 mg/L was observed at pH 6 [26]. At adverse pH condition, algae tend to divert their energy for the biosynthesis of lipids instead of biomass [16].

14.4 Biodiesel Properties of *S. abundans*

Even after producing sufficient quantities of lipids, the alga cannot be used as feedstock for biodiesel production until and unless its fatty acid-derived biodiesel properties are in accordance with various national and international biodiesel standards. There are certain biodiesel standards such as EN 14214, European biodiesel standards; IS 15607, Indian biodiesel standards; and ASTM D6751–08, American biodiesel standard that provide standard range of values for biodiesel properties. There are four important biodiesel properties such as degree of unsaturation (DU), cetane number (CN), iodine value (IV), and saponification value (SV). DU is the sum of monounsaturated (MUFA) and polyunsaturated fatty acids (PUFA) that influences

Table 14.2 Comparative biodiesel properties of *S. abundans* with other biodiesel feedstocks

Biodiesel feedstock	DU (wt. %)	CN	IV (g I_2/100 g oil)	SV (mg KOH/g oil)	References
S. abundans	84.5	52.15	94.06	202.02	[25]
Aphanothece sp.	70.6	55.8	65.4	225.1	[13]
Soybean	143.8	49	128	–	[39]
Sunflower	152.2	50	132	–	,,
Peanut	113.1	53	97	–	,,
EN 14214	–		Min. 51	Max. 120	–
IS 15607	–		Min. 51	Max. 120	–
ASTM D6751-08	–		Min. 47	–	–

DU degree of unsaturation, *CN* cetane number, *IV* iodine value, *SV* saponification value
EN 14214, European biodiesel standards; IS 15607, Indian biodiesel standards; ASTM D6751–08, American biodiesel standards

the oxidative stability of biodiesel. PUFA contain large number of reactive sites that are susceptible to free radial attack; therefore, having higher PUFA content negatively affects the oxidative stability of biodiesel. Fatty acid profile having large quantity of long-chain saturated fatty acid (SFA) and MUFA improves oxidative stability of biodiesel [2, 7, 12, 49].

CN is another biodiesel property that influences the fuel combustion. Biodiesel having higher value of CN have low nitrous oxide (NO) emission, easy startup, and less knocking of engine. Higher value of SFA and MUFA results in higher CN value. IV indicates unsaturation of biodiesel oil; it increases with increase in double bonds. It is the amount of iodine required to saturate 100 g of oil. Biodiesel with high IV results in engine deposit. SV is another biodiesel property which is defined as the amount (mg) of potassium hydroxide (KOH) required to saponify 1 gram of oil sample. SV is also used to calculate the CN value of biodiesel [19, 36].

Table 14.2 shows the comparative biodiesel properties of *S. abundans* with other biodiesel feedstocks. Study carried out by Mandotra et al. [25] on *S. abundans* reported the presence of 6.4% of linolenic acid and complete absence of fatty acid ⩾4 double bond. For an ideal biodiesel, linolenic acid and PUFA with ⩾4 double bond should not increase 12% and 1%, respectively [15, 33].

14.5 Conclusion

The present chapter reviews the biodiesel potential of freshwater green microalga *S. abundans*. It is evident by the present data that the alga possesses higher biomass and lipid content naturally. However, the biomass and lipid content could be further enhanced by slight modification in the growth conditions. Fatty acid-derived biodiesel properties of *S. abundans* were also reviewed and found in accordance with various national and international biodiesel standards (EN 14214, IS 15607, and ASTM D 6751–08). As a result, it could be concluded that the alga could be explored for large-scale commercial biodiesel feedstock for quality biofuel production.

References

1. Alaswad A, Dassisti M, Prescott T, Olabi AG (2015) Technologies and developments of third generation biofuel production. Renew Sust Energ Rev 51:1446–1460
2. Arias-Peñaranda MT, Cristiani-Urbina E, Montes-Horcasitas C, Esparza-Garcıa F, Torzillo G, Cañizares-Villanueva RO (2013) Scenedesmus incrassatulus CLHE-Si01: a potential source of renewable lipid for high quality biodiesel production. Bioresour Technol 140:158–164
3. Bajhaiya AK, MandotraS K, AnsoliaA, Barsana A (2017) Recent advances in improving eco-physiology of microalgae for biofuels. In: Algal biofuels. Springer International Publishing, pp 141–162
4. Baunillo KE, Tan RS, Barros HR, Luque R (2012) Investigations on microalgal oil production from Arthrospira platensis: towards more sustainable biodiesel production. RSC Adv 2(30):11267–11272
5. Becker EW (1994) Microalgae: biotechnology and microbiology, vol 10. Cambridge University Press, Cambridge
6. Chellamboli C, Perumalsamy M (2014) Application of response surface methodology for optimization of growth and lipids in Scenedesmus abundans using batch culture system. RSC Adv 4(42):22129–22140
7. Chen L, Liu T, Zhang W, Chen X, Wang J (2012) Biodiesel production from algae oil high in free fatty acids by two-step catalytic conversion. Bioresour Technol 111:208–214
8. Chisti Y (2007) Biodiesel from microalgae. Biotechnol Adv 25(3):294–306
9. Chiu SY, Kao CY, Tsai MT, Ong SC, Chen CH, Lin CS (2009) Lipid accumulation and CO 2 utilization of Nannochloropsis oculata in response to CO 2 aeration. Bioresour Technol 100(2):833–838
10. Damiani MC, Popovich CA, Constenla D, Leonardi PI (2010) Lipid analysis in Haematococcus pluvialis to assess its potential use as a biodiesel feedstock. Bioresour Technol 101(11):3801–3807
11. Dasgupta CN, Suseela MR, Mandotra SK, Kumar P, Pandey MK, Toppo K, Lone JA (2015) Dual uses of microalgal biomass: an integrative approach for biohydrogen and biodiesel production. Appl Energy 146:202–208
12. Demirbaş A (2008) Production of biodiesel from algae oils. Energy Sources Part A 31(2):163–168
13. Francisco ÉC, Neves DB, Jacob-Lopes E, Franco TT, (2010) Microalgae as feedstock for biodiesel production: carbon dioxide sequestration, lipid production and biofuel quality. J Chem Technol Biot 85(3):395–403
14. González-Garcinuño Á, Tabernero A, Sánchez-Álvarez JM, del Valle EMM, Galán MA (2014) Effect of nitrogen source on growth and lipid accumulation in Scenedesmus abundans and Chlorella ellipsoidea. Bioresour Technol 173:334–341
15. Gouveia L, Oliveira AC (2009) Microalgae as a raw material for biofuels production. J Ind Microbiol Biotechnol 36(2):269–274
16. Guckert JB, Cooksey KE (1990) Triglyceride accumulation and fatty acid profile changes in Chlorella (Chlorophyta) during high pH-induced cell cycle inhibition. J Phycol 26(1):72–79
17. Ho SH, Chen CY, Chang JS (2012) Effect of light intensity and nitrogen starvation on CO 2 fixation and lipid/carbohydrate production of an indigenous microalga Scenedesmus obliquus CNW-N. Bioresour Technol 113:244–252
18. Hu Q, Sommerfeld M, Jarvis E, Ghirardi M, Posewitz M, Seibert M, Darzins A (2008) Microalgal triacylglycerols as feedstocks for biofuel production: perspectives and advances. Plant J 54(4):621–639
19. Knothe G (2012) Fuel properties of highly polyunsaturated fatty acid methyl esters. Prediction of fuel properties of algal biodiesel. Energy Fuel 26(8):5265–5273
20. Kumar K, Ghosh S, Angelidaki I, Holdt SL, Karakashev DB, Morales MA, Das D (2016a) Recent developments on biofuels production from microalgae and macroalgae. Renew Sust Energ Rev 65:235–249

21. Kumar P, Mandotra SK, Suseela MR, Toppo K, Joshi P (2016b) Characterization and transesterification of fresh water microalgal oil. Energy Sources Part A 38(6):857–864
22. Li Y, Horsman M, Wang B, Wu N, Lan CQ (2008) Effects of nitrogen sources on cell growth and lipid accumulation of green alga Neochloris oleoabundans. Appl Microbiol Biotechnol 81(4):629–636
23. Liang K, Zhang Q, Gu M, Cong W (2013) Effect of phosphorus on lipid accumulation in freshwater microalga Chlorella sp. J Appl Phycol 25(1):311–318
24. Maity JP, Bundschuh J, Chen CY, Bhattacharya P (2014) Microalgae for third generation biofuel production, mitigation of greenhouse gas emissions and wastewater treatment: present and future perspectives–a mini review. Energy 78:104–113
25. Mandotra SK, Kumar P, Suseela MR, Ramteke PW (2014) Fresh water green microalga Scenedesmus abundans: a potential feedstock for high quality biodiesel production. Bioresour Technol 156:42–47
26. Mandotra SK, Kumar P, Suseela MR, Nayaka S, Ramteke PW (2016) Evaluation of fatty acid profile and biodiesel properties of microalga Scenedesmus abundans under the influence of phosphorus, pH and light intensities. Bioresour Technol 201:222–229
27. Mata TM, Martins AA, Caetano NS (2010) Microalgae for biodiesel production and other applications: a review. Renew Sust Energ Rev 14(1):217–232
28. Minowa T, Yokoyama SY, Kishimoto M, Okakura T (1995) Oil production from algal cells of Dunaliella tertiolecta by direct thermochemical liquefaction. Fuel 74(12):1735–1738
29. Muthuraj M, Palabhanvi B, Misra S, Kumar V, Sivalingavasu K, Das D (2013) Flux balance analysis of Chlorella sp. FC2 IITG under photoautotrophic and heterotrophic growth conditions. Photosynth Res 118(1–2):167–179
30. Natrah FMI, Yusoff FM, Shariff M, Abas F, Mariana NS (2007) Screening of Malaysian indigenous microalgae for antioxidant properties and nutritional value. J Appl Phycol 19(6):711–718
31. Nigam PS, Singh A (2011) Production of liquid biofuels from renewable resources. Prog Energy Combust Sci 37(1):52–68
32. Nigam S, Rai MP, Sharma R (2011) Effect of nitrogen on growth and lipid content of Chlorella pyrenoidosa. Am J Biochem Biotechnol 7(3):124–129
33. Pereira H, Barreira L, Custódio L, Alrokayan S, Mouffouk F, Varela J, Abu-Salah KM, Ben-Hamadou R (2013) Isolation and fatty acid profile of selected microalgae strains from the Red Sea for biofuel production. Energies 6(6):2773–2783
34. Philipose MT (1967) Chlorococcales, vol 8. Indian Council of Agricultural Research, New Delhi
35. Pokoo-Aikins G, Nadim A, El-Halwagi MM, Mahalec V (2010) Design and analysis of biodiesel production from algae grown through carbon sequestration. Clean Techn Environ Policy 12(3):239–254
36. Predojević Z, Škrbić B, Đurišić-Mladenović N (2012) Transesterification of linoleic and oleic sunflower oils to biodiesel using CaO as a solid base catalyst. J Serb Chem Soc 77(6):815–832
37. Rabbani S, Beyer P, Lintig JV, Hugueney P, Kleinig H (1998) Induced β-carotene synthesis driven by triacylglycerol deposition in the unicellular alga Dunaliella bardawil. Plant Physiol 116(4):1239–1248
38. Raghothama KG (2000) Phosphate transport and signaling. Curr Opin Plant Biol 3(3):182–187
39. Ramos MJ, Fernández CM, Casas A, Rodríguez L, Pérez Á (2009) Influence of fatty acid composition of raw materials on biodiesel properties. Bioresour Technol 100(1):261–268
40. Richmond A (ed) (2008) Handbook of microalgal culture: biotechnology and applied phycology. Wiley, New York
41. Rodolfi L, Chini Zittelli G, Bassi N, Padovani G, Biondi N, Bonini G, Tredici MR (2009) Microalgae for oil: strain selection, induction of lipid synthesis and outdoor mass cultivation in a low-cost photobioreactor. Biotechnol Bioeng 102(1):100–112
42. Ruangsomboon S (2012) Effect of light, nutrient, cultivation time and salinity on lipid production of newly isolated strain of the green microalga, Botryococcus braunii KMITL 2. Bioresour Technol 109:261–265

43. Schnurr PJ, Espie GS, Allen DG (2013) Algae biofilm growth and the potential to stimulate lipid accumulation through nutrient starvation. Bioresour Technol 136:337–344
44. Scott SA, Davey MP, Dennis JS, Horst I, Howe CJ, Lea-Smith DJ, Smith AG (2010) Biodiesel from algae: challenges and prospects. Curr Opin Biotechnol 21(3):277–286
45. Sims RE, Mabee W, Saddler JN, Taylor M (2010) An overview of second generation biofuel technologies. Bioresour Technol 101(6):1570–1580
46. Singh A, Olsen SI (2011) A critical review of biochemical conversion, sustainability and life cycle assessment of algal biofuels. Appl Energy 88(10):3548–3555
47. Solovchenko AE, Khozin-Goldberg I, Didi-Cohen S, Cohen Z, Merzlyak MN (2008) Effects of light intensity and nitrogen starvation on growth, total fatty acids and arachidonic acid in the green microalga Parietochloris incisa. J Appl Phycol 20(3):245–251
48. Spijkerman E, Wacker A (2011) Interactions between P-limitation and different C conditions on the fatty acid composition of an extremophile microalga. Extremophiles 15(5):597
49. Wahlen BD, Morgan MR, McCurdy AT, Willis RM, Morgan MD, Dye DJ, Bugbee B, Wood BD, Seefeldt LC (2012) Biodiesel from microalgae, yeast, and bacteria: engine performance and exhaust emissions. Energy Fuel 27(1):220–228
50. Williams PJLB, Laurens LM (2010) Microalgae as biodiesel & biomass feedstocks: review & analysis of the biochemistry, energetics & economics. Energy Environ Sci 3(5):554–590

Exploring the Diversity of Marine Planktonic Cyanobacterial Assemblages in a Mangrove Ecosystem: Integration of Uncultured and Cultured Approaches

15

Tarkeshwar Singh and Punyasloke Bhadury

Abstract

Mangrove ecosystems represent an unique ecotone between land and ocean boundary and therefore harbour rich biodiversity. The estuarine waters within a mangrove ecosystem are highly productive and sustain rich fisheries that support livelihood of millions of population globally. Phytoplankton communities hold the key to sustenance of mangrove fisheries by contributing to aquatic primary production. However, community composition of marine planktonic cyanobacteria, a key constituent of phytoplankton assemblages, and their resulting importance in primary production are not very well understood across mangroves, in particular from South and Southeast Asia. In this book chapter, snapshot of marine planktonic cyanobacterial community structure along seasonal scales has been highlighted using Sundarbans mangrove ecosystem as an example. The importance of integrating taxonomy with molecular tools is paramount towards understanding structure and function of marine planktonic cyanobacterial communities in coastal ecosystems such as mangroves and thus have consequences for sustainence of rich coastal fisheries.

15.1 Introduction

Coastal ecosystems contribute substantially to primary production amounting to 0.7 gigatons of carbon annually. Cyanobacterial communities account for 50% of all aquatic primary production or approximately 25% of total global primary

T. Singh · P. Bhadury (✉)
Integrative Taxonomy and Microbial Ecology Research Group, Department of Biological Sciences and Centre for Climate and Environmental Studies, Indian Institute of Science Education and Research Kolkata, Mohanpur, Nadia, West Bengal, India
e-mail: pbhadury@iiserkol.ac.in

L. B. Sukla et al. (eds.), *The Role of Microalgae in Wastewater Treatment*,
https://doi.org/10.1007/978-981-13-1586-2_15

production [1]. In land-ocean boundary, planktonic cyanobacteria constitute a key component of phytoplankton assemblages (e.g. [2]). Ubiquitous picoplanktonic cyanobacterial populations represented mainly by two genera, namely, *Synechococcus* and *Prochlorococcus*, are key players of aquatic primary production in coastal waters globally (e.g. [3, 4]). *Synechococcus* is more widespread from mesotrophic to eutrophic coastal ecosystems [3, 5], while *Prochlorococcus* is more abundant in open ocean environments [6]. On a latitudinal scale, both these picocyanobacterial genera have distribution ranging from tropics to high latitudes [7].

Among coastal ecosystems, mangroves represent an ecotone and spread across tropical to subtropical regions covering approximately 75% of the coastlines [8]. The distribution of mangroves globally indicates a tropical dominance with latitudinal limits linked to major ocean currents. Biogeographically diverse regions such as the Indo-West Pacific, Indonesia, Australia, Brazil and Nigeria are represented by 43% of the world's mangrove forests. Mangroves are more widespread in South and Southeast Asia [9]. Mangroves serve as the nursery ground for rich coastal fisheries and thus serve livelihood of millions of people globally. For example, mangrove-related fish and crab species account for 32% of the small-scale fisheries landings in the Gulf of California [10]. The sustenance of rich fisheries in mangroves is directly or indirectly influenced by phytoplankton communities present in the water. Phytoplankton communities thus play a key role in energy transfer through trophic food webs including in fishes [9]. Hence, it is extremely important to elucidate the structure of phytoplankton communities at temporal and spatial scales, in particular for planktonic marine cyanobacteria, which can constitute an important component of overall phytoplankton assemblages in mangrove ecosystems.

15.2 Studies on Planktonic Marine Cyanobacterial Communities in Mangrove Ecosystems

Several studies have been undertaken to elucidate cyanobacterial communities in mangrove ecosystems based on microscopy. Majority of these studies have looked into cyanobacterial communities in pneumatophores of mangrove plants, rhizosphere, sediment, and partly from water. Based on bright field microscopy and cultured approaches, 15 marine planktonic cyanobacterial genera belonging to Chroococcales were reported from red mangroves in Brazil [11], whereas 10 benthic cyanobacterial genera including *Lyngbya*, *Oscillatoria* and *Anabaena* were reported from mangrove sediments of Zanzibar in Kenya [12]. In a recent study, it has been shown that mangrove sediment can harbour novel cyanobacterial species [13]. Based on published literature, 73 genera of cyanobacteria represented by 276 species have been reported (see review by [14]). Therefore, there is very limited knowledge on planktonic marine cyanobacterial communities in mangrove ecosystems compared to other phytoplankton groups at the taxonomic and functional levels, and thus their exact contribution to aquatic primary production seasonally and spatially remains largely unknown from this type of ecosystem. Moreover, many of these planktonic cyanobacteria could ultimately turn out to be extremely important in translational research such as in the field of biotechnology such as bioremediation or generation of biofuels.

15.3 Bridging Taxonomy with Molecular Tools

Given that 41% of the world's mangrove occurs in South and Southeast Asia, it has become imperative to accelerate our understanding of cyanobacterial communities such as planktonic forms from this particular ecosystem. Morphological identification using microscopy exclusively relies on observation of morphological features present within a cell. However, there is paucity of experienced taxonomists globally in the field of cyanobacterial taxonomy. Morpho-taxonomy may be difficult at times and thus can lead to underestimation of marine planktonic cyanobacterial diversity across various types of marine ecosystems including mangroves [2, 15]. Moreover, available identification manuals on cyanobacterial taxonomy include only a small fraction of genuine marine cyanobacterial taxa. This can be further compounded by the fact that coastal ecosystems in tropical and subtropical regions have remained largely unexplored from the viewpoint of cyanobacterial community composition.

The advent of molecular techniques has revolutionized biology over the last three decades. Techniques such as polymerase chain reaction (PCR), DNA sequencing, terminal-restriction fragment length polymorphism (T-RFLP) and in recent times the application of next-generation sequencing (NGS) are more increasingly used in biological research [16, 17]. The field of marine cyanobacterial taxonomy has been also benefited with the increasing application of molecular techniques. For example, molecular tools based on small subunit ribosomal RNA (16S rRNA marker) have been used to study planktonic cyanobacterial assemblages from various types of marine environment including coastal ecosystems (e.g. [18–21]). The 16S rRNA operons are known to be present in more than one copy per genome within a prokaryotic cell compared to protein coding genes. Thus this molecule has the ability to increase the possibility towards discovery of rare microbial taxa from the environment. It is also the most preferred molecular marker in prokaryotic studies including for cyanobacteria [19, 20]. One of the most widely used approaches for assessing prokaryotic diversity is based on 16S rRNA clone library and sequencing approach. Information obtained from clone library and sequencing approach can provide key information on novel prokaryotic species that are widespread across marine environments (e.g. [22]). Moreover, in clone libraries cyanobacterial diversity can be also studied at microhabitat levels and thus allow improved phylogenetic resolution along with detection of uncultured cyanobacteria [23]. Therefore, integration of taxonomy with molecular tools can help to unravel marine planktonic cyanobacterial assemblages from lesser known coastal ecosystems such as mangroves.

In this chapter, we have discussed how molecular tools can be integrated with taxonomy to unravel information on the marine planktonic cyanobacterial assemblages in a mangrove ecosystem using Sundarbans mangroves as an example.

15.4 Sundarbans Mangrove Ecosystem

The mangroves in the Ganga-Brahmaputra-Meghna Delta shared between India and Bangladesh, popularly known as the Sundarbans, are the only contiguous and largest mangrove ecosystem in the world. The Sundarbans covers an area of approximately one million hectare which is greater than the combined area of Wadden Sea shared between Denmark, Germany and the Netherlands [24, 25]. The Sundarbans (21°32′ and 22°40′N; 88°05′ and 89°E), part of the world's largest deltaic mangrove ecosystem located at the apex of Bay of Bengal, encompasses over 102 islands with a network of innumerable rivers, rivulets and creeks [26]. This vast deltaic region covers a total area of approximately 10,000 km^2 and is strongly influenced by coastal water entering from the Bay of Bengal. There is huge freshwater flow from GBM riverine systems (42,000 m^3/s), in particular during monsoon, along with saline water influences, which leads to variability in salinity across different parts of Sundarbans mangrove. In 1987, UNESCO declared the core mangrove forest of Indian Sundarbans (2585 km^2) as Sundarbans Biosphere Reserve (SBR) for protection and conservation of mangrove flora and fauna including the Royal Bengal Tiger from anthropogenic disturbances. One of the important attributes of this mangrove ecosystem is the high load of suspended particulate matter in water column compared to other coastal ecosystems (e.g. [27, 28]).

Eukaryotic phytoplankton communities have been relatively well studied from the Sundarbans. Diatoms are the major contributor to aquatic primary production in Sundarbans, and they show temporal and spatial variability in link with prevailing environmental conditions [28–32]. The high fish productivity in the Sundarbans region is also dependent on aquatic primary production driven by phytoplankton communities [33]. However, knowledge on the marine planktonic cyanobacterial communities in estuarine or coastal water of Sundarbans is not very well understood. Moreover, given the area of this ecosystem being vast with topographic and seasonal hydrological differences, planktonic cyanobacterial community structure can also vary at temporal and spatial scales.

Therefore, to disentangle complexity of marine planktonic cyanobacterial communities and to understand their exact contribution to aquatic primary production in coastal ecosystems such as Sundarbans, it is necessary to integrate morphology-based taxonomy with molecular tools.

15.4.1 Sundarbans Biological Observatory Time Series

Since many of the mangrove ecosystems, especially in South and Southeast Asia, cover huge area, therefore in order to get an in-depth understanding of marine planktonic cyanobacteria, there is a need to establish time series focusing on a part within a mangrove ecosystem. The time series can help to understand temporal and spatial trends of biological communities, such as marine planktonic cyanobacteria, and ultimately link their importance to ecosystem functioning. Globally, there are numerous coastal time series and most notable among them include the Western

Channel Observatory and San Pedro Ocean Time Series. In case of Sundarbans, such a time series has been established in 2010 in its western part. This is known as Sundarbans Biological Observatory Time Series (SBOTS). SBOTS is located in Sagar Island, the largest island of Sundarbans which is approximately 6.7 m above sea level, and mean tidal amplitude of this island is 3.5–6.0 m throughout the year. Sagar Island is surrounded by Muri Ganga River in the east and Hooghly River in the north and west, while the southern part faces the coastal Bay of Bengal. SBOTS comprises three stations, namely, Stn1 (21°40′44.4″, 88°08′49.5″ E), Stn2 (21°40′59.3′′N; 88°09′13.1′′E) and Stn3 (21°40′40.6″ N, 88°09′19.2″ E), which are routinely monitored on a monthly basis to disentangle complexity of biological communities such as marine planktonic cyanobacterial communities (Fig. 15.1). Stn1 is located upstream of the Chemaguri creek located in southeast of Sagar Island, while Stn2 is located at the mouth of creek facing the Mooriganga River. On the other hand, Stn3 is located on the Mooriganga River and directly influenced by coastal water entering from the Bay of Bengal on a diurnal basis. All these stations

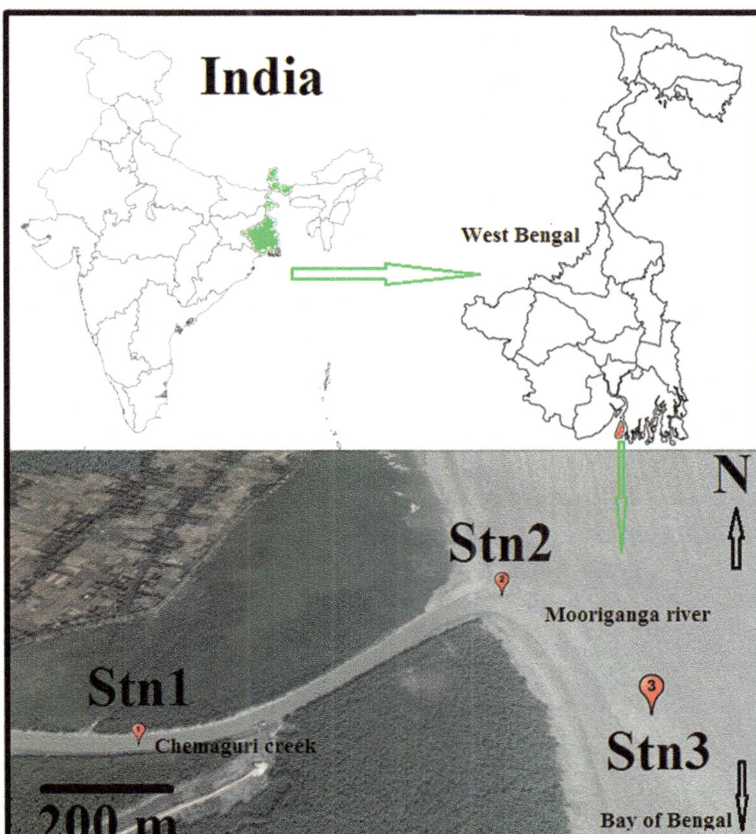

Fig. 15.1 Map of the study area as part of SBOTS (Source image modified from Google Earth)

have shallow water depth. Interestingly, salinity in Stn1 is much lower compared to the other two stations of SBOTS [28, 34].

15.5 Monitoring Marine Planktonic Cyanobacterial Communities in SBOTS

In order to get an in-depth understanding of marine planktonic cyanobacterial communities and also their overall contribution to total phytoplankton assemblages in coastal ecosystems, there is an increasing need to study and unravel functional complexity in biological groups such as cyanobacteria either at the temporal or spatial scales. In case of mangrove ecosystems, this approach is more relevant given the vast geographical expanse along with heterogeneity in topography and prevailing environmental conditions. In SBOTS, marine planktonic cyanobacterial community structure was elucidated at the temporal scale, in particular seasonal scale, by applying both microscopy and molecular tools.

In Sundarbans mangroves, monsoon and post-monsoon seasons strongly influences ecosystem level functioning [34, 35]. During monsoon time, seasonal precipitation considerably increases freshwater flow from major rivers into this ecosystem. The increased freshwater flow is also accompanied with nutrient run-offs from land boundaries. Although tidal influences in this ecosystem during monsoon persist, nevertheless huge freshwater inflow lowers salinity during this season. Such changes also alter the aquatic primary production in this ecoregion. During post-monsoon season, the increased freshwater flow continues, and there is an in-depth increase in water column. This leads to improved light penetration in the water column and an overall increase in aquatic primary production in the region. Therefore seasons can influence the structure and functioning of biological communities in Sundarbans mangrove ecosystem including in marine planktonic cyanobacterial communities.

Time series can be particularly useful to understand seasonal influences on biological communities and also to unravel complexity of these communities such as for phytoplankton. As part of SBOTS, the seasonal influence on structure of marine planktonic cyanobacterial communities has been looked into, in addition to elucidation of the community structure. Since salinity in Stn1 is much lower compared to Stn3 of SBOTS, therefore, these two stations were selected because the influence of monsoon and post-monsoon seasons on marine planktonic cyanobacterial communities will be evident. In 2012, marine planktonic cyanobacterial communities from Stn1 and Stn3 of SBOTS were monitored during monsoon (July to September) and post-monsoon (October to December) seasons. Briefly, surface water was collected on a monthly basis from Stn1 and Stn3 in above seasons and then subjected to PCR clone library and sequencing by targeting cyanobacteria-specific 16S rRNA region ([19]; see protocols for clone library approach in [32]). At the time of collection, *in situ* environmental parameters were recorded in addition to collection of surface water for dissolved nutrient analysis and also for undertaking bright field microscopy.

The sequenced 222 clones showed varying identity (90–100%) at the nucleotide level with published uncultured and cultured 16S rRNA sequences belonging to *Cyanobacteria*. Interestingly, 160 out of 222 clones showed 95–100% identity with published 16S rRNA sequences available in nucleotide databases (GenBank/DDBJ/ENA) belonging to 17 strains of cultured marine *Synechococcus*. From seasonal viewpoint, in monsoon, 105 sequences were *Synechococcus*-like, while in post-monsoon 55 sequences were *Synechococcus*-like. From SBOTS stations perspective, 69 sequences were *Synechococcus*-like in Stn1, while 91 sequences were *Synechococcus*-like in Stn3. In Stn1, the sequences showed significant identity with 15 cultured strains of *Synechococcus*, while in case of Stn3, identity was observed with 12 cultured strains of *Synechococcus* (see relative abundances in Fig. 15.2). Most importantly, out of 160 *Synechococcus*-like sequences, 71 sequences showed 95–100% identity at the nucleotide level with published 16S rRNA sequence of cultured *Synechococcus* sp. KORDI-78 strain (Acc. no. FJ497748) isolated previously from the coastal water of East China Sea. The KORDI-78-like sequences were found to be highly abundant in monsoon and post-monsoon seasons of SBOTS. Moreover, a new species of marine planktonic cyanobacterium belonging to the genus *Synechococcus* has been successfully isolated from SBOTS, and using polyphasic taxonomic approaches, it has been also confirmed that this species contains unique signatures of fatty acids indicative of ecophysiological adaptation in this ecosystem.

Although *Synechococcus*-like sequences dominated SBOTS, nevertheless, sequences representative of other cyanobacterial taxa, although in low numbers, were also encountered from clone library approach. For example, some sequences showed identity with published 16S rRNA sequences belonging to *Hapalosiphon* sp., *Fischerella* sp., *Chroococcidiopsis* sp. and *Trichocoleus* sp. The detection of *Hapalosiphon*- and *Fischerella*-like sequences, albeit in low numbers from Stn1 of study area only during post-monsoon, indicates the importance of resuspension of

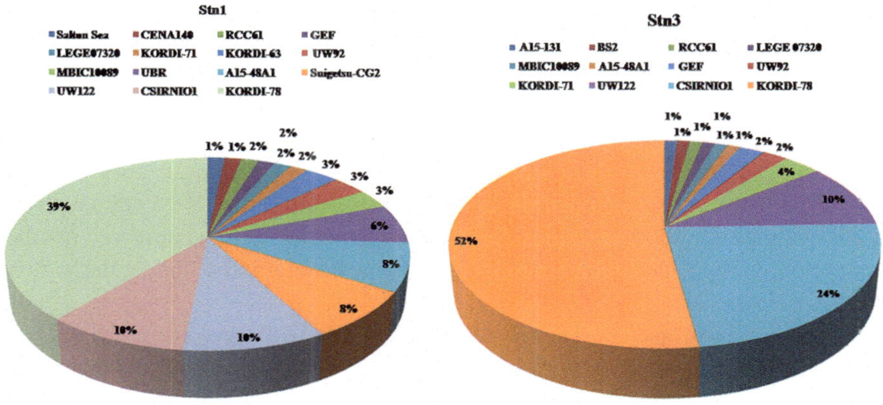

Fig. 15.2 Relative abundance of sequences showing identity with cultured *Synechococcus* strains in Stn1 and Stn3 of SBOTS

benthic cyanobacterial mats in surface water of Sundarbans which can be due to occasional vertical mixing as a result of increased freshwater flow especially during end of monsoon season. The resuspension of benthic cyanobacterial mats has been also reported from this ecosystem [28].

Sequences representing four major orders of planktonic cyanobacteria, namely, Chroococcales, Pleurocapsales, Oscillatoriales and Stigonematales, were encountered from SBOTS. Some of these orders have been also reported from other mangrove ecosystems (e.g. [11, 12, 36]). Sequences representing the orders Pleurocapsales, Oscillatoriales, and Stigonematales were only found in Stn1 for the months of July and December of 2012. The Stn3 throughout monsoon and post-monsoon seasons was represented by sequences belonging only to Chroococcales.

The bright field microscopy approach adopted during elucidating marine planktonic cyanobacterial communities from both stations of SBOTS also showed the presence of cells belonging to *Synechococcus* sp., *Chroococcidiopsis* sp., *Phormidium* sp., *Trichocoleus* sp. and *Trichodesmium* sp. Interestingly, *Synechococcus* cells were observed from July to December in both stations, while cells of *Trichodesmium* sp. were encountered only in Stn3 during October and November 2012. Generally, there was congruency between bright field microscopy and molecular tools while elucidating marine planktonic cyanobacterial assemblages from Sundarbans mangrove ecosystem.

The cyanobacterial 16S rRNA sequences generated from SBOTS stations were subjected to deep phylogeny to get an in-depth understanding of their ecophysiological adaptation in this ecosystem. It was found that several novel clades of marine planktonic cyanobacteria exist in Sundarbans and that salinity could be an important factor that could control distribution of these clades in estuarine waters of this ecosystem. For example, salinity-specific *Synechococcus* clades have been reported from other marine environments [37, 38]. The identification of these novel clades highlights that the number of undiscovered marine planktonic cyanobacteria in mangrove ecosystem such as Sundarbans is still potentially visited and integrative taxonomic approaches can help unravel the vast species richness of marine planktonic cyanobacteria. The existence of undiscovered cyanobacteria in coastal ecosystems has been also highlighted before and considered as an emerging challenge in cyanobacterial taxonomy [2].

At the same time, it was found that prevailing environmental parameters including their variability, in particular salinity and dissolved nutrients (e.g. orthophosphate concentration), along seasonal scales such as in monsoon or post-monsoon can also influence the composition as well as seasonal variability of marine planktonic cyanobacterial communities. Such trends are expected to be observed in other mangrove ecosystems which are influenced by freshwater influx, particularly those found in South Asia.

Interestingly, from SBOTS *Prochlorococcus*-like sequences were not encountered either during monsoon or post-monsoon, and the absence of this genus was also confirmed by bright field microscopy. This finding with respect to Sundarbans is significant, although implications of such in coastal primary production warrant further investigation. Moreover, this also opens up avenue for undertaking in-depth research to look into the distribution of *Prochlorococcus* populations in other mangroves of South as well as Southeast Asia.

15.5.1 Way Forward

The findings from SBOTS of Sundarbans indicate that populations of *Synechococcus* dominate this mangrove ecosystem across seasonal scales. Besides *Synechococcus*, other cyanobacterial genera, some of which are terrestrial and found in the estuarine water due to run-off as a result of freshwater flow, were also encountered. The importance of environmental parameters such as salinity is pivotal in controlling structure of marine planktonic cyanobacterial communities, especially in mangroves such as Sundarbans which is influenced by freshwater inflow. The detection of novel clades once again reconfirms that unexplored marine cyanobacterial diversity is unusually high in coastal ecosystems such as mangrove and this type of ecosystem is a 'hotspot' of cyanobacterial diversity including for marine planktonic cyanobacteria. Many of these novel sequences can be ultimately represented in the form of establishment of marine planktonic cyanobacterial cultures, and some of these could be potentially applied in the field of biotechnology (e.g. [39]). The unexplored marine planktonic cyanobacterial diversity holds key to broader understanding of biogeochemical cycling including carbon and nitrogen cycles in mangrove ecosystems such as Sundarbans and also their role in sustaining the rich coastal fisheries.

References

1. Kirchman DL (2012) Processes in microbial ecology. Oxford University Press, Oxford
2. Hoffman L (1992) Marine cyanobacteria in tropical regions: diversity and ecology. Eur J Phycol 34:371–379
3. Lomas MW, Moran SB (2011) Evidence for aggregation and export of cyanobacteria and nano-eukaryotes from the Sargasso Sea euphotic zone. Biogeosciences 8:203–216
4. Palenik B, Ren Q, Dupont CL et al (2006) Genome sequence of *Synechococcus* CC9311: insights into adaptation to a coastal environment. Proc Natl Acad Sci U S A 103:13555–13559
5. Liu HB, Nolla HA, Campbell L (1997) *Prochlorococcus* growth rate and contribution to primary production in the equatorial and subtropical North Pacific Ocean. Aquat Microb Ecol 12:39–47
6. Partensky F, Garczarek L (2010) *Prochlorococcus*: advantages and limits of minimalism. Annu Rev Mar Sci 2:305–331
7. Huang S, Wilhelm SW, Harvey HR, Taylor K, Jiao N, Chen F (2012) Novel lineages of *Prochlorococcus* and *Synechococcus* in the global oceans. ISME J 6(2):285–297. https://doi.org/10.1038/ismej.2011.106
8. Lalli CM, Parsons TR (1997) Biological oceanography: an introduction. The Open University oceanography series, 2nd edn. Butterworth-Heinemann, Oxford
9. Alongi DM (2002) Present state and future of the world's mangrove forests. Environ Conserv 29:331–349
10. Aburto-Oropeza O, Ezcurra E, Danemann G, Valdez V, Murray J, Sala E (2008) Mangroves in the Gulf of California increase fishery yields. Proc Natl Acad Sci U S A 105:10456–10459

11. Nogueira NMC, Ferreira-Correia MM (2001) Cyanophyceae/cyanobacteria in red mangrove forest at Mosquitos and Coqueiros estuaries, São luís, state of Maranhão, Brazil. Braz J Biol 61:347–356
12. Kyaruzi JJ, Kyewalyanga MS, Muruke MHS (2003) Cyanobacteria composition and impact of seasonality on their *in situ* nitrogen fixation rate in a mangrove ecosystem adjacent to Zanzibar Town. WIO J Mar Sci 2:35–44
13. Debnath M, Singh T, Bhadury P (2017) New records on Cyanobacterial morphotypes with *Leptolyngbya indica* sp. nov. from terrestrial biofilms of the lower Gangetic plain, India. Phytotaxa 316:101–120
14. Alvarenga DO, Rigonato J, Branco LHZ, Fiore MF (2015) Cyanobacteria in mangrove ecosystems. Biodivers Conserv 2401:799–817
15. Komárek J (2008) Cyanoprokaryota. 1. Teil/1st part: Chroococcales. In: Büdel B, Gärdner G, Krienitz L, Schagerl M (eds) Süsswasserflora von Mitteleuropa, vol 19/1. Elsevier, München, pp 1–548
16. Ishak HD, Plowes R, Sen R, Kellner K, Meyer E, Estrada DA, Dowd SE, Mueller UG (2011) Bacterial diversity in *Solenopsis invicta* and *Solenopsis geminata* ant colonies characterized by 16S amplicon 454 pyrosequencing. Microb Ecol 61:821–831
17. Sogin ML, Morrison HG, Huber JA, Welch DM, Huse SM, Neal PR, Arrieta JM, Herndl GH (2006) Microbial diversity in the deep sea and the underexplored "rare biosphere". Proc Natl Acad Sci U S A 103:12115–12120
18. Bolhuis H, Stal LJ (2011) Analysis of bacterial and archaeal diversity in coastal microbial mats using massive parallel 16S rRNA gene tag sequencing. ISME J 5(11):1702–1712. https://doi.org/10.1038/ismej.2011.5201
19. Mühling M, Allen JW, Murrell JC, Joint I (2008) Improved group-specific PCR primers for denaturing gradient gel electrophoresis analysis of the genetic diversity of complex microbial communities. ISME J 2:379–392
20. Nübel U, Garcia-Pichel F, Muyzer G (1997) PCR primers to amplify 16S rRNA genes from cyanobacteria. Appl Environ Microbiol 63:3327–3333
21. Singh T, Bhadury P (2018) Biogeographic patterns of marine planktonic cyanobacterial assemblages in marine marginal habitats using 16S rRNA phylogeny. Phycol Res 66:189–198
22. Giovannoni S, Rappé M (2000) Evolution, diversity and molecular ecology of marine prokaryotes. In: Kirchman DL (ed) Microbial ecology of the oceans. Wiley-Liss, New York, pp 47–84
23. Li L (2014) Infectious microecology: theory and application. Springer, Berlin/Heidelberg, p 650
24. Boggs S (2006) Principles of sedimentology and stratigraphy, 4th edn. Pearson/Prentice Hall, Upper Saddle River
25. Gopal B, Chauhan M (2006) Biodiversity and its conservation in the Sundarban mangrove ecosystem. Aquat Sci 68:338–354
26. Spalding MD, Blasco E, Field C (1997) World mangrove atlas. International Society for Mangrove Ecosystems, Okinawa
27. Chaudhuri K, Manna S, Sarma KS, Naskar P, Bhattacharyya S, Bhattacharyya M (2012) Physicochemical and biological factors controlling water column metabolism in Sundarbans estuary, India. Aquat Biosyst 8(1):1–16. https://doi.org/10.1186/2046-9063-8-26
28. Bhattacharjee D, Samanta B, Danda AA, Bhadury P (2013) Temporal succession of phytoplankton assemblages in a tidal creek system of the Sundarbans mangroves – an integrated approach. Int J Biodivers 2013:1–15
29. Biswas H, Mukhopadhyay SK, De TK, Sen S, Jana TK (2004) Biogenic controls on the air-water carbon dioxide exchange in the Sundarban mangrove environment, northeast coast of Bay of Bengal, India. Limnol Oceanogr 49:95–101
30. Biswas H, Dey M, Ganguly D, De TK, Ghosh S, Jana TK (2010) Comparative analysis of phytoplankton composition and abundance over a two decade period at the land-ocean boundary of a tropical mangrove ecosystem. Estuar Coasts 33:384–394
31. Manna S, Chaudhuri K, Bhattacharyya S, Bhattacharyya M (2010) Dynamics of estuarine ecosystem: eutrophication induced threat to mangroves. Saline Syst 68:1–16

32. Samanta B, Bhadury P (2014) Analysis of diversity of chromophytic phytoplankton in a mangrove ecosystem using *rbc*L gene sequencing. J Phycol 50:328–340
33. Mukherjee S, Chaudhuri A, Kundu N, Mitra S, Homechaudhuri S (2013) Comprehensive analysis of fish assemblages in relation to seasonal environmental variables in an estuarine river of Indian Sundarbans. Estuar Coasts 36:192–202
34. Choudhury AK, Das M, Philip P, Bhadury P (2015) An assessment of the implications of seasonal precipitation and anthropogenic influences on a mangrove ecosystem using phytoplankton as proxies. Estuar Coasts 38:854–8723
35. Choudhury AK, Bhadury P (2015) Relationship between N:4 P: Si ratio and phytoplankton community composition in a tropical estuarine mangrove ecosystem. Biogeosci Discuss 12:2307–2355
36. Branco LHZ, Célia Leite S'A, Paiva Axevedo de MT, Sormus L (1997) Cyanophyte flora from Cardoso Island mangroves, São Paulo State, Brazil. 2. Oscillatoriales. Algol Stud 84:39–520
37. Jing H, Liu H (2012) Phylogenetic composition of *Prochlorococcus* and *Synechococcus* in cold eddies of the South China Sea. Aquat Microb Ecol 65:207–219
38. Xia X, Vidyarathna NK, Palenik B, Lee P, Liu H (2015) Comparison of the seasonal variations of *Synechococcus* assemblage structures in estuarine waters and coastal waters of Hong Kong. Appl Environ Microbiol 81:7644–7655
39. Saha M, Goecke F, Bhadury P (2017) Minireview: algal natural compounds and extracts as antifoulants. J Appl Phycol 30(3):1859–1874. https://doi.org/10.1007/s10811-017-1322-0

Metabolic Engineering Prospects for Enhanced Green Fuel Production by Microalgae

16

Rajesh Kumar Sahoo and Enketeswara Subudhi

Abstract

The world has been entering into a phase of acute energy crisis due to the logarithmic increase in global population and the continuous depletion of finite fossil fuel resources. Generation of greenhouse gasses due to combustion of fossil fuels adds further menace to the environment bringing about global warming. There is an urgent need to search for an alternative fuel source which is economic and environmental friendly. Microalgae are the photosynthetic microorganisms which have the potential to convert light energy into biofuel through series of biochemical reactions. However, the major drawbacks for algae-based biofuel production are the time-consuming processes. According to a report published by US Department of Energy, there are 3000 different microalgae having potential to produce TAG as main precursor of biofuel. There are several genes which encode enzymes for the lipid metabolism responsible for enhancing the lipid content in microalgae. Development of molecular tools and techniques and synthetic biology may give more insight into the synthesis of triacylglycerides (TAG) in lipid-accumulating microalgae. In this article, we review the possibility of development of metabolic engineering for lipid synthesis in microalgae and discuss the various strategies such as single gene expression, carbon assimilation, expression of transcription factors and flux balance analysis which increase the lipid accumulation via metabolic engineering.

R. K. Sahoo
Biomics & Biodiversity Laboratory, Centre for Biotechnology,
Siksha 'O' Anusandhan (Deemed to be University), Bhubaneswar, Odisha, India
e-mail: rajeshkumarsahoo@soa.ac.in

E. Subudhi (✉)
Centre for Biotechnology, Siksha 'O' Anusandhan (Deemed to be University),
Bhubaneswar, Odisha, India
e-mail: enketeswarasubudhi@soa.ac.in

L. B. Sukla et al. (eds.), *The Role of Microalgae in Wastewater Treatment*,
https://doi.org/10.1007/978-981-13-1586-2_16

16.1 Introduction

Nowadays, the world is facing two major problems, i.e. energy crisis and global warming, due to increased transportation, population, industrialization and use of fossil fuels. In last three decades, the serious environmental problems are occuring in worlwide due to increasing consumption of fossil fuels. According to the reports of International Energy agency (https://data.worldbank.org/indicator/EG.USE.COMM.FO.ZS), more than 80% energy from fossil fuel are consumed in world wide. At present, staggering rate of consumption, available world fossil oil reserves will be exhausted in less than 50 years [36]. In recent times, many of the countries are extracting energy from different sources such as geothermal, agricultural waste, solar, wind and water, which can be used as alternatives to fossil-based fuels [25]. According to the International Energy Agency (IEA) report, the energy derived from waste biomass has higher potential for alternative fuel in comparison with other renewable energy sources [20]. Similarly, microalgal lipids are apprehended as a best source for biofuel as compared to other renewable energy sources and waste of both household and industries. Third-generation biofuels are derived from microalgae, which are less disadvantageous than first- and second-generation biofuels. Biofuels from agricultural and animal products such as soya bean oil, rapeseed oil, palm oil and animal fats are called first-generation biofuel, but it has created negative effect on global food market [12, 13]. Biofuels from nonedible oil plants such as *Madhuca indica*, *Jatropha curcas*, *Pongamia pinnata* and *Simarouba glauca* and lignocellulose biomass are called second-generation biofuel, but these plants require huge area of land for cultivation, and also due to less advanced technologies and knowledge for commercial utilization [31], these are not feasible.

Microalgae are group of unicellular eukaryotic organisms and found in large and diverse aquatic environment such as fresh water, saline water and sea water. Most of the microalgae are photoautotrophic and convert solar energy to chemical energy. There are 40,000–70,000 species belonging to nine phyla, and some researcher's prophesize that there could be species that are undiscovered or unclassified [15]. Microalgae have the potential for biofuel production because they are capable of producing useful quantities of polysaccharides and triglycerides which are considered as raw materials for bioethanol and biodiesel and also have the ability to fix huge amount of CO_2 from environment. They are also producing valuable proteins and compounds which can be beneficial source for animal feed and pharmaceutical industries. But the extraction of biofuel from microalgae is not commercially feasible due to the low yield and cost of the downstream processing.

The growth rate of microalgae varies from species to species, and many researchers reported that microalgae are growing faster but producing less amount of lipid under environmentally favourable conditions. Under physiological (pH, temperature, salinity and nutrient availability) and chemical stress (heavy metals) conditions, microalgae accumulate lipid and carbohydrate [6, 19]. But, these conditions could be inhibitory for growth, and lower biomass production possibility of contamination may increase [19]. For commercial feasibility of microalgae-based biofuel, is needed for better understanding of lipid biosynthesis pathway.

Metabolic pathway engineering is a powerful approach to enhance biofuel as well as biomass production by the microalgae. Bioengineering and recombinant DNA technology can be used for the alteration of cellular metabolism through the insertion deletion and/or modification of metabolic pathways to enhance biofuel production. So, biofuel from metabolically engineered microalgae/microorganism can be referred to as fourth-generation biofuel [26]. Now, the researchers are focusing on the development of high lipid content microalgae and cultivated in large-scale open pond by using metabolic engineering approach [5]. In this study, we discuss about metabolic engineering strategy for over production of lipid in microalgae and major challenges faced in commercial production.

16.2 Biochemistry of Algal Lipid

Microalgae are autotrophic organisms and producing rich reservoir of lipid, protein, vitamins, long-chain polyunsaturated fatty acids and bioactive compounds by using light energy and inorganic nutrients [29]. The production of lipids and secondary metabolites can be enriched under specific environmental stress condition like nitrogen and phosphate restriction [1, 7, 19]. Hence, controlling the environmental stress condition (pH, temperature, carbon, nitrogen, phosphate, ammonium, nitrate, light and some heavy metals) and microalgal strain selections are the common approaches used for enhancing the lipid accumulation. The complete knowledge on lipid metabolism and inducing factors for algal growths are the two important approaches for maximizing the production of biofuel.

The CO_2 is converted to glycerol-3-phosphate (G3P) through photosynthetic reaction. This G3P molecule is an important precursor for the synthesis of polysaccharides and lipid. The G3p molecule is converted to pyruvate and thereafter acetyl-CoA in the presence of pyruvate dehydrogenase. The acetyl-CoA and bicarbonate are converted to malonyl-CoA, which is catalysed by acetyl-CoA carboxylase (Fig. 16.1). Malonyl-ACP (acetyl carrier protein) is produced from malonyl-CoA and catalysed by malonyltransferase. After that, the acetyl group is condensed with malonyl-ACP to produce ketobutyryl-ACP in the presence of 3-ketoacyl-ACP synthase (KAS). This ketobutyryl-ACP is converted to fatty acyl-ACP through sequential reactions, and the fatty acyl-ACP thioesterase is catalysed to release fatty acids from ACP.

Triacylglycerides (TAG) are produced by the sequential acylation of glycerol-3-phosphate (G3P) with three acyl-CoA and fatty acids catalysed by a group of enzymes such as acyltransferases. The G3P is converted to lysophosphatidic acid (LPA) by the action of glycerol-3-phosphate acyltransferase, and further LPA acylated by lysophosphatidic acid acyltransferase (LPAT) produces phosphatidic acid (PA). Phosphate group removed from PA to produce diacylglycerol (DAG) and the reaction catalysed by phosphatidic acid phosphatase (PAP). The TAG is produced from DAG and catalysed by DGAT (diacylglycerol acyltransferase). TAG is stored

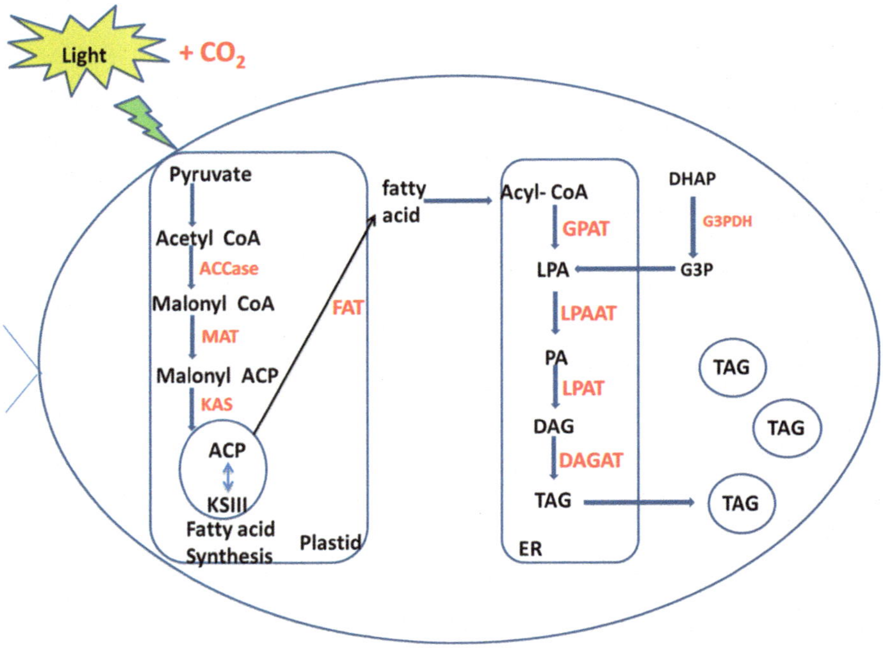

Fig. 16.1 Microalgal lipid biosynthesis. *ACCase* acetyl-CoA carboxylase, *MAT* malonyl-ACP transacylase, *KAS* 3-ketoacyl-ACP synthase, *ACP* acyl carrier protein, *KSIII* 3-ketoacyl-acyl carrier protein synthase III, *FAT* fatty acyl-ACP thioesterase, *GPAT* glycerol-3-phosphate acetyltransferase, *LPA* lysophosphatidic acid, *LPAAT* lysophosphatidic acid acyltransferase, *PA* phosphatidic acid, *LPAT* lysophosphatidylcholine acyltransferase, *DAGAT* diacylglycerol acyltransferase, *DHAP* dihydroxyacetone phosphate, *G3P* glyceraldehyde-3-phosphate, *G3PDH* glycerol-3-phosphate dehydrogenase, *TAG* triacylglycerides

in the form of lipid bodies of the algal cell. The accumulation of TAG is not only resource of carbon and energy but also serves as important physiological function in algal cell. This algal lipid is transesterified to produce biofuel.

16.3 Conventional Approach to Enhance Lipid Content

In conventional approach, hyperaccumulation of lipid bodies in microalgal cell is only due to effect of physiological stresses (pH, light intensity, photo-oxidative stress and temperature(s)) and chemical parameters (nutrient deprivation and salinity) [21, 38, 41], which are increasing the activity of several enzymes responsible for synthesis of lipid bodies. Many of the researchers found that TAG accumulation in different class of microalgae are induced by nutrient deficiency specifically nitrate, phosphate and iron. Lacking of nitrogen and phosphorus in growth medium which results stimulating the various acyltransferase and phosphorus transporter systems and converting acyl-CoA to TAG. De Bhowmick and co-authors [8] gave three possible reasons for lipid accumulation under nitrogen deficiency, i.e.

thylakoid membrane content, acyl hydrolase activation and phospholipid hydrolysis stimulation [38, 43].

Other key enzymes like phosphogluconate dehydrogenase (PGD), glucose-6-phosphate-1-dehydrogenase (GPD) and pyruvate decarboxylase (PDC) are expressed under deficiency of nitrogen [32]. Diacylglycerol trimethylhomoserine (DGTS) and digalactosyldiacylglycerol (DGDG) are inducing in the absence of phosphorus [22]. Malic enzyme is an important key player for providing NADPH for intracellular fatty acid synthesis and highly expressed in starvation of all nutrients in *Chlorella pyrenoidosa* [11]. Similarly, physiological factors such as temperature, pH, CO_2 and heavy metals are also responsible for lipid accumulation. The growth temperature will vary depending upon the microalgae species. In daytime, the growth rate is higher which tends higher photosynthetic activity and also results higher biomass production. Similarly, higher concentration of CO_2 results in higher biomass production, but it will decrease the pH [2, 24]. From this conventional approach, stress and starvation are the key players to increase the lipid bodies, and it is employed after attaining a significant quantity of algal biomass. The major disadvantage of conventional approaches are low photosynthetic activity and slow growth rate. However, the combination of bioengineering technique and insight of metabolic pathway and biochemical stresses could be possible to develop a new strain for greater lipid productivities.

16.4 Metabolic Engineering Approach to Enhance Lipid Content

Metabolic engineering approach is the genetic modification of biosynthesis pathway in a cell that triggers the synthesis of target molecule. Several methodologies are implemented in enhancing the lipid accumulations in microalgae such as engineering of gene towards biosynthesis of lipid, engineering of transcription factor, improvement of photosynthetic efficiency, modification of the carbon assimilation pathway, identification of the rate-limiting enzyme and flux balance analysis [2].

Several researchers have overexpressed the key enzymes which are involved in synthesis of TAG. Among these, acetyl-CoA carboxylase (ACCase) catalyses the carboxylation of acetyl-CoA to form malonyl-CoA, the first committed step for fatty acid synthesis. Dunahay et al. [10] overexpressed of ACCase in *Cyclotella cryptic* and *Navicula saprophila*, and found only the increase of ACCase activity by two- to threefold but no effects on fatty acid accumulation. Other enzymes like 3-ketoacyl carrier protein synthase III (KASIII) (catalyses the initial condensing reaction of fatty acid synthesis) from spinach (*Spinacia oleracea*) were overexpressed in tobacco, *Arabidopsis* and *Brassica*; lipid content in seed was not increased [9]. From the above study, the cloning and expression of the single genes related to fatty acid synthesis did not increase the fatty acid content as well as lipid. Ma and co-authors [27] reported that *Phaeodactylum tricornutum* was producing 82% total neutral lipid. They are using an antisense C-DNA which knocked down the regulation of pyruvate carboxylase kinase (PDK). PDK is an enzyme which catalyses to deactivate the pyruvate dehydrogenase complex (PDC) through phosphorylation. Similarly, malic enzyme was overexpressed in *Phaeodactylum tricornutum*, which resulted 2.5-fold increase of total lipid accumulation as compared to control [15]. Niu et al. [34] cloned and overexpressed the type 2 DGAT in heterologous host *Phaeodactylum tricornutum* producing 35% increased TAG accumulation as compared to control. Overexpression of individual genes like GPAT, LPAT, DAGAT and G3PDH (Fig. 16.1) of lipid biosynthesis pathway had partial effect on lipid synthesis, whereas all the five genes are cloned in single construct and overexpressed twofold of total lipid content [17, 23].

In addition to that lipid accumulation of microalgae is not sufficient, refining of lipid is also desirable. Biodiesel are the methyl esters of fatty acids, which need to contain correct chain length (C10–C18). Thioesterase is an enzyme, which catalyse to termination of fatty acid chain elongation. Genetic engineering of this enzyme is an alternative strategy to alter the fatty acid content of the biodiesel. This enzyme was successfully cloned and expressed in plant species for altering the fatty acid composition [39]. In 2011, Gong et al. [16] applied the same approach in microalgae, but it did not alter fatty acid carbon chain. Blatti et al. [4] applied the same approach with some modification in *Chlamydomonas reinhardtii*; they did the protein-protein interaction between fatty acid acyl carrier protein (ACP) and thioesterase (TE) that regulates fatty acid synthesis. They found increased level of short-chain fatty acids within a chloroplast of *Chlamydomonas reinhardtii*. This review shows that fatty acid synthesis, altered fatty acid chain length and hyperaccumulation of TAG can be possible through metabolic engineering. Blocking of the β-oxidation pathway can induce the lipid accumulation in microalgae [8]. This type of gene could be suppressed by random mutagenesis or RNA silencing [35]. Some reports suggested that lipid degrading enzymes like lipase, phospholipase and acyltransferase increased the fatty acid synthesis without affecting the growth of *Thalassiosira pseudonana* [40].

In photosynthetic microalgae, Rubisco is a key enzyme which catalyses fixation of CO_2 into ribulose-1,5-biphosphate (RuBP) to form 3-phosphoglycerate in the Calvin-Benson-Bassham cycle. Several reports have shown that Rubisco is the major enzyme in CBB cycle, when CO_2 is absent in the medium or under high light intensity or high temperature. Engineering of Rubisco, can be improved the photosynthetic CO_2 assimilation remains a matter of debate. Whitney and Andrews [42] successfully replaced the Rubisco from Tabaco plant with its counterpart from *Rhodospirillum rubrum*, which has a naturally highly CO_2 specificity. They found that CO_2 specificity of genetically engineered plant was very low as compared to wild plants. This study showed that expression of Rubisco and inaccurate folding of protein in heterologous host might be responsible for low specificity of CO_2 assimilation. In 2010, Genkov and co-authors [14] transformed the gene encoding the small subunit of Rubisco of *Arabidopsis* and sunflower into a Rubisco gene-deficient strain of *Chlamydomonas*. They observed 11% increased CO_2 specificity in vitro as compared to control.

In every living cell, the regulation of growth, development, cell cycle progression, physiological and metabolic acclimation in variable environments, are the results of the regulation of gene expression. The regulation (how much of DNA is transcribed to RNA) is controlled by DNA binding transcription factors. Similarly very few transcription factors were identified in microalgae which are responsible for lipid enhancement. The overexpression of transcription factors which induced 50% in lipid production [37]. Zhang et al. [44] isolated soybean GmDoF4 transcription factor and overexpressed in *Chlorella ellipsoidea*, which results in significant enhancement of lipid content as compared to control. Similarly, other transcription factors such as wrinkled 1 and LEC2 of *Arabidopsis* were overexpressed which resulted in crease of fatty acid and lipid accumulation, respectively. The overexpression of Wrinkled 1 transcription factor genes regulates acyl carrier protein (ACP1), acetyl-CoA carboxylase and ketoacyl-acyl carrier protein synthase (KAS1) [28]. LEC2 transcription factor is involved to regulate the lipid metabolism during seed maturation [8]. Hu et al. [18] identified 11 transcription factors in *Nannochloropsis* which have regulatory role in lipid metabolism. Engineering of microalgal transcription factor can be utilized as a multigene targeted approach, which has the ability to overproduce lipid in microalgae.

Flux balance analysis (FBA) is a mathematical method for studying the flow of metabolites in particular genome scale metabolic network reconstruction that have been built in the last decade. This metabolic network reconstruction contains complete information about the genes which encode the enzyme regulating the metabolic reaction and detailing of metabolic pathway of an organism. It calculates metabolite flow through metabolic network and also predicts growth rate of an organism or production rate of industrially important metabolites.

Mathematically, metabolic flux represents stoichiometric model, i.e. $S.v = 0$, where S is stoichiometric matrix and v is the flux vector (represents relationship between metabolites and products) [30]. Stoichiometric matrix S is m x n matrix (m, number of metabolites; n, the number chemical reaction or fluxes within that of metabolic network). FBA can be used in microalgae metabolism to maximize the TAG accumulation. Recently, Baroukh et al. [3] have attempted a dynamic flux balance analysis to study the carbon storage in microalgae under diurnal light cycle. dFBA is one of the FBA, which can be used to study the metabolism of microalgae during the light-dark transition cycle [33]. The combination of metabolic modelling and in silico analysis of microalgae can be a useful strategy for the development of lipid overproducing strain.

16.5 Conclusion

Algae are the largest photosynthetic organisms in the world which converts the greenhouse gas into carbohydrate and lipid. These lipid bodies are used as biodiesel, which has major advantage over the plant-based biofuel. But they have some limitation to produce quantitative amount of lipid. The metabolic engineering of microalgae is an approach for regulating multiple enzymes which control the lipid metabolism and generate significant amount of lipid.

Acknowledgement The authors are grateful to Prof (Dr.) S. C. Si, Dean, Centre of Biotechnology, and Prof (Dr.) M. R. Nayak, President, Siksha 'O' Anusandhan University, for providing financial support and encouragement throughout.

References

1. Amaro HM, Guedes AC, Malcata FX (2011) Advances and perspectives in using microalgae to produce biodiesel. Appl Energy 88:3402–3410. https://doi.org/10.1016/j.apenergy.2010.12.014
2. Banerjee C, Dubey KK, Shukla P (2016) Metabolic engineering of microalgal based biofuel production: prospects and challenges. Front Microbiol 7:–432. https://doi.org/10.3389/fmicb.2016.00432
3. Baroukh C, Steyer JP, Bernard O, Chachuat B (2016) Dynamic flux balance analysis of the metabolism of microalgae under a diurnal light cycle. IFAC-PapersOnLine 49:791–796. https://doi.org/10.1016/j.ifacol.2016.07.285
4. Blatti JL, Beld J, Behnke CA et al (2012) Manipulating fatty acid biosynthesis in microalgae for biofuel through protein-protein interactions. PLoS One 7(9):e42949. https://doi.org/10.1371/journal.pone.0042949
5. Brennan L, Owende P (2010) Biofuels from microalgae – a review of technologies for production processing and extractions of biofuels and co-products. Renew Sust Energ Rev 14:557–577. https://doi.org/10.1016/j.rser.2009.10.009
6. Chen CY, Zhao XQ, Yen HW et al (2013) Microalgae-based carbohydrates for biofuel production. Biochem Eng J 78:1–10. https://doi.org/10.1016/j.bej.2013.03.006
7. Courchesne NMD, Parisien A, Wang B, Lan CQ (2009) Enhancement of lipid production using biochemical, genetic and transcription factor engineering approaches. J Biotechnol 141(1–2):31–41. https://doi.org/10.1016/j.jbiotec.2009.02.018
8. De Bhowmick G, Koduru L, Sen R (2015) Metabolic pathway engineering towards enhancing microalgal lipid biosynthesis for biofuel application—a review. Renew Sust Energ Rev 50:1239–1253. https://doi.org/10.1016/j.rser.2015.04.131
9. Dehesh K, Jones A, Knutzon DS, Voelker TA (1996) Production of high levels of 8:0 and 10:0 fatty acids in transgenic canola by overexpression of Ch FatB2, a thioesterase cDNA from Cuphea hookeriana. Plant J 9:167–172. https://doi.org/10.1046/j.1365-313X.1996.09020167.x
10. Dunahay TG, Jarvis EE, Dais SS, Roessler PG (1996) Manipulation of microalgal lipid production using genetic engineering. Appl Biochem Biotechnol 57/58:223–231. https://doi.org/10.1007/BF02941703
11. Fan J, Cui Y, Wan M et al (2014) Lipid accumulation and biosynthesis genes response of the oleaginous Chlorella pyrenoidosa under three nutrition stressors. Biotechnol Biofuels 7:17. https://doi.org/10.1186/1754-6834-7-17
12. FAO (2007) Sustainable bioenergy: a framework for decision makers
13. FAO (Food and Agriculture Organization of the United Nation) (2008) The state of food and agriculture
14. Genkov T, Meyer M, Griffiths H, Spreitzer RJ (2010) Functional hybrid rubisco enzymes with plant small subunits and algal large subunits: engineered rbcS cDNA for expression in chlamydomonas. J Biol Chem 285:19833–19841. https://doi.org/10.1074/jbc.M110.124230
15. Gimpel JA, Henríquez V, Mayfield SP (2015) In metabolic engineering of eukaryotic microalgae: potential and challenges come with great diversity. Front Microbiol 6:1376. https://doi.org/10.3389/fmicb.2015.01376
16. Gong Y, Guo X, Wan X et al (2011) Characterization of a novel thioesterase (PtTE) from Phaeodactylum tricornutum. J Basic Microbiol 51:666–672. https://doi.org/10.1002/jobm.201000520
17. Hsieh HJ, Su CH, Chien LJ (2012) Accumulation of lipid production in Chlorella minutissima by triacylglycerol biosynthesis-related genes cloned from Saccharomyces cerevisiae and Yarrowia lipolytica. J Microbiol 50:526–534. https://doi.org/10.1007/s12275-012-2041-5

18. Hu J, Wang D, Li J et al (2015) Genome-wide identification of transcription factors and transcription-factor binding sites in oleaginous microalgae Nannochloropsis. Sci Rep 4:5454. https://doi.org/10.1038/srep05454
19. Hu Q, Sommerfeld M, Jarvis E et al (2008) Microalgal triacylglycerols as feedstocks for biofuel production: perspectives and advances. Plant J 54:621–639. https://doi.org/10.1111/j.1365-313X.2008.03492.x
20. IEA (2010) Key world energy statistics. Paris
21. Karemore A, Pal R, Sen R (2013) Strategic enhancement of algal biomass and lipid in Chlorococcum infusionum as bioenergy feedstock. Algal Res 2:113–121. https://doi.org/10.1016/j.algal.2013.01.005
22. Khozin-Goldberg I, Cohen Z (2006) The effect of phosphate starvation on the lipid and fatty acid composition of the fresh water eustigmatophyte *Monodus subterraneus*. Phytochemistry 67:696–701. https://doi.org/10.1016/j.phytochem.2006.01.010
23. Klok A, Lamers P, Martens D et al (2014) Edible oils from microalgae: insights in TAG accumulation. Trends Biotechnol 32:521–528
24. Kumar A, Ergas S, Yuan X et al (2010) Enhanced CO2 fixation and biofuel production via microalgae: recent developments and future directions. Trends Biotechnol 28:371–380. https://doi.org/10.1016/j.tibtech.2010.04.004
25. Lam MK, Lee KT (2012) Microalgae biofuels: a critical review of issues, problems and the way forward. Biotechnol Adv 30:673–690. https://doi.org/10.1016/j.biotechadv.2011.11.008
26. Lü J, Sheahan C, Fu P (2011) Metabolic engineering of algae for fourth generation biofuels production. Energy Environ Sci 4:2451. https://doi.org/10.1039/c0ee00593b
27. Ma Y, Wang X, Niu Y, et al (2014) Antisense knockdown of pyruvate dehydrogenase kinase promotes the neutral lipid accumulation in the diatom Phaeodactylum tricornutum. 1–9. https://doi.org/10.1186/s12934-014-0100-9
28. Maeo K, Tokuda T, Ayame A et al (2009) An AP2-type transcription factor, WRINKLED1, of Arabidopsis thaliana binds to the AW-box sequence conserved among proximal upstream regions of genes involved in fatty acid synthesis. Plant J 60:476–487. https://doi.org/10.1111/j.1365-313X.2009.03967.x
29. Markou G, Nerantzis E (2013) Microalgae for high-value compounds and biofuels production a review with focus on cultivation under stress conditions. Biotechnol Adv 31:1532–1542 https://doi.org/10.1016/j.biotechadv.2013.07.011
30. May P, Christian J-O, Kempa S, Walther D (2009) ChlamyCyc: an integrative systems biology database and web-portal for Chlamydomonas reinhardtii. BMC Genomics 10:209. https://doi.org/10.1186/1471-2164-10-209
31. Medipally SR, Yusoff FM, Banerjee S, Shariff M (2015) Microalgae as sustainable renewable energy feedstock for biofuel production. Biomed Res Int 2015:519513 https://doi.org/10.1155/2015/519513
32. Miller R, Wu G, Deshpande RR et al (2010) Changes in transcript abundance in Chlamydomonas reinhardtii following nitrogen deprivation predict diversion of metabolism. Plant Physiol 154:1737–1752. https://doi.org/10.1104/pp.110.165159
33. Muthuraj M, Palabhanvi B, Misra S et al (2013) Flux balance analysis of Chlorella sp. FC2 IITG under photoautotrophic and heterotrophic growth conditions. Photosynth Res 118:167–179. https://doi.org/10.1007/s11120-013-9943-x
34. Niu Y-F, Zhang M-H, Li D-W et al (2013) Improvement of neutral lipid and polyunsaturated fatty acid biosynthesis by overexpressing a type 2 diacylglycerol acyltransferase in marine diatom Phaeodactylum tricornutum. Mar Drugs 11:4558–4569. https://doi.org/10.3390/md11114558
35. Radakovits R, Jinkerson RE, Darzins A, Posewitz MC (2010) Genetic engineering of algae for enhanced biofuel production. Eukaryot Cell 9:486–501. https://doi.org/10.1128/EC.00364-09
36. Rodolfi L, Zittelli GC, Bassi N et al (2009) Microalgae for oil: strain selection, induction of lipid synthesis and outdoor mass cultivation in a low-cost photobioreactor. Biotechnol Bioeng 102:100–112. https://doi.org/10.1002/bit.22033
37. Sapphire Energy I (2011) Stress-induced lipid trigger

38. Takagi M, Karseno, Yoshida T (2006) Effect of salt concentration on intracellular accumulation of lipids and triacylglyceride in marine microalgae Dunaliella cells. J Biosci Bioeng 101:223–226. https://doi.org/10.1263/jbb.101.223
39. Thelen JJ, Ohlrogge JB (2002) Metabolic engineering of fatty acid biosynthesis in plants. Metab Eng 4:12–21. https://doi.org/10.1006/mben.2001.0204
40. Trentacoste EM, Shrestha RP, Smith SR et al (2013) Metabolic engineering of lipid catabolism increases microalgal lipid accumulation without compromising growth. Proc Natl Acad Sci 110:19748–19753. https://doi.org/10.1073/pnas.1309299110
41. Verma N, Mehrotra S, Shukla A, Mishra B (2010) Prospective of biodiesel production utilizing microalgae as the cell factories: a comprehensive discussion. Afr J Biotechnol 9:1402–1411. https://doi.org/10.5897/AJBx09.071
42. Whitney SM, Andrews TJ (2001) Plastome-encoded bacterial ribulose-1,5-bisphosphate carboxylase/oxygenase (RubisCO) supports photosynthesis and growth in tobacco. Proc Natl Acad Sci U S A 98:14738–14743. https://doi.org/10.1073/pnas.261417298
43. Xin L, Hong-ying H, Ke G, Ying-xue S (2010) Effects of different nitrogen and phosphorus concentrations on the growth, nutrient uptake, and lipid accumulation of a freshwater microalga Scenedesmus sp. Bioresour Technol 101:5494–5500. https://doi.org/10.1016/j.biortech.2010.02.016
44. Zhang J, Hao Q, Bai L et al (2014) Overexpression of the soybean transcription factor GmDof4 significantly enhances the lipid content of Chlorella ellipsoidea. Biotechnol Biofuels 7:128. https://doi.org/10.1186/s13068-014-0128-4

Microalgae: Gizmo to Heavy Metals Removal

17

Sradhanjali Singh, Debabrata Pradhan, and Lala Behari Sukla

Abstract

The memento of environmental pollution is mounting so fast that if we don't kill it, it will kill us. In the race to prove our supremacy, we have forgotten the rules of "Mother Nature," thereby deteriorating the nature's beauty and precious resources. We are living in the planet as if we have another option to go. Now we have to be a part of solution, not a part of pollution, and need to take stringent control measures to save the lives on earth. Several environmental management plans and policy have been implemented to seize pollution to some extent. However, one such application which has caught the attention of the world is use of microalgae as an indicator for removal of environmental pollutant. They can be used for biotransformation, bioaccumulation, and biodegradation of specific pollutants from wastewater. Nonetheless, the relevance of microalgae makes them a jewel for reducing pollution. Along with this, they have the ability to reduce biological oxygen demand, remove N and P, and suppress the growth of coliforms bacteria. Being a photosynthetic organism, they convert solar energy and CO_2 into biomass enriched with N and P. They capture CO_2 gas from the atmosphere during the basic photosynthesis process resulting in reducing greenhouse gases. This algal-based system can help in removing pollutant from wastewater by separating the biomass which is supersaturated with metal content from the medium resulting in high-quality reusable effluent water. Notably, another approach can also be made by immobilizing the algal components that can act as an absorbent and studying the adsorption capacity for uptaking of toxic metals

S. Singh
CSIR-National Environmental and Engineering Research Institute, Delhi Zonal Center, New Delhi, India

D. Pradhan (✉) · L. B. Sukla (✉)
Biofuels and Bioprocessing Research Center, Institute of Technical Education and Research, Siksha 'O' Anusandhan (Deemed to be University), Bhubaneswar, Odisha, India
e-mail: debabratapradhan@soa.ac.in; lalabeharisukla@soa.ac.in

L. B. Sukla et al. (eds.), *The Role of Microalgae in Wastewater Treatment*,
https://doi.org/10.1007/978-981-13-1586-2_17

such as lead, cadmium, mercury, scandium, tin, arsenic, and bromine from wastewater. This will minimize the disposal cost and eliminate the generation of secondary pollutions, which would be a perfect replacement to the conventional technologies.

17.1 Introduction

The use of potential microorganisms and/or their aggregates is practically beneficial in the wastewater treatment process compared to different conventional physical and chemical processes. Microorganisms, such as bacteria, fungi, and microalgae, are competent to remove different targeted pollutants from the wastewater. The biological processes are operated either by the direct mixing of free microorganisms with the wastewater or by immobilizing/encapsulating the microbial cells within a matrix [1]. However, use of immobilized/encapsulated cells over free cells is more effective in the bio-treatment process due to different advantages such as higher biodegradation rates, easier solid-liquid separation, higher biomass loading, better operation stability, greater protection from toxic substances, and increased plasmid stability of immobilized cells.

The choice of right kind of microbial species is essential for the pilot plant scale-up of the bio-treatment process. In this context, microalgae prove to be efficient as they can grow very fast compared to the terrestrial plants which enormously fulfill the need of our present environmental scenario (www.allaboutalgae.com). In recent years, use of microalgae has not been limited to only food or feed, but rather they have equal potential in different bioremediation processes. Interestingly, it can create jobs in various sectors ranging from construction to farming, research and development to engineering, and marketing to financial services [23, 26, 30, 37].

Another interesting use of the biomass of microalgae is the role of active adsorbent in the biosorption of different pollutants from the wastewater. Similar to biological species like bacteria, fungi, and other species, the biomass of microalgae has sufficient active surface area to bind different metal ions present in the wastewater. Biosorption of heavy metal ions depends on different functional groups present in the microalgae biomass. Also the possibility of the bioaccumulation of different heavy metals has been accounted during the growth of microalgae in the active interaction, unlike the biosorption process which is the passive interactions between heavy metals and biomass (https://www.witpress.com/elibrary/wit-transactions-on-ecology-and-the-environment/78).

17.2 Mechanism Involved in Biosorption and Bioaccumulation

Uptake of metal ions from the solution on the surface of either live or dead biological cells by means of physical or chemical forces, like electrostatic interactions, van der Waals bonding, covalent bonding, and ion exchange, is called biosorption. However, from the solution metal ions entering to cells of different living organisms through the nutrient channels followed by their absorption within the cells due to the natural cellular metabolism is referred to as bioaccumulation [3, 8, 29]. Bioaccumulation renders toxicity and propagation of diseases in the organisms [37]. Biomass used as adsorbent in the biosorption can be regenerated absolutely; however, it is difficult in the case of bioaccumulation [13, 14, 41]. In the former case, the metal ions deposited can be removed from the cell surface by acid or alkali treatment. Biomass of several species, such as fungi, cyanobacteria, bacteria, and algae, has been used for both biosorption and bioaccumulation process of metal ions from the solution. Biosorption of metal ions is a rapid process due to the structures, compositions, and presence of functional groups of the cell wall [6, 7, 18, 39]. The above properties of cell wall define the metal ion-binding mechanisms, affinities, and nature of the biosorption process [19]. However, bioaccumulation is a slow process as it depends on the natural cellular growth of the living organisms.

Although the mechanism is understood well and several reports are available so as to study the detail mechanisms involved in the bioprocesses, a pictorial representation showing the bacterial cell is shown in Fig. 17.1 in order to explain different biosorption mechanisms. Interactions that occur between dissolved metal ions in the solution and cellular components of the living or dead organisms are either of

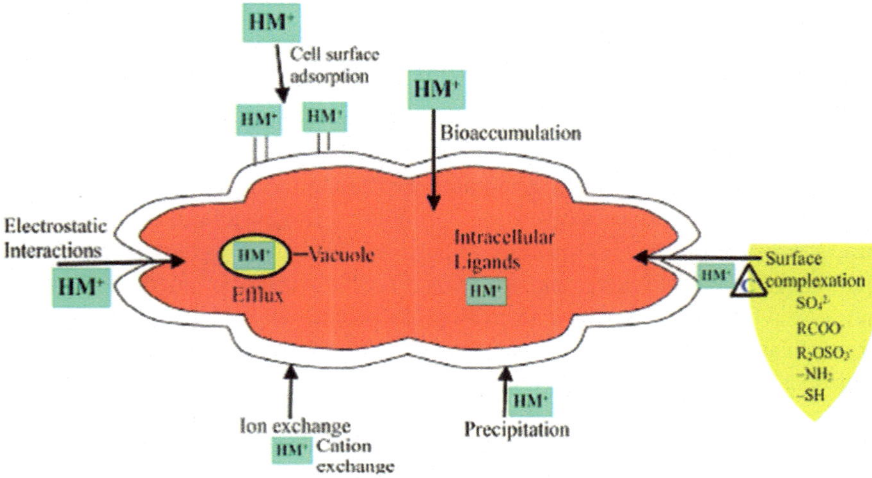

Fig. 17.1 Schematic diagram showing the different mechanisms of bacterial biosorption (Adapted from Ref. [5])

physical adsorption, ion exchange, complexation, coordination, microprecipitation, chelations, crystallization, and diffusion [16, 17, 42, 44, 45, 48].

There are two mechanisms such as binding of metal ions to different components of the cell surface and intracellular accumulation of metal ions through the cellular metabolism by living microorganisms. Whereas, Malik [25] pointed that the intracellular accumulations is slow and mostly dependent on nutrients and environmental conditions. Nonetheless, biosorption is a passive process [33, 45]. Akhtar et al. [2] suggested that there were two phases of adsorption phenomenon, e.g., first one related to physical sorption and second due to both structural change and surface transformation, involved in the biosorption of uranium in live *Trichoderma harzianum*. In another study of copper bioaccumulation by the live yeast, *Saccharomyces cerevisiae* was found to be biphasic [22]. It consisted of an initial and rapid surface binding and then followed by a second slow intracellular uptake.

Muraleedharan and Venkobachar [31] reported that different metal ions may be transported into the cells and may react to form a product that remains within the cells. Furthermore, there may be formation of chelating complex when metal cations mixed with the negatively charged sugar units at the end of polysaccharides chain [40]. Besides, they cited that the extracellular polyphosphate groups were associated with the sugar metabolism and complex metal ions by chelating through negatively charged oxygen atoms.

The ion exchange reaction is an important mechanism involved in the biosorption process, where the cell walls of biomaterials act as the ion exchanger similar to different ion exchange resins. The cell walls contain a mixture of monovalent and divalent cations, and the substitution within the cations supports the formation of ligand complexes which act as the metal binding resins [4].

17.3 Algae

Algae are the diverse group of photosynthetic organisms and found in different aquatic habitats. Their size ranges from unicellular microalgae to multicellular giant kelp. The species biodiversity of microalgae is more than two million; however, only a tiny fraction of the total species biodiversity has been isolated and characterized [35]. They play a key role as initiator of different food webs in the biospheres. In addition, they produce oxygen gas during their photosynthesis which is a key process in restoring the atmospheric composition. They use different water resources like brackish and freshwater and wastewater for the purpose of micro- and macronutrients. Other algae can grow heterotrophically in the absence of light using sugar or starch as nutrients. Some other algae species can grow both autotrophic and heterotrophic modes which is called mixotrophic growth. Contrasting to the terrestrial plants, they have no true roots, leaves, and stems. They contain high levels of protein, minerals, vitamins, and trace elements like iodine, calcium, and iron. They contain low amount of fat but high amount of fiber. Owing to several advantages, algal species are considered to be the best possible weapon to deal with different

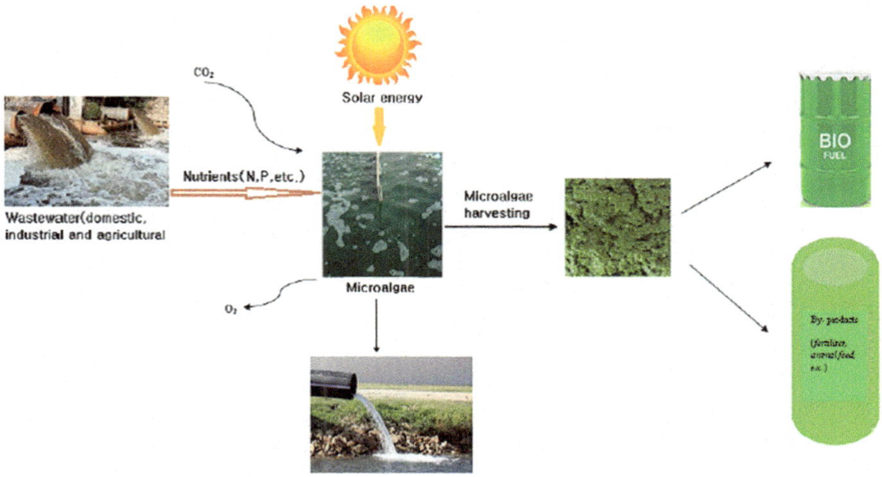

Fig. 17.2 Wastewater treatment through microalgae. (Adapted from Ref. [34])

pollution issues [9–12, 15, 24, 28, 32, 36, 38, 43, 47]. Figure 17.2 shows a simple wastewater treatment process using microalgae.

17.4 Immobilization Techniques

Immobilization technique confines either of the enzymes and live/dead cells on or within an inert support for their stability and functional reuse. There are four principal methods available for immobilizing the enzymes, such as adsorption, covalent binding, entrapment, and encapsulation, as shown in Fig. 17.3.

There are different conventional treatment processes, such as filtration, sedimentation, flocculation, chlorination, and activated sludge process, which have been used to remove different contaminants from the wastewater [27]. These methods are based on multiple steps conducted for separation followed by elimination of different pollutants from the wastewater, which results in an expensive process. Therefore, search for a cost-effective and integrative wastewater treatment process is necessary for the complete removal of different pollutants (e.g., organic and inorganic) from the wastewater. In this context, the immobilization technique is effective for the complete eradication of different nutrients and mineral components from the wastewater.

17.5 Future Prospective

Biosorption of heavy metals is a self-driven process for enhanced heavy removal from industrial wastes and wastewater. For sustainable approach, development of an appropriate biosorbent is highly essential. The properties of algal biomass, such as

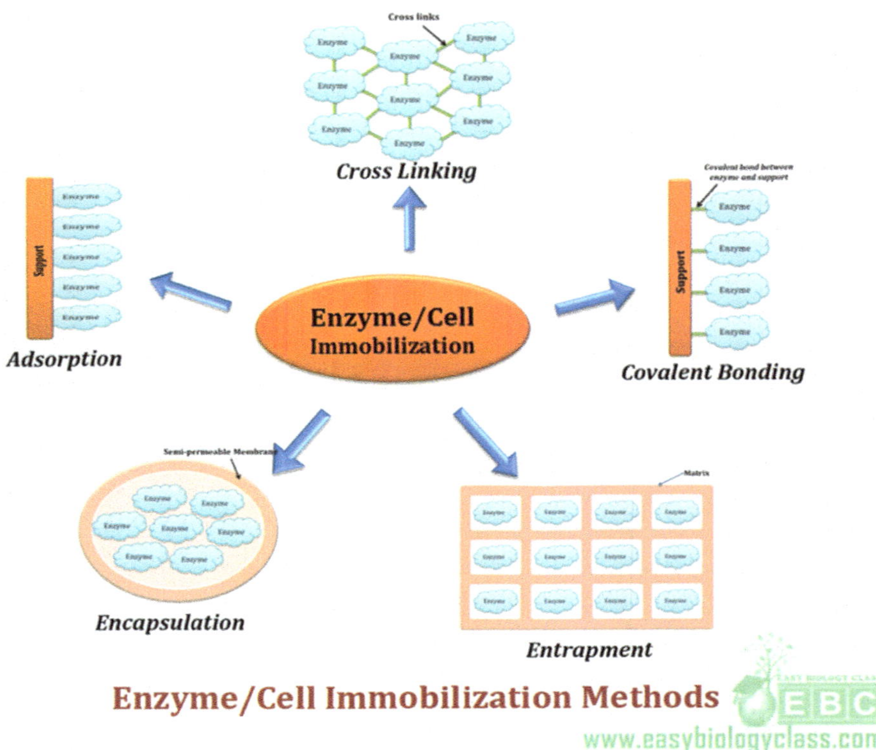

Fig. 17.3 Different methods of immobilization technique. (Source: http://www.easybiologyclass.com/enzyme-cell-immobilization-techniques)

cellular structure, polysaccharide contents, exposure of cell wall, presence of metal binding functional groups, and extracellular polysaccharides, determine the effective biosorption of metal ions. The algal species such as *Chlorella* and *Dunaliella* were used for biomass production and treating wastewater for more than 75 years. Although the process and mechanism are well understood, its commercial applications are a matter of concern. Beside the wastewater treatment, algae have other benefits like they can be used in pharmaceuticals, animal feed, composting, agriculture, biofuels, aquaculture, etc. The genetically engineered products of algal biomass have numerous health benefits such as antibacterial, antiviral, antitumor/anticancer, and antihistamine. Probably some more beneficial properties of microalgae are yet to be explored (e.g., some microalgae can tolerate high level ionization radiations; perhaps there is a long way to go with microalgae).

References

1. Adholeya A, Das M (2015) Potential uses of immobilized Bacteria, Fungi, algae, and their aggregates for treatment of organic and inorganic pollutants in wastewater, water challenges and solutions on a global scale. ACS symposium series. Am Chem Soc 1206:319–337
2. Akhtar K, Akhtar MW, Khalid AM (2007) Removal and recovery of uranium from aqueous solutions by *Trichoderma harzianum*. Water Res 41:1366–1378
3. Aksu Z, Gönen F, Demircan Z (2002) Biosorption of chromium(VI) ions by Mowital® B3CH resin immobilized activated sludge in a packed bed: comparison with granular activated carbon. Process Biochem 38(2):175–186
4. Al-Asheh S, Duvnjak Z (1997) Sorption of cadmium and other heavy metals by pine bark. J Hazard Mater 56:35–51
5. Ayangbenro AS, Babalola OO (2017) A new strategy for heavy metal polluted environments: a review of microbial biosorbents. Int J Environ Res Public Health 14:94
6. Barkley NP (1991) Extraction of mercury from groundwater using immobilized algae. J Air Waste Manage Assoc 41(10):1387–1393
7. Bayramoğlu G, Tuzun I, Celik G, Yilmaz M, Arica MY (2006) Biosorption of mercury(II), cadmium(II) and lead(II) ions from aqueous system by microalgae Chlamydomonas reinhardtii immobilized in alginate beads. Int J Miner Process 81(1):35–43
8. Brady D, Duncan J (1994) Bioaccumulation of metal cations by Saccharomyces cerevisiae. Appl Microbiol Biotechnol 41(1):149–154
9. Crist RH, Martin JR, Crist DR (1999) Interaction of metal ions with acid sites of biosorbents peat moss and *Vaucheria* and model substances alginic and humic acids. Environ Sci Technol 33:2252–2256
10. Crist RH, Oberholser K, Shank N, Nguyen M (1981) Nature of binding between metallic ions and algal cell wells. Environ Sci Technol 15:1212–1217
11. Das N, Vimala R, Karthika P (2008) Biosorption of heavy metals – an overview. Indian J Biotechnol 7:159–169
12. De Philippis R, Colica G, Mecarozzi P (2011) Exopolysaccharide-producing cyanobacteria in heavy metal removal from water: molecular basis and practical applicability of the biosorption process. Appl Microbiol Biotechnol 92:697–708
13. de-Bashan LE, Bashan Y (2010) Immobilized microalgae for removing pollutants: review of practical aspects. Bioresour Technol 101(6):1611–1627
14. Dönmez G, Aksu Z (2002) Removal of chromium (VI) from saline wastewaters by Dunaliella species. Process Biochem 38(5):751–762
15. Eccles H (1999) Treatment of metal-contaminated wastes: why select a biological process? Trends Biotechnol 17:462–465
16. Fourest E, Roux JC (1992) Heavy metal biosorption by fungal mycelial by-products: mechanisms and influence of pH. Appl Microbiol Biotechnol 37:399–403
17. Gadd GM (1998) Accumulation of metal by microorganisms and algae. In: Rehm H (ed) Biotechnology: a complete treatise. vol. 6B, special microbial processes vol 4. Verlagsgesellschaft. VCH, Weinheim, pp 401–430
18. Gekeler W, Grill E, Winnacker E-L, Zenk MH (1988) Algae sequester heavy metals via synthesis of phytochelatin complexes. Arch Microbiol 150(2):197–202
19. Godlewska-Żyłkiewicz B (2003) Biosorption of platinum and palladium for their separation/preconcentration prior to graphite furnace atomic absorption spectrometric determination. Spectrochim Acta B 58(8):1531–1540
20. http://www.easybiologyclass.com/enzyme-cell-immobilization-techniques
21. https://www.witpress.com/elibrary/wit-transactions-on-ecology-and-the-environment/78
22. Huang C, Huang CP (1996) Application of *Aspergillus oryzae* and *Rhizopus oryzae* for Cu (II) removal. Water Res 30:1985–1990
23. Johnson EA, An GH (1991) Astaxanthin from microbial sources. Crit Rev Biotechnol 11(4):297–326

24. Kaplan D (1998) Algal polysaccharides as natural metal Chelators. BARD & North Carolina Biotechnology Centre, Beaufort, pp 51–55
25. Malik A (2004) Metal bioremediation through growing cells. Environ Int 30:261–278
26. Mallick N (2002) Biotechnological potential of immobilized algae for wastewater N, P and metal removal: a review. Biometals 15(4):377–390
27. Metcalf EH (2003) Wastewater engineering: treatment and reuse. McGraw-Hill, New York
28. Mishra A, Kavita K, Jha B (2011) Characterization of extracellular polymeric substances produced by micro-algae *Dunaliella salina*. Carbohydr Polym 83:852–857
29. Moreno-Garrido I, Blasco J, Gonzalez-Delvalle M, Lubián L (1998) Differences in copper accumulation by the marine microalga Nannochloropsis gaditana, submitted to two different thermal treatments. Ecotoxicol Environ Restor 1:43–47
30. Mulbry W, Westhead EK, Pizarro C, Sikora L (2005) Recycling of manure nutrients: use of algal biomass from dairy manure treatment as a slow release fertilizer. Bioresour Technol 96(4):451–458
31. Muraleedharan TR, Venkobachar C (1990) Mechanism of biosorption of copper (I) by *Ganoderma lucidum*. Biotechnol Bioeng 35:320–325
32. Naja G, Volesky B (2011) The mechanism of metal cation and anion biosorption. In: Kotrba P, Mackova M, Macek T (eds) Microbial biosorption of metals. Springer Science+Business Media B.V, Dordrecht, pp 19–58
33. Pardo R, Herguedas M, Barrado E, Vega M (2003) Biosorption of cadmium, copper, lead and zinc by inactive biomass of *Pseudomonas putida*. Anal Bioanal Chem 376:26–32
34. Pogaku R, Hardinge BS, Vuthaluru H, Amir HA (2016) Production of bio-oil from oil palm empty fruit bunch by catalytic fast pyrolysis: a review. Biofuels 7(6):647–660
35. Pradhan D, Singh S, Sukla LB (2017a) Microalgae for carbon sequestration vis-à-vis bio-fuels production. Inglomayor 13:56–67
36. Pradhan D, Sukla LB, Acevedo R (2017b) Microalgae for future biotechnology industries. Inglomayor 13:40–55
37. Pradhan D, Sukla LB, Sawyer M, Rahman PKSM (2017c) Recent bioreduction of hexavalent chromium in wastewater treatment: a review. J Ind Eng Chem 55:1–20
38. Roy D, Greenlaw PN, Shane BS (1993) Adsorption of heavy metals by green algae and ground rice hulls. J Environ Sci Health (Part A) 28:37–50
39. Schiewer S, Volesky B (2000) Biosorption processes for heavy metal removal. In: Lovley DR (ed) Environmental microbe–metal interactions. ASM Press, Washington, DC, pp 329–362
40. Shumate H, Strandberg G, Parrott J (1978) Biological removal of metal ions from aqueous process streams. Biotechnol Bioeng Symp 8:12–20
41. Velásquez L, Dussan J (2009) Biosorption and bioaccumulation of heavy metals on dead and living biomass of Bacillus sphaericus. J Hazard Mater 167(1):713–716
42. Vijayaraghavan K, Yun Y-S (2008) Bacterial biosorbents and biosorption. Biotechnol Adv 26:266–291
43. Volesky B (1990) Biosorption of heavy metals. CRC Press, Boca Raton, p 396
44. Volesky B (2003) Sorption and biosorption. BV Sorbex, Inc. Montreal-St. Lambert, Quebec, Canada
45. Volesky B, Holan ZR (1995) Biosorption of heavy metals. Biotechnol Prog 11:235–250
46. www.allaboutalgae.com
47. Xue HB, Stumm W, Sigg L (1988) The binding of heavy metals to algae surfaces. Water Res 22:917–926
48. Yan G, Viraraghavan T (2000) Effect of pretreatment on the bioadsorption of heavy metals on Mucor rouxii. Water SA 26:119–124

Characterization of Extracellular Proteins to Explore Their Role in Bio-Flocculation for Harvesting Algal Biomass for Wastewater Treatment

18

Surajit Debnath

Abstract

Bioprocess technology aims to production of high-value end products from natural materials in an eco-friendly method. While doing so it can also solve environmental and industrial problems in addition to product yield. Bio-flocculation is a dynamic phase in most of the bioprocess such as wastewater treatment, harvesting of biofuels, bioremediation of activated sludge, and yielding of bio-materials from a bioreactor. In algae-based bioprocess technology, harvesting of algal biomass is enormously energy-intensive step. This alone is the main constraint on commercial development of numerous conceivable methodologies of environmental management through algal systems worldwide.

Several strategies are currently investigated in order to enhance auto-flocculation in a regulated way to avoid energy demanding centrifugation and successive processing. A successful master plan in this domain would lead to potentially low-cost harvesting technique. Some approaches that are under scrutiny involve co-culture of bio-flocculent producing organisms. However, the bottleneck of biomass harvesting at minimal cost is yet to be circumvented. In this study, the flocculation enhancing proteins of *Saccharomyces cerevisiae* ie, *Flo 1, Flo 5* and *Flo 9* have been analyzed using computation biology tools to evaluate their structural and functional characteristics to assess dynamic behavior and flocculating properties. It is followed by a dry run of molecular biology intervention. Bioinformatics simulations such as molecular dynamics, normal mode analysis, characterization of protein active sites, and protein network interaction are some low cost yet efficient tools that are used in this chapter for their near precise prediction on a protein behavior. This study is novel and aids to the ongoing brainstorming of the bioprocess biotechnology fraternity to establish an economical harvesting protocol for algal biomass for wastewater treatment.

S. Debnath (✉)
Department of Medical Laboratory Technology, Women's Polytechnic,
Government of Tripura, Agartala 799130, Tripura, India

L. B. Sukla et al. (eds.), *The Role of Microalgae in Wastewater Treatment*,
https://doi.org/10.1007/978-981-13-1586-2_18

18.1 Introduction

The challenge of today's world in the changing scenario of extensive industrialization is the sustainability of very basic livelihood resources such as drinking water. If the living world is considered as a single unit in the biosphere, then we can appreciate the role of water more logically. All physiological reactions that drive the living world take place in an aqueous environment. Therefore optimization of bioprocess that recycles water from its waste counterpart is of great importance. Total bioremediation of wastewater would be a highly acclaimed bioprocess technology that would be sustainable and eco-friendly. With suitable inputs and some scientific breakthrough this can become the most impactful discovery of mankind. Wastewater treatment by microalgal system is the current hot topic towards achieving total bioremediation of waste water and sustainable environmental management. Microalgae harvesting for wastewater treatment for purposes like drinking water supply and environmental management of inland water [43, 56, 108] are being explored. Recycling of wastewater via microalgal systems have an added advantage of coproduction of industrially important biomass. It was observed that the synergistic coupling of microalgae propagation with carbon sequestration and wastewater treatment can be potentially used to mitigate environmental impacts of high consumption of energy and fuel [10]. Historically it was Oswald and Gotaas [69] who first reported the use of wastewater ponds to cultivate algae that may lead to recycling. Over the years it has been established that wastewater treatment is one of the favorable applications of algal systems [6–8, 52, 82, 88]. In spite of the obvious advantages widespread application of algal biofilm-based treatment of municipal, industrial, and agricultural waste streams has been limited so far [44]. The scientific deadlock that is limiting this promising bioprocess of algal system of wastewater treatment has been the lack of efficient separation technique of the biomass from the resultant product. So the bottleneck of sustainable downstream processing after the bioremediation process still persists. For a review on applications of microalgae in various sustainable process including wastewater treatment the literature [37] can be followed.

Recycling of wastewater in a sustainable way is a real challenge that should be addressed to mitigate the obvious impacts of climate change.

In order to separate the treated water and the algal biomass, several processes were under inspection by the scientists. Assisted flocculation for separation of biomass after recycling of wastewater was able to generate much attention due to several advantages discussed later in the chapter. However, in the post-genomic era, molecular biology and genetic engineering can be a promising tool to overcome scientific hurdle of the bioprocess. In this chapter computational biology is harnessed to predict suitability of flocculation enhancing proteins from *Saccharomyces* which can be targeted for enhanced flocculation of algal biomass through molecular biology and genetic engineering platforms. FLO1, FLO5, and FLO9 gene products ie *Flo1*, *Flo5* & *Flo9* proteins were analyzed by structural bioinformatics so that the best suitable candidate can be identified for transformation in a suitable algal host that may be genetically modified for enhanced flocculation. Along with this the molecular basis of flocculation is reviewed and perspective strategy for genetic modification of algal system for enhanced flocculation is critically discussed in this chapter.

18.2 Background

Figure 18.1 gives the glimpses of the wastewater recycling by algal systems and the scientific bottleneck we are facing today. The fresh water undergoes severe physico-chemical modification through the process of industrialization that includes pollution caused by agrochemicals, mining, etc. The conservative mode of recycling the water is capital intensive and not viable for developing economies to sustain for long. For example, conventional technologies for the treatment of large volumes of wastewater, such as reverse osmosis, ion exchange, precipitation, etc., are less efficient and commercially no viable [27].

On the other hand, using microalgal systems is a sustainable option once we can able to optimize the downstream processing of removing the accumulated algal mass from the treated water. The added advantage of the process is that, we can scale up the process to produce industrially important biomass as a by-product with no or minimal extra cost. The wastewater itself provide nutrients to sustain biomass yield. Several inorganic and organic compounds generally present in wastewater make it suitable as a substrate for biomass yield [14, 54, 76, 105, 109]. After the recycling of wastewater and formation of biomass, the products (recycled water and the biomass) must be separated from each other for their anticipated utilization. For this process a sustainable protocol is yet to be in place.

The promising sky of algal bioprocess for wastewater treatment is yet to be cleared from the dark cloud of the scientific bottleneck that hinders separation of products after the bioprocess.

The concept of wastewater treatment by microalgae is promising but possesses the hurdle of biomass separation after microalgae treatment.

18.2.1 Separation of Biomass from Algal Bioprocess for Wastewater: An Energy-Intensive Method

Currently, removing the biomass from treated water is conventionally done by energy-intensive methods. As mentioned already, the techniques used for harvesting microalgal cells such as centrifugation, filtration, flocculation, gravity sedimentation and flotation [33, 35, 61, 75, 87, 98], all have some major drawbacks. Large-scale harvesting and separation of biomass from algal bioprocess for wastewater remain an energy-intensive method mainly due to three microalgal features, which

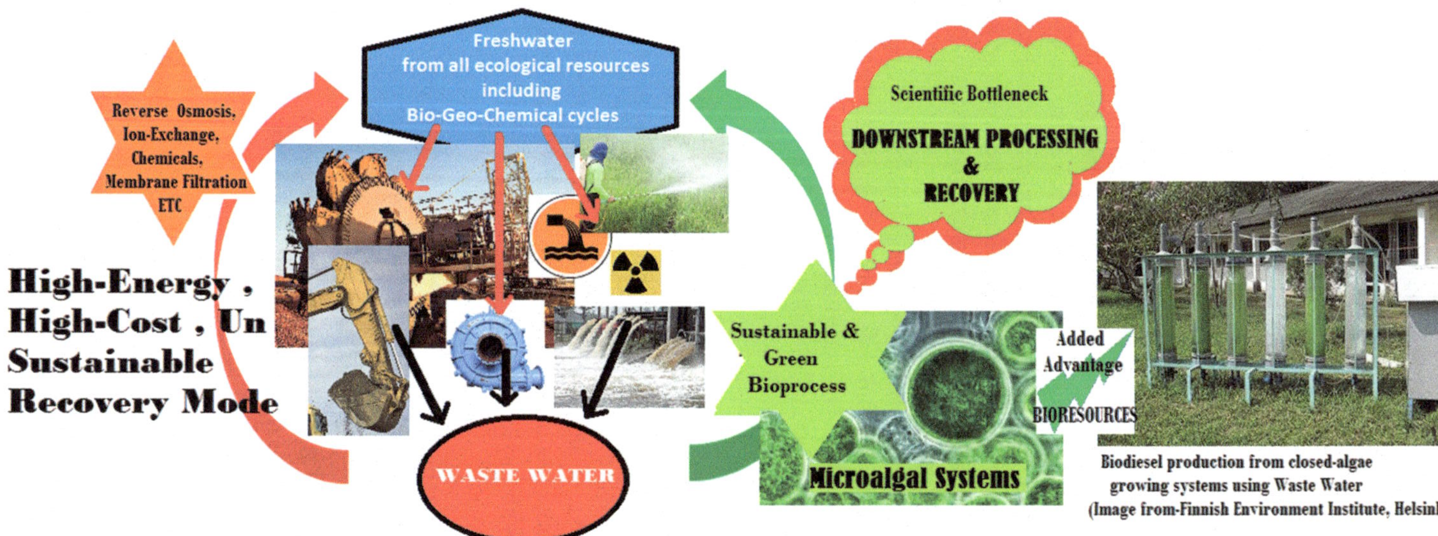

Fig. 18.1 Wastewater recycling by algal systems waiting for the breakthrough

are the small size of cells (5,30 mm), low concentration of the yield (0.02–0.05% DW), and negative surface charge [104] on the cells. Extraction of biomass after recovery of wastewater is also very costly due to expensive solvents and high electricity consumption [64, 65]. Slower growth rate on account of only 0.03–0.06% of atmospheric CO_2 is another problem with microalgae culture [50] in normal condition due to mass transfer limitation [57].

18.2.2 Bio-flocculation: A Promising Method for Separation of Biomass from Algal Bioprocess for Wastewater

To reduce the cost of removing the accumulated biomass from the treated water, bio-flocculation seems to be the most promising one among several options that have been studied with much detail. Flocculation, a process by which the algae forms a 3D structure or clumps called flocs which can be easily segregated, is a method of choice for removing algal biomass post-treatment of wastewater. However under normal growth conditions flock formation is inhibited by the negatively charged microalgal surfaces preventing their self-flocculation [30, 73, 83, 84]. Therefore various approaches are harnessed to address this by altering the surface charges [98]. For example, chemical or biological flocculants (inorganic and organic) for neutralizing or reducing microalgal negative surface charge can be used [98]. But these methodologies, however, are not universally successful and do not work for all microalgae strains [48, 49, 98]. However, bio-flocculation methods due to their high efficiency and low energy input are becoming increasingly popular [14, 70]. Among biological organisms that promote flocculation are bacteria and fungi [48, 49, 54, 74] that enable the algae to form flocks. A review of current strategies for flocculation methods [101] can be read.

18.2.3 Fungal-Assisted Algal Bio-flocculation

Co-cultivation of fungal and microalgal cells or fungal-assisted bio-flocculation is getting high attention because of low energy inputs, high efficiency of bio-flocculation of microalgal cells, and no requirement of added chemicals [66, 104]. Concentration of microalgal cells within fungal filaments can be achieved by bio-flocculation via hydrogen bonds, electrostatic interactions, and/or using a matrix of extracellular polymeric substance (EPS) secreted by fungal and algal cells (that are vital for removing the biomass after treatment of the wastewater) [62] (Fig. 18.2).

Bio-flocculation of microalgae after wastewater treatment is a wonderful method but needs optimization.

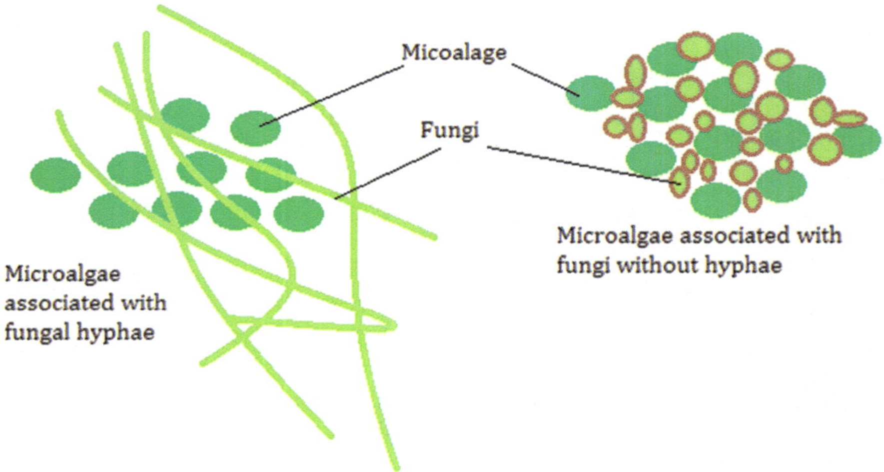

Fig. 18.2 Algae-fungi cocultivation for enhanced separation

Cocultivation of microalgae and fungus showed an additive effect on wastewater treatment efficiency, as per previous studies [42, 104], and several species of fungi [4, 110] have been used for efficient wastewater treatment in experimental setups.

18.3 A Novel Strategy for Enhanced Flocculation Phenotype in Microalgae for Wastewater Treatment

In the above background, it is understood that bio-flocculation is an effective mean that may aid to separate the algal biomass after the algae has treated the wastewater by its much celebrated curative properties through accumulation of toxic components from the wastewater. In order to enhance the flocculation efficiency of algae, a strategic approach of molecular biology-genetic engineering intervention is proposed here which is fortified by the results of computational biology analysis described in the chapter (Fig. 18.3). To identify the candidate algal host, the suitable organisms with already proven track record of enhanced flocculation are also reviewed here. The focus is to pinpoint candidate proteins that may enhance flocculation features. A thorough practical analysis of perspective protein candidates enabled identification of most suitable proteins for further possible experiments using wet lab facilities. Results from the computational simulation of the study are at parity with published reports on the identified proteins. Thus the identified candidate proteins may be introduced in the suitable algal vector with established transformation and expression characteristics. A possible work plan of the molecular biology-genetic engineering intervention is also described in this chapter.

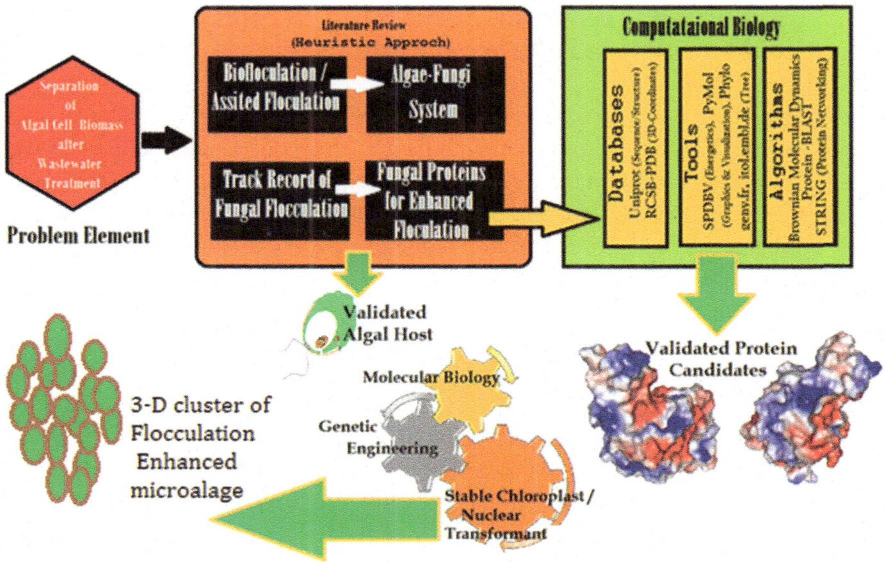

Fig. 18.3 Schema of a novel strategy for enhanced flocculation of algal systems

In the post-genomic era, molecular biology-genetic engineering interventions can revolutionize wastewater treatment by microalgae.

18.4 Search for a Well-Understood Molecular Mechanism of Flocculation

Among fungi-based co-culture of microalgae, yeast is a suitable candidate since the physiology, biochemistry, and molecular biology of yeast flocculation are well understood. It has a track record of efficient flocculation, and there is a scope that molecular biology and genetic engineering is harnessed in this regard.

Cell flocculation is used for industrial wastewater treatment and harvesting yeast (*Saccharomyces cerevisiae*) biomass from the fermentation broth as cheap and simple method already [102]. The reason behind this is its economy and easy availability and potential accumulation of heavy metals irrespective of external conditions [89].

Saccharomyces cerevisiae cells are able to flocculate in Zn^{2+}, Cu^{2+}, Cd^{2+}, Ni^{2+}, or Cr^{3+} containing ionic solutions [53, 96]. As per studies carried out, the advantages of yeast flocculation in a bioreactor are threefold: (1) ability to function in high-throughput bioreactors that lead to high yield in a short processing time [24], (2) retention of biomass in a variety of suspended biomass reactor configurations [25], and (3) high metabolism that ensures lower risk of external contamination [51]. Another important advantage for using yeast for assisted bio-flocculation is well-understood molecular and genetic basis of yeast flocculation along with the understanding of the intricate control of its flocculation mechanism [15, 71, 111].

18.5 Molecular Mechanism of Yeast Bio-flocculation and Role of Flocculation Enhancing Proteins (*Flo*)

The hallmark of yeast flocculation is formation of aggregates through interaction of flocculins (FLO, lectin-like proteins) with their receptors on a neighboring cell wall [16, 99]. The classical lectin theory of yeast flocculation is schematically represented in Fig. 18.4. *Flo* proteins bind to Mannans (Carbohydrate residues) on the surface of neighboring cells leading to the cross binding of cells. Cross binding leads to flock formatuion. The resultant flocks thus produced would sediment much faster relative to the free cells due to reduced surface to volume ratio of the aggregated cells. CDC.dk).

The regulation of flocculation in yeast through Flocculation enhancing proteins is a complex process. And several FLO proteins that are involed in the complex orchestra has been identified. Among these, some FLO proteins are transcriptional regulators such as FLO8 and the rest are cell wall surface glycoproteins. Cell wall charge and hydrophobicity also play an important role in Flocculation. It is beacuse expression of several proteins (Flo1, Flo5, Flo9, Flo10, and Flo11p) in yeast cell wall is correlated to increased surface hydrophobicity of yeast [32, 36, 63]. FLO1, FLO5, FLO9, and FLO10 of the FLO gene family are in close proximity in the genome [60]. Although different FLO genes display different degree of flocculation and sugar sensitiveness, overexpression of these four genes induces flocculation [36, 63]. Several environmental factors such as cations, pH, temperature, and aeration also impose regulation of flocculation via FLO genes [27, 45, 90, 100]. For a review on the yeast flocculation, the review can be read [27].

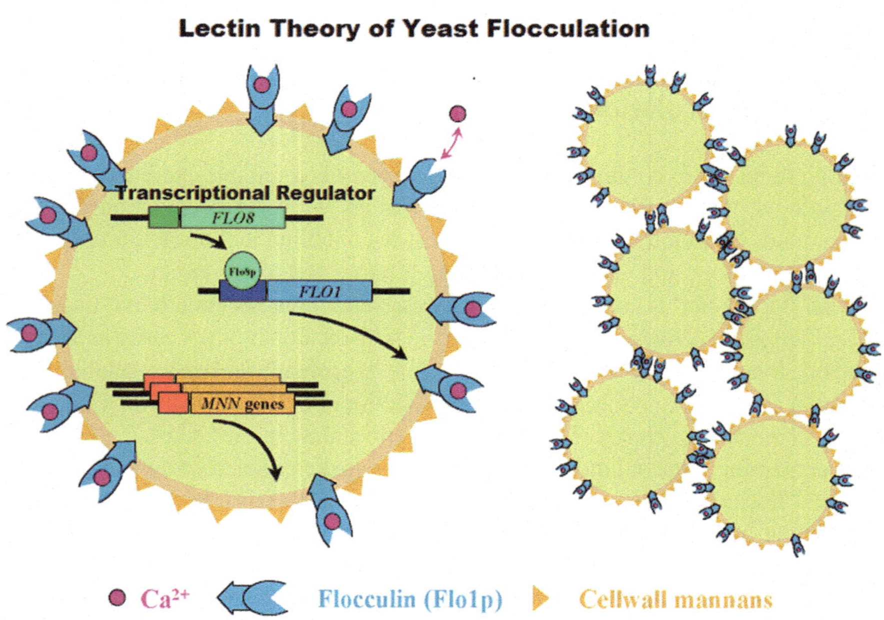

Fig. 18.4 Lectin theory of yeast flocculation. (After CDC.dk)

Yeast can be a source of motivation for transgenic intervention for bio-flocculation since the mechanism and regulation of yeast flocculation is well understood.

18.6 Rationale Behind Choosing FLO1, FLO5, and FLO9 genes of Yeast for Evaluation in Assisted Flocculation for Wastewater Treatment by Microalgal Systems

A heuristic literature review indicated that the FLO1, FLO5, and FLO9 of yeast are best suitable candidates for molecular biology and genetic engineering-based intervention for genetically modified flocculation-enhanced phenotype in suitable microalgae. Initially to assess the molecular aspect of the *Flo* proteins of yeast, *Flo1, Flo5* and *Flo9* of yeast were subjected to extensive phylogenetic analysis that is discussed in detail later in the chapter. From the comparative analysis from generated dendrogram (Fig. 18.5) it was clear that the proteins share extensive homology.

Among the several *Flo* proteins in yeast, the *Flo1, Flo5* and *Flo9* are responsible for cell-cell adhesion that subsequently leads to flocculation, whereas *Flo11* is responsible for substrate adhesion [26]. According to studies conducted during the present decade, it was understood that methylation (addition of CH_3 Group) of a regulatory protein Set1 (COMPASS) could be a critical point. It is because, a mutation in COMPASS protein that renders the protein defective leads to induction of a flocculent behavior because of the increasing amounts of FLO1, FLO5, and FLO9 mRNA transcripts [27] of the corresponding genes.

Other than that, in terms of the role of *Flo1*, it is observed that widely exposed flocculins are present on *Flo1*-expressing cell surface, which is the site of adhesion [31]. Figure 18.6 depicts the role of *Flo1* in flocculation. It is also seen that multiple weak lectins interact together with strong unfolding forces (both of which are associated with *Flo1* molecules) lead to cell-cell adhesion bonds. Microscale cell adhesion behavior correlates with single-molecule and single-cell data that suggests *Flo1* mechanics to be critical for yeast flocculation [31].

On the other hand, the importance of FLO5 gene in the development of high flocculent characteristics of yeasts is also well understood [97]. Therefore it was clear that among the perspective candidates for molecular biology and genetic engineering, intervention of *Flo* proteins, *Flo1* and *Flo5* would be the target proteins for further study. The computational biology analysis later discussed in the chapter also indicate that *Flo1* and *Flo5* proteins are indeed the target proteins for generating genetically engineered enhanced flocculation phenotype in microalgal host for wastewater treatment.

Flo proteins may be excellent target for transgene expression for enhanced flocculation phenotype in microalgae for wastewater treatment.

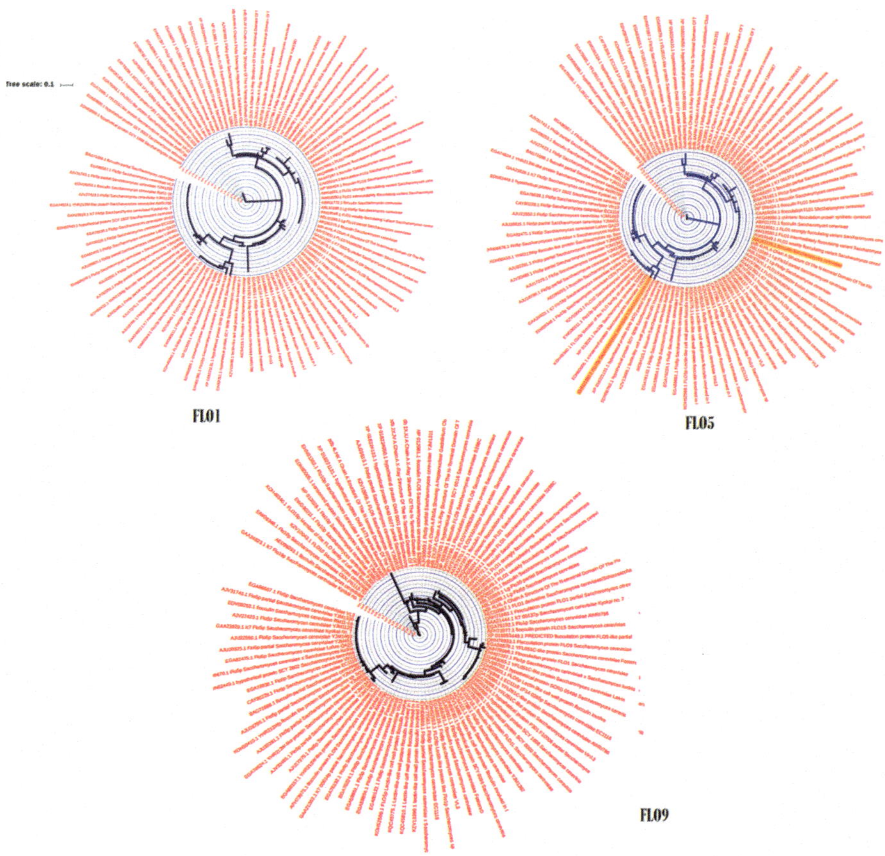

Fig. 18.5 Homology of yeast *Flo1, Flo5* and *Flo9* phylogeny

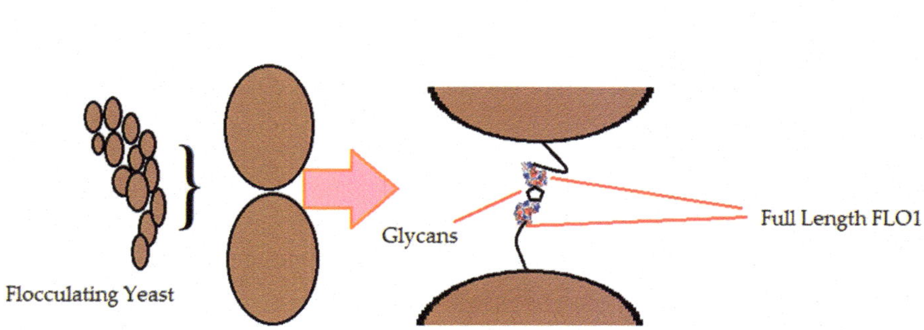

Fig. 18.6 Role of *Flo1* in cell-cell adhesion

18.7 Analysis of *Flo* Proteins Using Computational Biology

Computational simulations in the microprocessor environment become a method of choice due to near precise prediction capabilities of a bimolecular phenomenon and its economic operational costs. To examine the behavior of *Flo* proteins, a systematic bioinformatics analysis was done. The method described here is easy to perform and requires minimal investments in terms of infrastructure but sensitive enough to differentiate intricate differences among protein candidates that aids to prioritizing the suitable protein for molecular biology-genetic engineering experiments. Thus this process helps to minimize cost of the overall trial and error expenditure. In the following sections, the materials and method used in the computational biology analysis and the results are also described.

18.7.1 Sequence and Structure Retrieval of *Flo* Proteins

UniProt, a global database, was queried for *Flo* proteins for the genus *Saccharomyces*. Only the reviewed entries were selected for the test run. The reviewed entries for *Flo1, Flo5* and *Flo9* were searched for their 3D coordinates. Among the entries of available proteins in a database, identification of the suitable structure often becomes a tricky job since several aspects of the structure parameters come in play. A suitable structure file should comply with the following: (1) a sufficient length of the primary structure indicated by the residue count should be present. If a small portion/segment/domain of the desired protein is available, it may not represent the molecular behavior of the full-length peptide; (2) if a mutant of the original protein is available, the wild-type characteristics may be absent; so, the mutants may be excluded from study, (3) if experimentally induced ligands are present in the structure in large number, it may hinder desired study with the protein structure; (4) resolution of the protein is an important factor and retrieval of high-resolution protein structures are advised; and finally (5) the method through which the structure was generated is another important parameter. It is known that structures generated through X-ray crystallography are more suitable then NMR-generated coordinates. Therefore, prior to selection of a 3D structure co-ordinate of a protein consideration of theses 5 factors would lead to suitable retrival of data sources.

18.7.2 3D Coordinates of the *Flo* Proteins and their Stereochemical Quality

The *Flo1* in RCSB-PDB returned different entries of the protein for the genus *Saccharomyces*. The entry 4LHL.pdb was selected for the study. For the *Flo5* protein among the numerous available entries, 2XJP.pdb was selected for further studies on the basis of the above mentioned criteria for selection. However, for the *Flo9* protein, there was no entry for its 3D coordinates (Table 18.1).

Table 18.1 Structural features of the selected 3D coordinates

Protein	PDB entry	Details	Method of structure deduction	Resolution	Residue count
Flo1	4LHL. Pdb	Structure of the N-terminal domain of the Flo1 adhesin (N-Flo1p) from the yeast *Saccharomyces* Cerevisiae	X-ray diffraction	1.43 Å	232
Flo5	2XJP. Pdb	Structure of the N-terminal domain of the Flo5 from the yeast *Saccharomyces cerevisiae* in complex with calcium and mannose	X-ray diffraction	0.95 Å	258
Flo9	NA	NA	NA	NA	NA

In the absence of 3D coordinates of a protein, it is customary to design the protein using protein modeling algorithms. Among all known protein sequences, only a fraction have their structures determined experimentally by X-ray or NMR. Due to increased reliability of alignment algorithms and modeling programs, there is a way to predict 3D coordinates of proteins for which we only have knowledge of the primary sequence. Therefore protein sequences sharing certain homology with that of experimentally determined proteins can be modeled with high accuracy [18, 20].

Flo9 of *Saccharomyces* was modeled and validated following the protocol of [9] based on the primary structure from the reviewed UniProt entry. In this method, the algorithm uses multiple templates and analyzes them for suitability in terms of QMEAN Z-Score. The score is directly proportional to the model quality. The method returns results with respect to the modeled residue range, template used and its resolution, % of sequence identity, and E-value. The complete program uses layers of multiple algorithms with some benchmark, such as conservative BLAST search with restrictive parameters of *E*-value cutoff: 10^{-5} and 60% minimum sequence identity. The process also includes library search of hidden Markov models for SMTL using HHSearch.

The overall stereochemical quality of the protein models were assessed through the method of [13]. At both global and local levels, the model quality is evaluated. The functions of optimized hydrogen placement, all-atom contact analysis, and updated versions of covalent geometry and torsion-angle criteria are considered. A number of serious steric overlaps (>0.4 Å) per 1000 atoms are generated by all atoms clashscore. Analysis for the number and percentage of Ramachandran outliers and Ramachandran favored residues done by classical Ramachandran plot. It also gives the number and percentage of poor roamers, Cβdeviations >0.25 Å, bad backbone bonds, and angles [13] (Table 18.2).

Table 18.2 The stereochemical parameters of FLO proteins

		Flo1		*Flo5*		*Flo9*		
Parameters	Details	No.s	Percentage	No.s	Percentage	No.s	Percentage	Expected Range
All-Atom Contacts	Clashscore, all atoms:	2.49		16.39		2.19		99th percentile* (N=479, 1.43Å ± 0.25Å)
	Clashscore is the number of serious steric overlaps (> 0.4 Å) per 1000 atoms.							
Protein Geometry	Poor rotamers	2	0.96%	6	2.19%	2	0.95%	Goal: <0.3%
	Favored rotamers	201	96.63%	261	95.26%	203	96.67%	Goal: >98%
	Ramachandran outliers	0	0.00%	0	0.00%	0	0.00%	Goal: <0.05%
	Ramachandran favored	233	95.49%	244	95.31%	235	95.14%	Goal: >98%
	MolProbity score^	1.35		2.34		1.33		92nd percentile* (N=3404, 1.43Å ± 0.25Å)
	Cβ deviations >0.25Å	0	0.00%	4	1.34%	2	0.88%	Goal: 0
	Bad bonds:	3 / 1935	0.16%	1 / 2475	0.04%	0 / 1957	0.00%	Goal: 0%
	Bad angles:	0 / 2649	0.00%	17 / 3464	0.49%	7 / 2687	0.26%	Goal: <0.1%
Peptide Omegas	Cis Prolines:	0 / 17	0.00%	0 / 19	0.00%	0 / 19	0.00%	Expected: ≤1 per chain, or ≤5%
	Cis nonProlines:	2 / 229	0.87%	3 / 278	1.08%	2 / 229	0.87%	Goal: <0.05%

In the two column results, the left column gives the raw count, right column gives the percentage.

* 100th percentile is the best among structures of comparable resolution; 0th percentile is the worst. For clashscore the comparative set of structures was selected in 2004, for MolProbity score in 2006.

^ MolProbity score combines the clashscore, rotamer, and Ramachandran evaluations into a single score, normalized to be on the same scale as X-ray resolution.

Table 18.3 Domain analysis of *Flo* proteins

Protein	Domain	Start site (amino acid position)	End site (amino acid position)	E-value
Flo1	PA14	76	221	6.48e-25
Flo5	PA14	87	235	2.36e-27
Flo9	PA14	78	226	5.56e-24

PA14 domain indicates a domain for yeast adhesion

18.7.3 Phylogenetic and Evolutionary Analysis of *Flo* Proteins

The pattern of evolutionary changes that ultimately shape proteins characteristics are also footprints of the evolution that candidate the origin of the protein. To decipher the evolutionary strategy, the primary protein sequences were retrieved in FASTA format from the reviewed entries in UniProt. The FASTA sequences were subjected to protein BLAST in the NCBI tools [41] platform. For protein searches, the options are *blastp* (default), *PSI-BLAST*, and *PHI-BLAST*, but the default BLAST search is good enough to retrieve 100 sequences with sufficient match along with a matching score. The BLAST sequences were downloaded and subjected to phylogenetic tree algorithm Phylogeny.fr. The Newick format tree was then loaded into a visualization program itol.embl.de/. The phylogenetic trees indicate considerable homology in the evolutionary makeup of the *Flo* protein.

18.7.4 Analysis of Active Site Motifs and Energetics of *Flo* Proteins

The active site design of a functional protein determines its specificity and its uniqueness in a reaction. The *Flo* proteins interact with various carbohydrate residues as well as various ions in the environment for successful flocculation behavior. Therefore it is imperative that the active site architecture is well understood. The 3D coordinates of the proteins were subjected to multiple algorithms for analysis of the behavior of the active site and the energetics of the proteins.

The primary structural data was subjected to simple modular architecture research tool of embl-heidelberg.de for domain analysis. The algorithm is designed in such a way that it not only identifies conserved domains or motifs from a peptide primary structure but also indicate their outlier homologues and homologues of known structure including PFAM domains, signal peptides etc, including internal repeated elements. The tool uses the databases such as Swiss-Prot, SP-TrEMBL and stable Ensembl proteomes. In this method E-values are calculated using hidden Markov models that leads to more accurate estimates (Table 18.3).

The secondary structural elements are functional aspects that determine how a protein behaves in the proximity of its substrate. The residues that built up the active site motif also give specific detail of the interactions that bound the substrate with the active site. The protein structural data as well as the validated homology model is subjected to Pymol [21] window to visualize these in their secondary structure (helix-sheet-loop). The substrates/solvents/ions were selected and proximal residues within a molecular distance of 4 Å were identified. Since the volume and the area of the catalytic cleft are important factor because they accommodate a substrate or a substrate inhibitor [19]. The proximal residues in this site and their nature generally determine the overall behavior of the active site. Among the proximal residues, detail search was made to identify continuous stretches of amino acid residues that may form a domain with in this site.

In terms of Bioenergetics the vacuum electrostatics or protein surface contact potentials were assessed for possible indication of protein-protein contact or protein-ligand contact information. Electrostatic potential of protein was mapped to the molecular surface of *Flo* proteins by vacuum electrostatics. The overall energetics of the proteins and hydrophobic surface patches were deduced by SPDBv tools [34]. Multiple file loading and handling simultaneously is one of the advantages of this tool. Each of the loaded file appears in a separate layer, and each layer holds individual chains that are composed of groups of either amino acids, nucleotides, or heterogeneous groups (e.g. NAD, HEME). Among the loaded coordinates, the first set will be considered the reference structure. For H-bond detection, SPDBv assigns Kollman's atom types.

18.7.5 Molecular Dynamics Simulation of *Flo* Proteins

To predict the energy of the molecule as a function of its conformation molecular mechanics or force-field methods uses classical type models [47]. Mechanical properties such as equilibrium geometries, transition states, and relative energies between conformers or between different molecules can be predicted through simulations. For force-field analysis of selected proteins energy scores in terms of bond energy, non-bond energy, angles energy, torsion energy, improper energy, electrostatic energy and the total energy of the protein models in Kilo Jules/mole were estimated (GROMOS96). In this analysis the total energy is expressed as a sum of Taylor series expansions for every pair of bonded atoms. Additional potential energy terms contributed by bending, torsional energy, van der Walls energy, and electrostatics [85] are added. These energy parameters are compared for the three studied proteins that will indicate their relative stability.

Along with force-field calculation, normal mode analysis is another parameter that is used in this study to emphasize on protein structure stability. Normal mode analysis compares collective motions of a group of atoms with that of the minimum energy conformation of the same group. The minimum energy conformation is based on the harmonic approximation of the potential energy function [67]. In this method thermodynamic and mechanical properties of catalytic proteins that accommodate a substrate in its interior would be such that the channel gates of the catalytic cleft will vibrate in a normal mode [92]. For studying collective motions in macromolecules, normal mode calculations can provide an alternative to molecular dynamic simulations [94]. The collective motions of the three *Flo* proteins were recorded and compared with each other.

A constitutive approach for evaluation of structure and function relationship in proteins is high-end computational methods such as molecular dynamics (MD) simulation. Protein stability, flexibility, conformational behavior, etc. are different properties of a macromolecule that can be elucidated by the parameters used in molecular dynamic simulation studies.

The two approaches of molecular dynamics simulation are classical (i.e., Newtonian) mechanics and Brownian dynamics. Since macromolecules such as proteins are more flexible and dynamic in nature, their behavior can be analyzed by molecular dynamics simulation. External factors such as temperature, pH, charge, ion concentration, phosphorylation, or binding of a ligand can lead to conformational changes in a protein [67] so influence of these factors can also be examined via molecular dynamics simulation.

For doing so, analysis of root mean square deviation (RMSD) is an effective measure. Root mean square deviation (RMSD) is the deviation observed between two heavy atoms to predict the stability of a protein. RMSD of the protein explains the protein folding nature, where the RMSD of "*N*" set of atoms at time "*t*" is calculated using

$$\mathrm{RMSD}(t) = \sqrt{\frac{\pounds_{i=0}^{N} \left| \vec{r}_i(t) - \vec{r}_i^{\,0} \right|^2}{N}}.$$

The Brownian analysis was performed in terms of analysis of B factor per residue and the RMSD per residue analysis. The identified protein domains were plotted on the RMSD plot to observe the behavior of active site residues in a Brownian simulation.

During the simulation the program rectifies the alternate location for several parameters. These includes the amide assignment, improper chirality if any, residue insertion in gaps if any, unusual Cis-Trans configurations, steric clashes, polar donor clashes, and ionic positive/negative clashes etc. The protocol also assigns or rectifies ligands in the structures. After these rectifications in the structure file the force field is applied. For the present analysis coarse-grained simulation of Brownian molecular dynamics (C-alpha) was opted mainly because of its ease of operation and informative results. Simulation parameters were set as follows : time - 100 PS; output frequency steps - 10; force constant (kcal/mole* Å^2) - 40 and distance between alpha carbon atoms - 3.8 Å.

According to the observation on B-factor assessment of the proteins it was found that, the *Flo1* has a range of B-factor/residue of 0.0–16.0 units in Å^2. For *Flo5* this range (B-factor/residue) is 0.0–25.0 units in Å^2 and for *Flo9* this range is 0.0–14.0 units in Å^2. From the analysis of RMSD/residue, it was observed that for *Flo1* the range of RMSD/residue is 0.1–0.8 units in Å, for *Flo5* the range of RMSD/residue is 0.1–1.0 Å, and for *Flo9* the range of RMSD/residue is 0.1–0.7 units in Å. Another dynamic parameter, i.e., radius of gyration of proteins has also been calculated during the simulation run for the studied *Flo* proteins. Radius of gyration of a protein indicates the overall dimension of a protein structure in space as well as compactness of the structure.

18.7.6 Protein Network Analyses

Possible interaction of *Flo* proteins with other proteins and pathways STRING database version 9 [94] was explored. The protein networks are used in bioinformatics of proteins for fine-tuning of interaction of protein functions. The major utilizations have been to increase the statistical power in genetics [72] and in study of gene interaction. This also aids to close gaps in metabolic enzyme knowledge [39]. The functionality that is harnessed in this study mainly deals with prediction of phenotypes and gene functions [103]. The generated interactions include physical and functional associations derived from various knowledge bases. These includes genomic context, high-throughput experiments, co-expression and prior knowledge [3]. After studying the interactions of *Flo* proteins individually, a common network comprising *Flo1, Flo5 & Flo9* was designed. This would enable to identify similar interaction behavior for prioritizing the candidate proteins.

18.8 Insight on FLO Proteins from the Bioinformatics Analysis

For a successful experiment, the superior quality of study sample is an important prerequisite to ensure anticipated results. The *Flo* protein structures that are used in the study are qualitatively noble (Table 18.1 and Figs. 18.7). The retrieved coordinates of *Flo1* and *Flo5* are of high-resolution X-ray analysis and therefore of good quality. Neither the retrieved sequences of *Flo1* and *Flo5* nor the generated model of *Flo9* have Ramachandran outliers according to our stringent quality check, and 95%< of residues in all the proteins are Ramachandran favored. The active site domain analysis reveals significant similarities of *Flo1* and *Flo9* (Table 18.2) and high homology of PA14 domain among the two. Therefore it is of no advantage using Flo1 and Flo9 together in the genetic transformation. Vacuum electrostatic analysis revealed *Flo1* potential to be ±70.009, *Flo5* potential ±69.749 and *Flo9* potential to be ±60.007 (Figs. 18.7, 18.8, and 18.9). For evaluation of the electrostatic properties of biomolecules, elucidation of the electrostatic potential is a standard practice in molecular biophysics which is done through the Poisson-Boltzmann equation [17]. Vacuum electrostatics indicate protein contact potential and so an indicator of molecular interactions that influence various aspects of most biochemical reactions [18, 20]. In Figs. 18.7, 18.8, and 18.9 the red colored area shows negatively charged and blue shows positively charged regions in the proteins. Potential surface maps between the *Flo1* and *Flo5* exhibit qualitative electrostatic similarity. The analysis of active site residues that interact the ions and the carbohydrates indicates that various residues interact with the ligands (Figs. 18.7, 18.8, and 18.9). Hydrophobic topology is proposed to be involved in various structural features of a protein, such as ligand binding on a surface [11], stabilization of macromolecular structure and shielding effect from solvent [5] and protein ensemble of orientation [22], etc. Therefore, the analysis of hydrophobic patches on the *Flo* proteins were done in silico (Fig. 18.10).

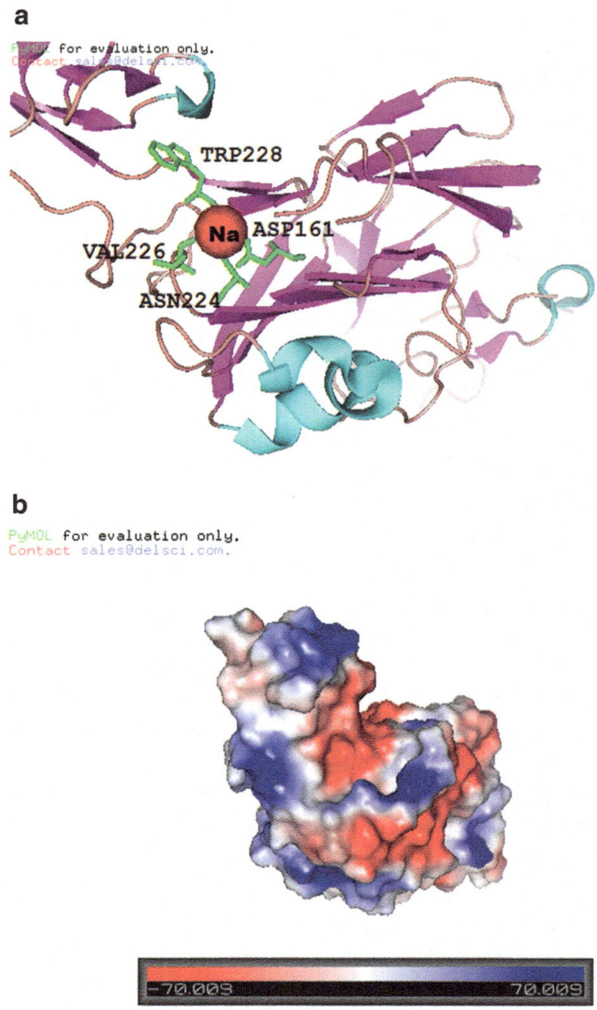

Fig. 18.7 (**a**) Ion-binding active site with vital amino acid residues of *Flo1*. (**b**) Electrostatic potential mapped to the molecular surface of *Flo1* by vacuum electrostatics

The bioenergetics analysis of *Flo* proteins (Table 18.4) indicates that the *Flo9* is extraordinarily stable, and it is markedly different from *Flo1* and *Flo5*. Similarly, the energy profiles of *Flo1* and *Flo5* are comparable.

Coarse-grained normal mode analysis (NMA) can predict experimentally observed functional motions of proteins [2]. In recent years NMA has reemerged as a powerful method for elucidating the structure-encoded dynamics of biomolecules. Information regarding the dynamics of molecular structures is required to establish the link between their structures and function because these entities are dynamic rather than static. To identify mobile regions and in some proteins to reproduce the direction of conformational change, normal modes can be used [23]. In this study highly mobile loops of *Flo1* and *Flo9* normal modes are seen, while *Flo5* structure is comparatively stable (Figs. 18.11, 18.12, 18.13).

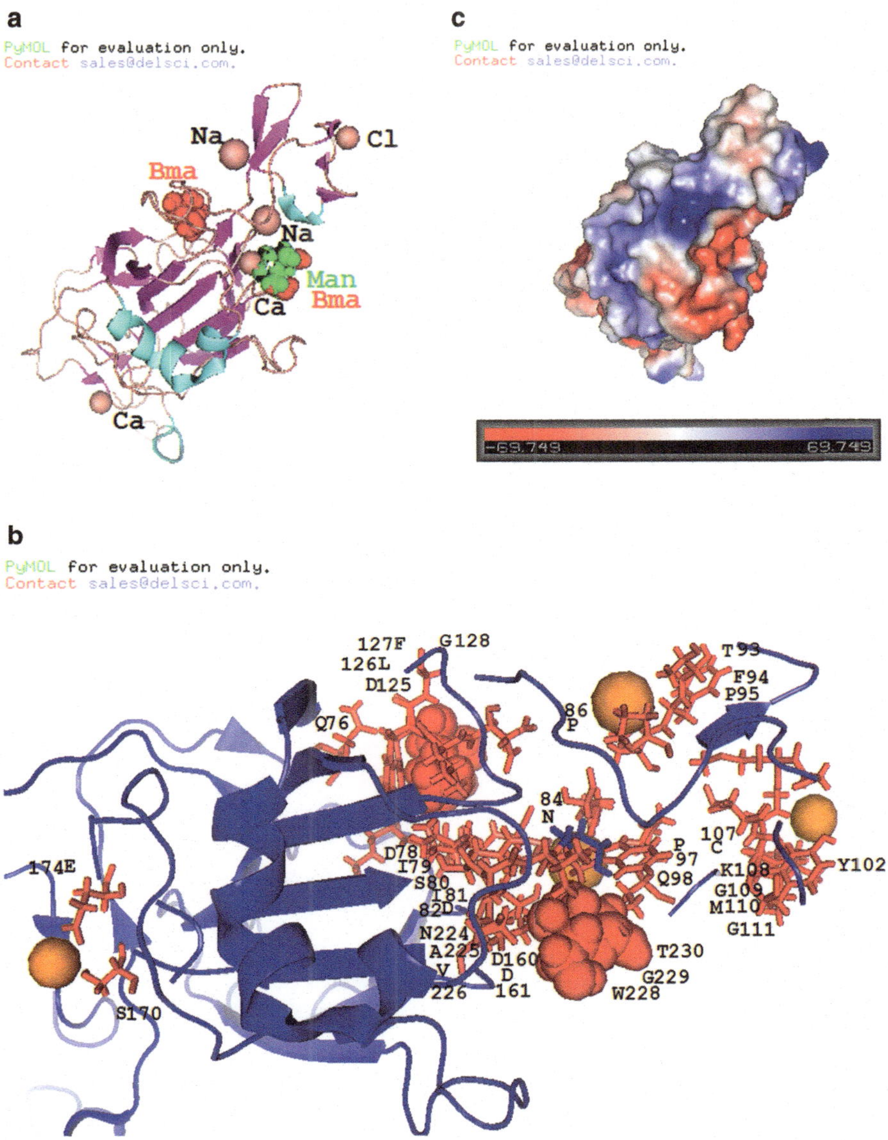

Fig. 18.8 (**a**) Ion and sugar binding sites of *Flo5*. (**b**) Ion and sugar binding active site with vital amino acid residues of *Flo5*. (**c**). Electrostatic potential mapped to the molecular surface of *Flo5* by vacuum electrostatics

B factors may be indices of dynamic and static disorder and displacements of the atoms from their mean positions as well. Temperature-dependent fluctuations in the atomic positions lead to dynamic disorder that has been observed to be the major contributor to the B factor. Therefore, B factor may accurately predict the backbone and side-chain dynamics [28]. From the B factor analysis of *Flo* protein residues

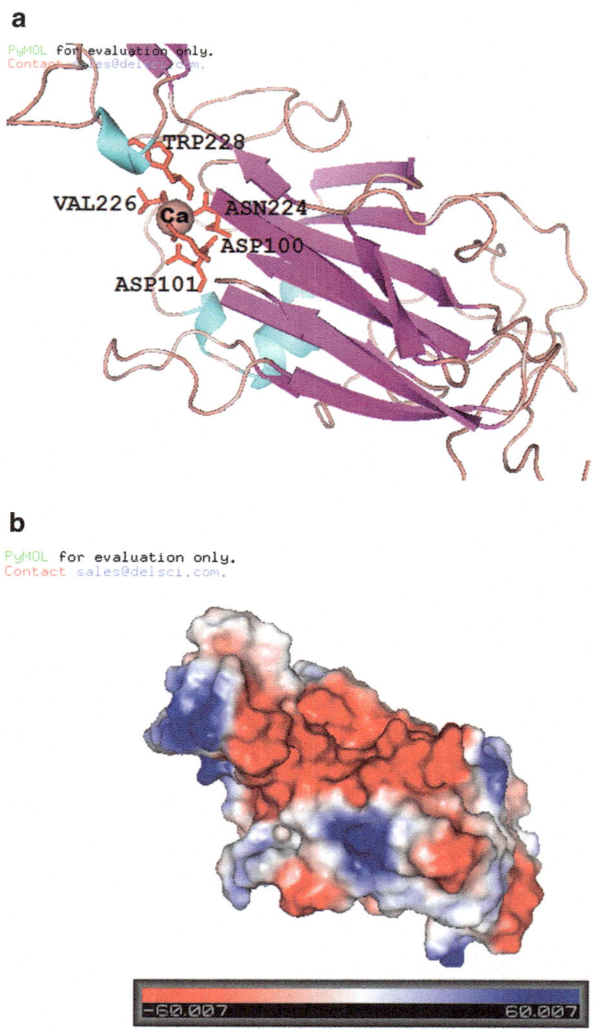

Fig. 18.9 (**a**) Ion-binding active site with vital amino acid residues of *Flo9*. (**b**) Electrostatic potential mapped to the molecular surface of *Flo9* by vacuum electrostatics

(Fig. 18.14), it is apparent that *Flo1* and *Flo9* share considerable similarity which is also reflected in the pattern of radius of gyration in proteins (Fig. 18.15). However, when the ligand-binding residues and domains are plotted on the RMSD/residue patterns, it reveals the uniqueness of *Flo5* interactions (Fig. 18.16). From the protein network analysis, several findings are worth mentioning in the context of *Flo* proteins. The vitality of *Flo1* is established from the network (Figs. 18.17 and 18.18). Seven interactions with factors associated with flocculation behavior are observed in *Flo1*, and in *Flo5* there were five interactions, whereas in *Flo9* only

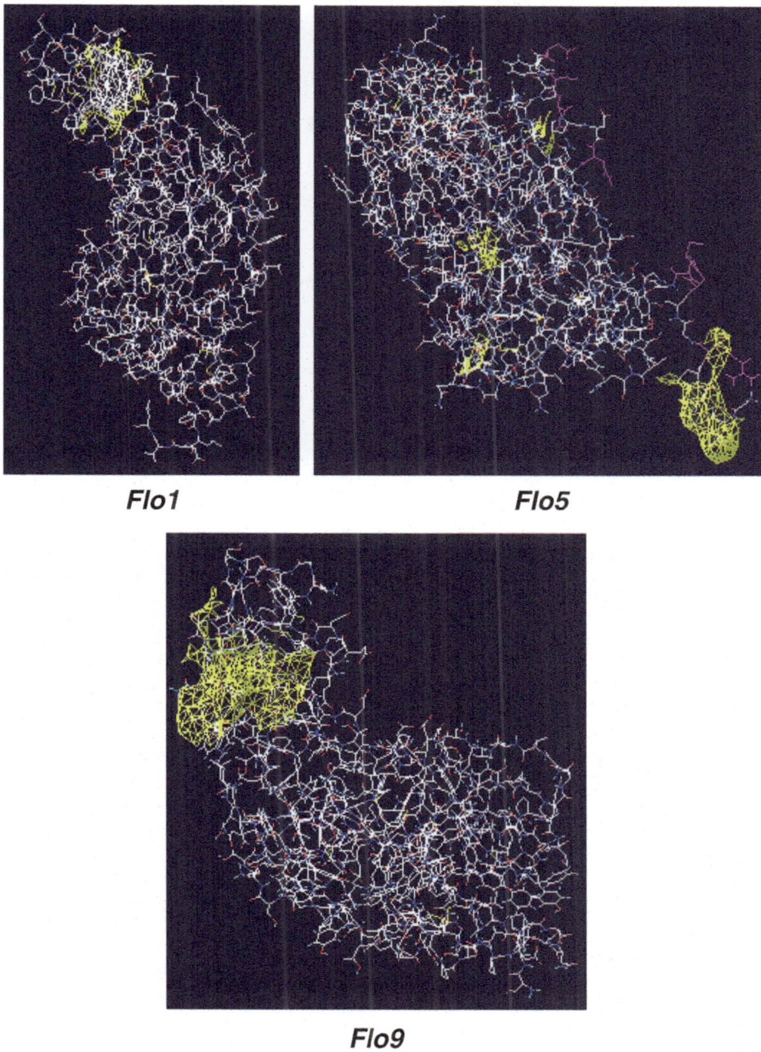

Fig. 18.10 Hydrophobic patches in *Flo* proteins

Table 18.4 Energetics comparison of FLO proteins

Proteins	Energy (KJ/mole)						
	Bonds	Angles	Torsion	Improper	Non-bonded	Electrostatic constraint	Total
Flo1	1932.878	1821.126	1336.378	408.048	−7685.32	−4720.17	−6907.063
Flo5	1039.343	1773.216	1359.668	318.787	−6420.53	−4615.82	−6545.334
Flo9	869.335	2074.848	1475.964	192.754	−7939.65	−5473.03	−8799.784

Energy computations are done in vacuum devoid of a reaction field implementing GROMOS96

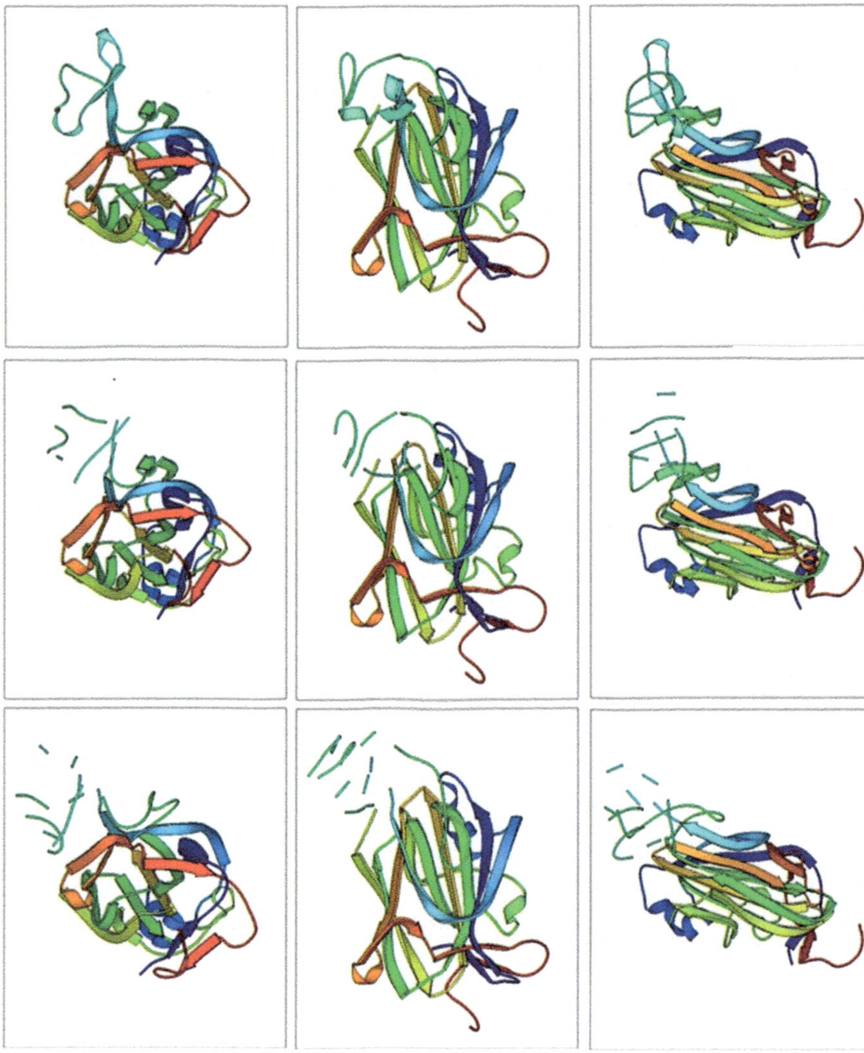

Fig. 18.11 Normal mode analysis of *Flo1* showing the vulnerable segments of the protein

three interactions were observed (Table 18.5). Therefore the analysis indicates that a transgene construct with *Flo1* and *Flo5* may promote flocculation in a transformed algae.

> *Transgenesis of Flo1 and Flo5 along with regulatory Flo8 in microalgae host would be the next big thing in wastewater treatment by microalgae.*

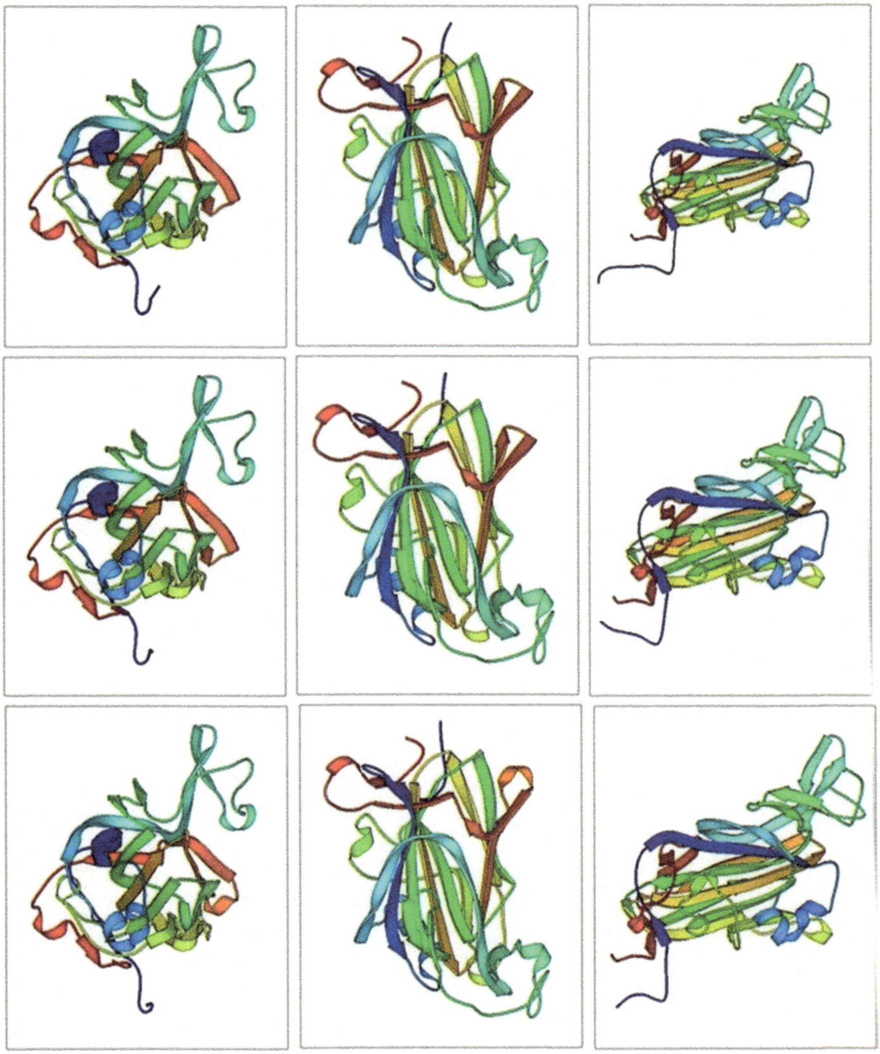

Fig. 18.12 Normal mode analysis of *Flo5* showing the vulnerable segments of the protein

18.9 Recombinant Gene Expressions for Enhanced Flocculation and Harvesting of Microalgae for Wastewater Treatment

18.9.1 The Basis for Transgenic Flocculation in Microalgae

Comprehension of the biochemical and genetic basis of cell flocculation becomes prerequisite to engineer the flocculation phenotype [101]. Therefore a thorough study of the published reports on the molecular basis of flocculation was done. This

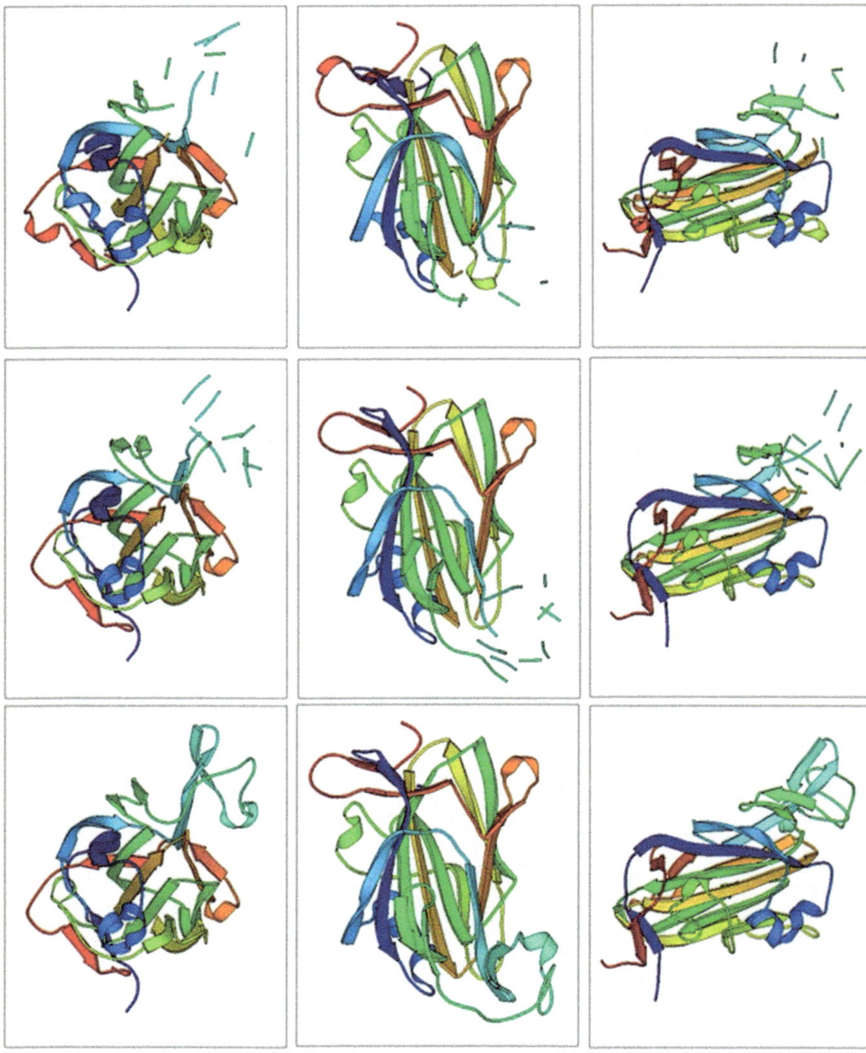

Fig. 18.13 Normal mode analysis of *Flo9* showing the vulnerable segments of the protein

was followed by subsequent bioinformatics analysis on the behavior of flocculation enhancing proteins. Nonklang et al. [68] is referred here among other similar works on the construction of flocculent *K. marxianus* by introduction of FLO genes from *Saccharomyces cerevisiae.* The strategy followed was overexpression of flocculation enhancing proteins via a transgene from a yeast species to another yeast species. Overexpression of a protein ensures increased molecular concentration of a gene product or ensures high-level production of the protein and therefore a noticeable change in the phenotype. In the experiment overexpression of *S. cerevisiae*, FLO genes were achieved by upstream insertion of constitutive TDH3 promoter,

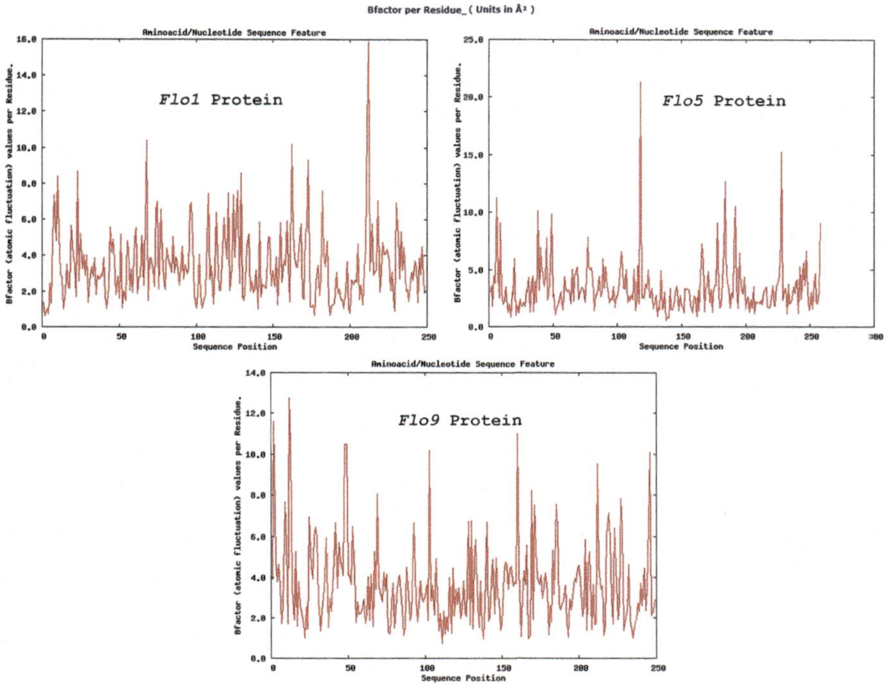

Fig. 18.14 B factor per residue in *Flo* proteins

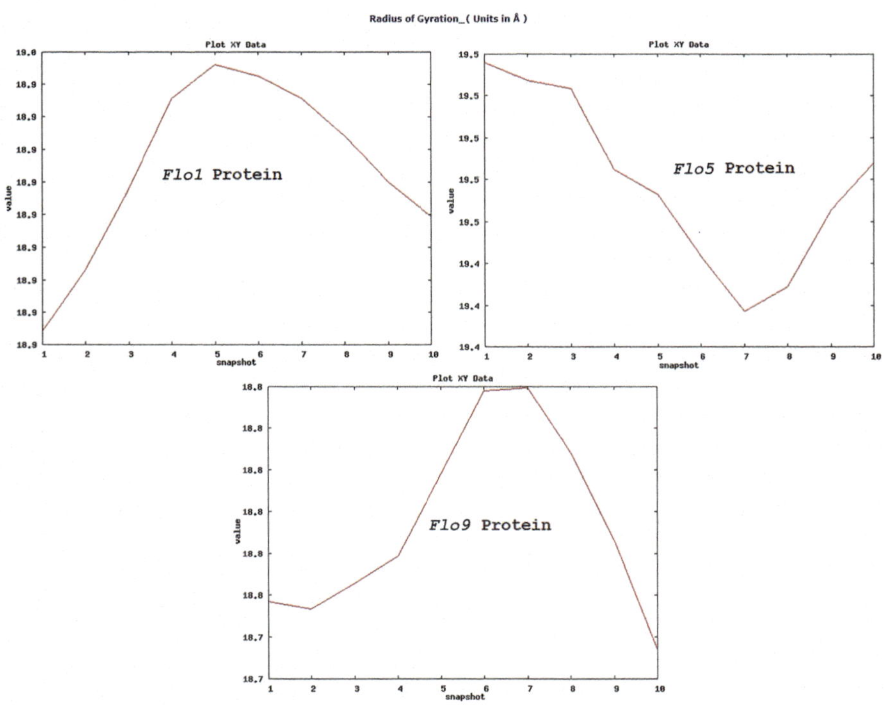

Fig. 18.15 Radius of gyration in *Flo* proteins

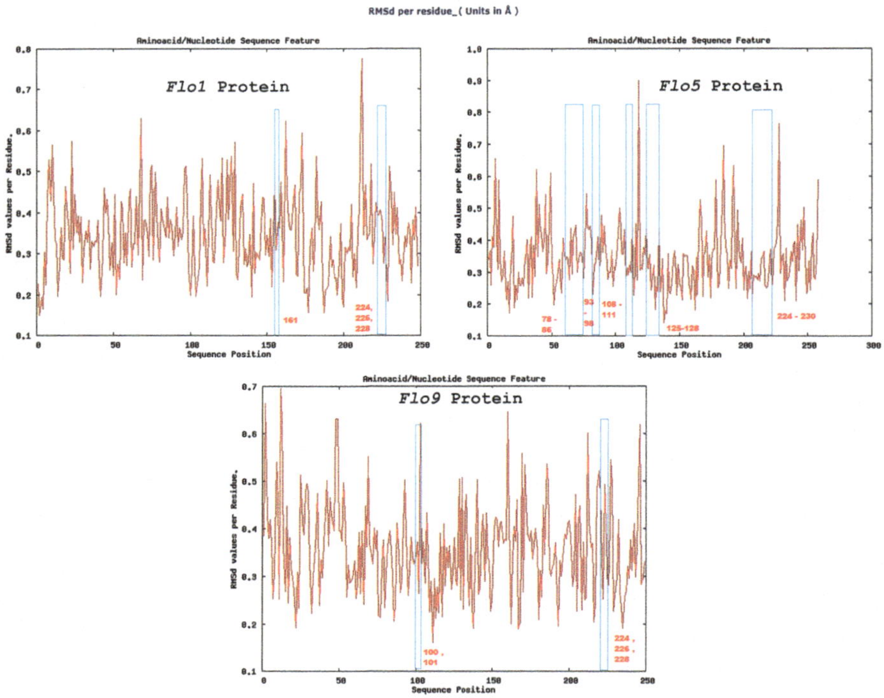

Fig. 18.16 RMSD/ residue and the effect of substrate binding in *Flo* proteins (bars indicate the interacting residues)

resulting in flocculent S. *cerevisiae* strains. These TDH3-FLO sequences were amplified by PCR and introduced directly into a *K. marxianus* strain which showed a flocculation phenotype. In another relevant experiment, controlled flocculation using ethanol-induced promoter in yeast strains was achieved [51].

On the basis of these molecular biology-genetic engineering experiments on transgenic flocculation phenotype, it is expected that the bottleneck of flocculation in microalgal systems can be effectively overcome. Transgenic expression of FLO 1 and FLO5 in a suitable algal host will revolutionize algal biomass harvesting in near future.

18.9.2 The Suitable Microalgae Host for Yeast FLO1 and FLO5 Transgene Expression

During last decades, mainly due to metabolic diversity, safety, sustainability, and scalability, biomanufacturing of recombinant proteins via microalga gained high advancement [95]. This includes cost-effective and easily scalable platform for production of medical and industrial recombinant proteins [79] .

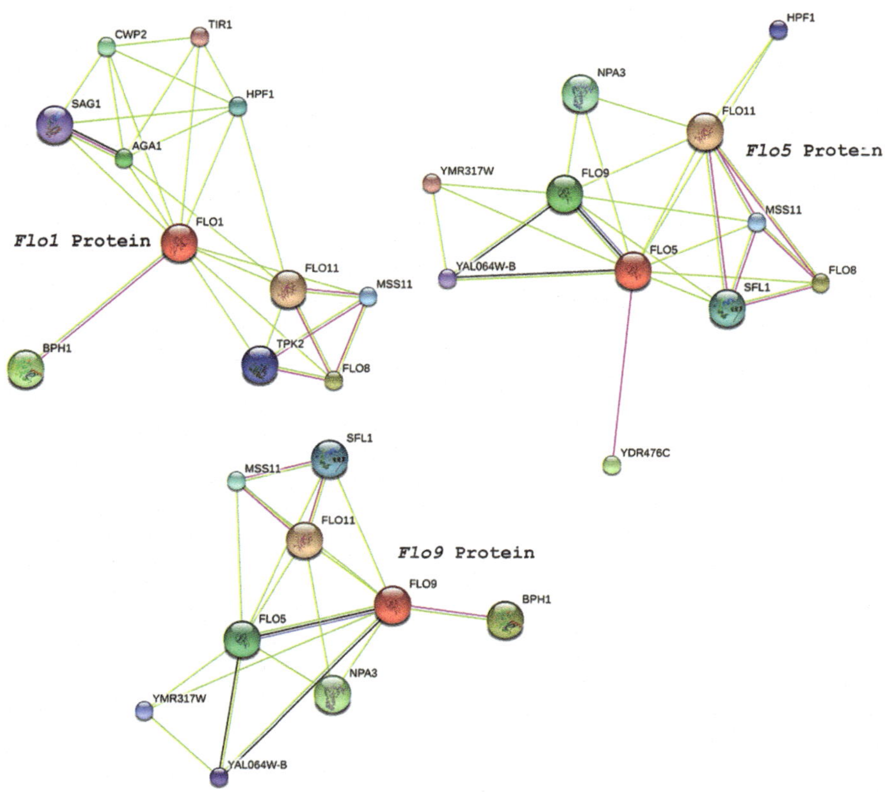

Fig. 18.17 Individual protein network of *Flo*

Among the most promising microalgal hosts, *Chlamydomonas reinhardtii* can serve as a potential expression platform for recombinant protein synthesis in vitro for applications including industrial biomaterials, bioenergy, therapeutics, and nutraceuticals [1].

Scarcity of existing infrastructure for commercial production and due to low product yield, commercial production of recombinant proteins from *C. reinhardtii* is hindered. Most of the recombinant proteins from *C. reinhardtii* are still in R&D phase [80].

The factors that made *Chlamydomonas reinhardtii* a model organism are its well-understood genetics, easy culturability, and simple life cycle. These factors are also responsible for most of the progress in the field of microalgae research through *C. reinhardtii*. Therefore, *C. reinhardtii* emerges as the most suitable host for transgene expression of FLO1 and FLO5 for enhanced flocculation phenotype for wastewater treatment.

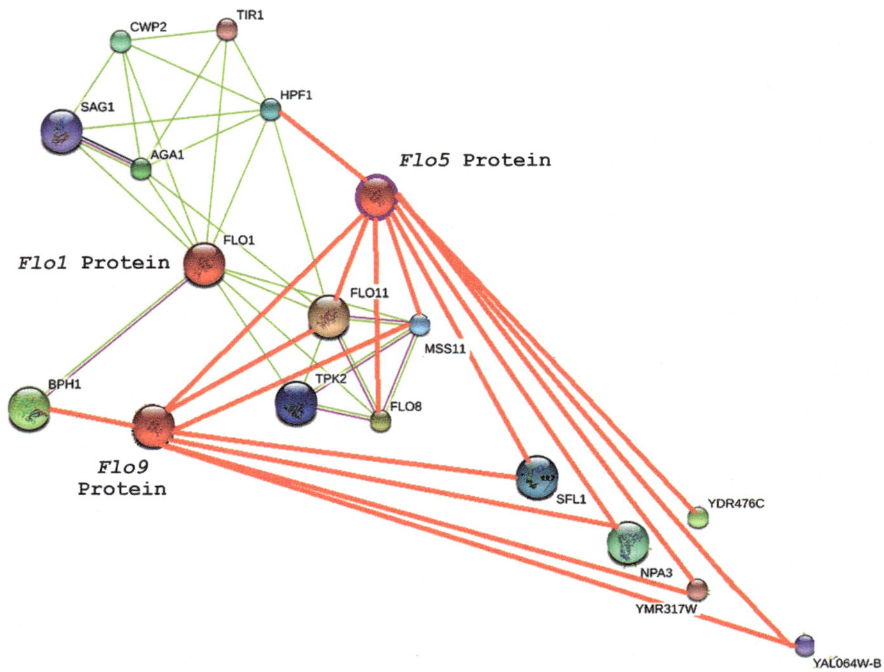

Fig. 18.18 Curated network of *Flo* proteins

18.9.3 Chloroplast or Nuclear Transformation of *C. reinhardtii*? Is the Question for Stable FLO1 and FLO5 Transgene Expression

Unlike the common bacterial transformation by a foreign transgene through bacterial nucleoid, recombinant proteins in *C. reinhardtii* can be expressed from either chloroplast [86] or nucleus [77]. Therefore from the most recent experiments on *C. reinhardtii* transgenesis, it is understood that the microalga to be a versatile host, but at the same time, this leads to a whole new direction in the genetic engineering of *C. reinhardtii* posing the inevitable question, whether stable & high-level expression is achieved from chloroplast or the nucleus?

However, according to published reports, *C. reinhardtii* chloroplast seems to be the site of choice due to the following reasons:

1. High copy number effect: Multiple chloroplasts in *C. reinhardtii*, which is a typical photosynthetic cell, ensure high transgene copy number, and so there would be no gene silencing, and multiple genes can be expressed in operons [40].
2. Chloroplast-specific molecular mechanism:
   (a) Due to little or no posttranslational modification in chloroplast-expressed proteins, they would contain the correct amino terminal sequence that would conserve the functional aspects of the expressed proteins [29].

Table 18.5 Networked proteins of *Flo*

Interacting proteins	Details of interaction	Score of interaction
Flo1 interactions		
FLO11[a]	GPI-anchored cell surface glycoprotein (flocculin); required for pseudohyphal formation, invasive growth, flocculation	0.892
FLO8[a]	Transcription factor required for flocculation, diploid filamentous growth, and haploid invasive growth; flocculation	0.821
BPH1	Protein homologous to human Chediak-Higashi syndrome and murine beige proteins, which are implicated in disease syndromes	0.654
AGA1[a]	Anchorage subunit of a-agglutinin of a-cells, highly O-glycosylated protein with N-terminal secretion signal and C-terminal signal	0.642
CWP2[a]	Covalently linked cell wall mannoprotein, major constituent of the cell wall; plays a role in stabilizing the cell wall	0.642
HPF1[a]	Haze-protective mannoprotein; reduces the particle size of aggregated proteins in white wines; involved in cell wall organization	0.627
MSS11[a]	Transcription factor involved in regulation of invasive growth and starch degradation; controls the activation of MUC1 and STA2	0.578
TPK2	cAMP-dependent protein kinase catalytic subunit; promotes vegetative growth in response to nutrients via the Ras-cAMP signal	0.540
SAG1	Alpha-agglutinin of alpha-cells, binds to Aga1p during agglutination, N-terminal half is homologous to the immunoglobulin superfamily	0.540
TIR1[a]	Cell wall mannoprotein of the Srp1p/Tip1p family of serine-alanine-rich proteins; expression is downregulated at acidic pH	0.540
Flo5 interactions		
FLO11[a]	GPI-anchored cell surface glycoprotein (flocculin); required for pseudohyphal formation, invasive growth, flocculation	0.862
FLO8[a]	Transcription factor required for flocculation, diploid filamentous growth, and haploid invasive growth; flocculation	0.800
YDR476C	Putative protein of unknown function; green fluorescent protein (GFP)-fusion protein localizes to the endoplasmic reticulum	0.704
FLO9[a]	Lectin-like protein with similarity to Flo1p, thought to be expressed and involved in flocculation	0.594
NPA3	Member of the conserved GPN-loop GTPase family; has a role in transport of RNA polymerase II to the nucleus	0.572
SFL1	Transcriptional repressor and activator; involved in repression of flocculation-related genes and activation of stress response	0.518
MSS11[a]	Transcription factor involved in regulation of invasive growth and starch degradation; controls the activation of MUC1 and STA2	0.509

(continued)

Table 18.5 (continued)

Interacting proteins	Details of interaction	Score of interaction
HPF1[a]	Haze-protective mannoprotein; reduces the particle size of aggregated proteins in white wines; involved in cell wall organization	0.449
YAL064W-B	Fungal-specific protein of unknown function	0.437
YMR317W	Putative protein of unknown function with some similarity to sialidase from *Trypanosoma*	0.418
Flo9 interactions		
FLO11[a]	GPI-anchored cell surface glycoprotein (flocculin); required for pseudohyphal formation, invasive growth, flocculation	0.791
BPH1	Protein homologous to human Chediak-Higashi syndrome and murine beige proteins, which are implicated in disease syndromes	0.673
NPA3	Member of the conserved GPN-loop GTPase family has a role in transport of RNA polymerase II to the nucleus	0.626
FLO5[a]	Lectin-like protein involved in flocculation	0.594
MSS11	Transcription factor involved in regulation of invasive growth and starch degradation; controls the activation of MUC1 and STA2	0.573
SFL1[a]	Transcriptional repressor and activator; involved in repression of flocculation-related genes and activation of stress responses	0.501
YAL064W-B	Fungal-specific protein of unknown function	0.441
YMR317W	Putative protein of unknown function with some similarity to sialidase from *Trypanosoma*	0.418

[a]Interactions associated with flocculation behavior

(b) Transgene expression in chloroplast has led to economically viable levels of protein accumulation in previous studies [55].
(c) Heterologous enzymes can be successfully targeted to the chloroplast [79].
(d) Lastly, eukaryotic transgene expression in *Chlamydomonas reinhardtii* chloroplasts ([55, 58, 81, 91, 106]) and chloroplast transformation in higher plants [38, 93] demonstrated exceptational integrity in terms of the desired traits of the protein expression.

Besides this, the protocol of chloroplast transformation seems to be technically non-exhaustive and is easily executable. High-yield secretion of recombinant proteins can be achieved through the microalga *Chlamydomonas reinhardtii* following protocol of Ramos et al. [78].

18.9.4 Transformation of *C. reinhardtii*

The protocol for transformation may be designed as per established reports on this method and can be summerized as follows: *Chlamydomonas reinhardtii* UVM4 strain may be transformed following the glass bead method [46]. This protocol uses 1 μg of a suitable vector (Expression vector with in which the transgene will be inserted) linearized with restriction endonuclease ScaI. Following transformation *Chlamydomonas reinhardtii* cells are to be harvested by centrifugation as pellet. The cell concentrate are to be plated for selection of transformed cells which are paromomycin resistant. Paromomycin-resistant colonies are to be further cultivated to screen gene integration through colony PCR as described by [12], using gene-specific primers GSP3 and GSP4 to amplify the full-length expression cassette. The following subsections below would complete the transformation for desired phenotype:

18.9.4.1 Immunoblotting

This method is employed to detect the expression of a transgene using a monoclonal antibody targeted to the transgene product. The marker antibody is generally tagged by a fluorescent marker. A positive signal (flourecence) indicate expression of the protein and vise-versa.

18.9.4.2 Fluorescence Localization Analysis

This is employed to localize the expression site of the protein. Monoclonal antibodies are again used which are fluorescent labeled. A fluorescence microscope is needed for the analysis. In general terms, presence of a gene can also be identified at nucleic acid level using fluorescent-labeled nucleotide probes. The method is generally termed as FISH (Fluorescent In Situ Hybridization).

18.9.4.3 Proteolytic Assay for Protein Stability

The expressed protein may be subjected to enzymatic, chemical, or physical (heat, pH change, ionic gradient, etc.) treatment which would lyse the protein. The degrees of lysis indicate relative stability of the expressed protein. Generally a spectrophotometer is employed in the test. For a review on chloroplast transformation, one can read Yusuke and Takashi [107] and Meyers et al. [59].

However, for expression of FLO1 and FLO5 transgene construct of yeast origin in *Chlamydomonas reinhardtii* (for enhanced flocculation phenotype for wastewater treatment) the nuclear transformation strategy also deserves a thorough evaluation. This is because either of the strategies, chloroplast or nuclear transformation, are yet to be experimentally validated for successful harvesting of algal biomass for wastewater treatment.

The considerations for nuclear transformation for expression of FLO1 and FLO5 transgene construct of yeast origin in *Chlamydomonas reinhardtii* are as follows:

1. The FLO1 and FLO5 proteins are lectin-like proteins, and therefore a glycosylation in the algal Golgi complex is easily achievable via nuclear transformation.
2. Secretion of FLO1 and FLO5 proteins to the cell wall surface is easily achievable via nuclear transformation.
3. Nuclear transformation is an established protocol in genetic engineering.

Therefore it is recommended that a thorough experimental evaluation is imperative for generation of successful microalgal transformation of FLO1 and FLO5 flocculation enhancing proteins for enhanced flocculation of microalgae for wastewater treatment.

18.10 Discussion

The above sections give an overview of the strategy that indicates that the transformation of fungal genes that are capable of enhanced flocculation into algal hosts would lead to enhanced flocculation phenotype. For this, transformation of chloroplast can be efficiently achieved [59]. The genetic engineering strategy, construction of vector, and possible outcome of the stretegy are depicted in Fig. 18.19. The in silico analysis described here confirms that FLO1 and FLO5 cassettes that can be inserted into a suitable host under regulation of FLO8 will eventually lead to expression of flocculation promoting proteins in *Chlamydomonas reinhardtii*.

In this chapter a detail review of the problems associated with wastewater treatment via algal systems and its solutions that may lead to their mitigation are described. Extensive review of literature indicates that molecular biology and genetic engineering-based interventions are needed for genesis of a sustainable

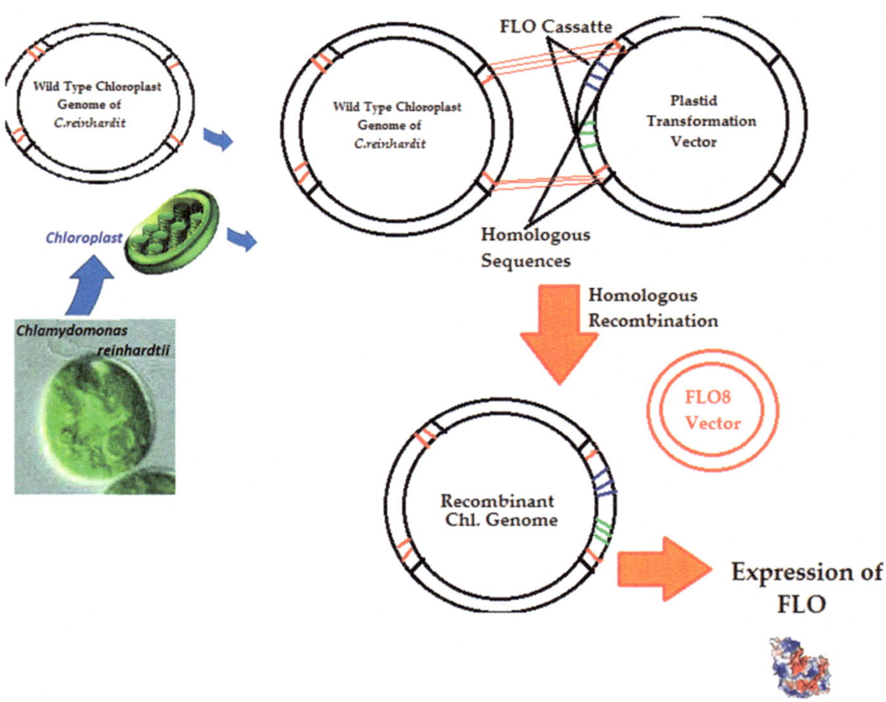

Fig. 18.19 Possible mechanism of chloroplast transformation for enhanced flocculation phenotype in microalgae for wastewater treatment

solution for this fundamental problem of algal biomass separation after wastewater treatment by algae. For preliminary experimentation *Chlamydomonas reinhardtii* will be the most promising host that will lead to substantial understanding toward outcome of the strategy. A thorough protocol that may be followed for the molecular biology experiment is also described; however a meticulous assessment and rectifications may be necessary by algal biotechnologists before finalizing the protocol. A detail methodology that may be employed to evaluate flocculation enhancing protein prioritization by computational tools is described. The methodology of computational simulation can be replicated for any other protein of industrial importance of any other origin. The schematic representations of concepts are easily comprehensible that may endorse brainstorming among the scientific fraternity currently focusing on algal biotechnology for wastewater treatment. Since the threats of climate change is real and it will impact the basic livelihood resources significantly, all sorts experimental and conceptual notions must be tested by the scientists in this high time. The priority of FLO1 and the FLO5 as flocculation enhancers is apparent from the study, and therefore a transformation experiment by the algal biotechnologists is highly anticipated by the author. The experimentations by computational tools such as domain analysis, molecular dynamics, protein structural and functional analysis, bioenergetics and protein network analysis, etc. employ the standard tools as per current norms. In this chapter several works, reviews, and papers by works of different fields are suggested which will aid the readers to have an in-depth understanding of the context that are discussed here. Therefore, the study is expected to aid in the understanding of bio-flocculation for wastewater treatment and other relevant sustainable utilization of microalgal systems.

Acknowledgment The author acknowledges Women's Polytechnic, Hapania, Government of Tripura for providing basic infrastructure for conducting the study.

References

1. Almaraz-Delgado AL, Flores-Uribe J, Perez-Espana VH, Salgado-Manjarrez E, Badillo-Corona JA (2014) Production of therapeutic proteins in the chloroplast of Chlamydomonas reinhardtii. AMB Express 4:57
2. Bahar I, Rader AJ (2005) Coarse-grained normal mode analysis in structural biology. Curr Opin Struct Biol 15(5):586–592
3. Baker NA, Sept D, Joseph S, Holst MJ, Andrew McCammon J (2001) Electrostatics of nanosystems: application to microtubules and the ribosome. Proc Natl Acad Sci 98(18):10037–10041
4. Bauer FF, Govender P, Bester MC (2010) Yeast flocculation and its biotechnological relevance. Appl Microbiol Biotechnol 88:31–39
5. Ben-nun-Shaul O, Bronfeld H, Reshef D, Schueler-Furman O, Oppenheim A (2009) The SV40 capsid is stabilized by a conserved pentapeptide hinge of the major capsid protein VP1. J Mol Biol 386(5):1382–1391
6. Boelee NC, Janssen M, Temmink H, Taparaviciute L, Khiewwijit R, Janoska A, Buisman CJN, Wijffels RH (2014) The effect of harvesting on biomass production and nutrient removal in phototrophic biofilm reactors for effluent polishing. J Appl Phycol 26(3):1439–1452

7. Boelee NC, Temmink H, Janssen M, Buisman CJN, Wijffels RH (2014) Balancing the organic load and light supply in symbiotic microalgal-bacterial biofilm reactors treating synthetic municipal wastewater. Ecol Eng 64:213–221
8. Boelee NC, Temmink H, Janssen M, Buisman CJN, Wijffels RH (2011) Nitrogen and phosphorus removal from municipal wastewater effluent using microalgal biofilms. Water Res 45(18):5925–5933
9. Bordoli L, Kiefer F, Arnold K, Benkert P, Battey J, Schwede T (2009) Protein structure homology modeling using SWISS-MODEL workspace. Nat Protoc 4(1):1–13
10. Brennan L, Philip O (2010) Biofuels from microalgae—a review of technologies for production, processing, and extractions of biofuels and co-products. Renew Sust Energ Rev 14(2):557–577
11. Buan NR, Escalante-Semerena JC (2005) Computer-assisted docking of flavodoxin with the ATP: Co (I) rrinoid adenosyltransferase (CobA) enzyme reveals residues critical for protein-protein interactions but not for catalysis. J Biol Chem 280(49):40948–40956
12. Cao M, Fu Y, Guo Y, Pan J (2009) Chlamydomonas (chlorophyceae) colony PCR. Protoplasma 235(1):107–110
13. Chen VB, Bryan Arendall W, Headd JJ, Keedy DA, Immormino RM, Kapral GJ, Murray LW, Richardson JS, Richardson DC (2009) MolProbity: all-atom structure validation for macromolecular crystallography. Acta Crystallogr D Biol Crystallogr 66(1):12–21
14. Christenson L, Sims R (2011) Production and harvesting of microalgae for wastewater treatment, biofuels, and bioproducts. Biotechnol Adv 29(6):686–702
15. Claro FB, Rijsbrack K, Soares EV (2007) Flocculation onset in Saccharomyces cerevisiae: effect of ethanol, heat and osmotic stress. J Appl Microbiol 102(3):693–700
16. Cunha AF, Missawa SK, Gomes LH, Reis SF, Pereira GA (2006) Control by sugar of Saccharomyces cerevisiae flocculation for industrial ethanol production. FEMS Yeast Res 6(2):280–287
17. Davis ME, Andrew McCammon J (1990) Electrostatics in biomolecular structure and dynamics. Chem Rev 90(3):509–521
18. Debnath, S., & Addya, S. (2015). In-silico modelling of SERAC1: protein involved in a developmental neural disorder MEGDEL syndrome characterized by 3-methyl glutaconic aciduria type IV with sensory-neural deafness, encephalopathy and Leigh-like syndrome. Int J Dev Neurosci, 47(Pt A):1–2
19. Debnath S, Addya S (2014) Structural basis for heterogeneous phenotype of ERG11 dependent azole resistance in C. albicans clinical isolates. Springerplus 3(1):660
20. Debnath S, Addya S (2015) Mis-sesnse mutations in Tafazzin (TAZ) that escort to mild clinical symptoms of Barth syndrome is owed to the minimal inhibitory effect of the mutations on the enzyme function: in-silico evidence. Interdiscip Sci Comput Life Sci 7(1):21
21. DeLano WL (2002) The PyMOL molecular graphics system
22. Díaz-Moreno I, Hulsker R, Skubak P, Foerster JM, Cavazzini D, Finiguerra MG, Díaz-Quintana A et al (2014) The dynamic complex of cytochrome c 6 and cytochrome f studied with paramagnetic NMR spectroscopy. Biochim Biophys Acta (BBA)-Bioenerg 1837(8):1305–1315
23. Dobbins SE, Lesk VI, Sternberg MJE (2008) Insights into protein flexibility: the relationship between normal modes and conformational change upon protein–protein docking. Proc Natl Acad Sci 105(30):10390–10395
24. Domingues L, Vicente AA, Lima N, Teixeira JA (2000a) Applications of yeast flocculation in biotechnological processes. Biotechnol Bioprocess Eng 5:288–305
25. Domingues L, Lima N, Teixeira JA (2000b) Contamination of a high-cell-density continuous bioreactor. Biotechnol Bioeng 68:584–587
26. El-Kirat-Chatel S, Beaussart A, Vincent SP, Flos MA, Hols P, Lipke PN, Dufrêne YF (2015) Forces in yeast flocculation. Nanoscale 7(5):1760–1767
27. Soares EV (2011) Flocculation in Saccharomyces cerevisiae: a review. J Appl Microbiol 110:1–18

28. Fenwick RB, van den Bedem H, Fraser JS, Wright PE (2014) Integrated description of protein dynamics from room-temperature X-ray crystallography and NMR. Proc Natl Acad Sci 111(4):E445–E454
29. Fischer R, Stoger E, Schillberg S, Christou P, Twyman RM (2004) Plant-based production of biopharmaceuticals. Curr Opin Plant Biol 7(2):152–158
30. Gonzalez-Fernandez C, Ballesteros M (2013) Microalgae autoflocculation: an alternative to highenergy consuming harvesting methods. J Appl Phycol 25:991–999
31. Goossens KVY, Ielasi FS, Nookaew I, Stals I, Alonso-Sarduy L, Daenen L, Sebastiaan E, Mulders V et al (2015) Molecular mechanism of flocculation self-recognition in yeast and its role in mating and survival. MBio 6(2):e00427–e00415
32. Govender P, Domingo JL, Bester MC, Pretorius IS, Bauer FF (2008) Controlled expression of the dominant flocculation genes FLO1, FLO5, and FLO11 in Saccharomyces cerevisiae. Appl Environ Microbiol 74:6041–6052
33. Gressel J, van der Vlugt CJB, Bergmans HEN (2013) Environmental risks of large scale cultivation of microalgae: mitigation of spills. Algal Res 2:286–298
34. Guex N, Peitsch MC (1997) SWISS-MODEL and the Swiss-PdbViewer: an environment for comparative protein modeling. Electrophoresis 18:2714–2723
35. Gultom SO, Hu B (2013) Review of microalgae harvesting via co-pelletization with filamentous fungus. Energies 6:5921–5939
36. Guo B, Styles CA, Feng Q, Fink G (2000) A Saccharomyces gene family involved in invasive growth, cell–cell adhesion, and mating. Proc Natl Acad Sci U S A 97:12158–12163
37. Harun R, Singh M, Forde GM, Danquah MK (2010) Bioprocess engineering of microalgae to produce a variety of consumer products. Renew Sust Energ Rev 14(3):1037–1047
38. He DM, Qian KX, Shen GF, Zhang ZF, Li YN et al (2007) Recombination and expression of classical swine fever virus (CSFV) structural protein E2 gene in Chlamydomonas reinhardtii chroloplasts. Colloids Surf B Biointerfaces 55:26–30
39. Janga SC, Diaz-Mejia JJ, Moreno-Hagelsieb G (2010) Network-based function prediction and interactomics: the case for metabolic enzymes. Metab Eng 13:1–10
40. Jinkerson RE, Jonikas MC (2015) Molecular techniques to interrogate and edit the Chlamydomonas nuclear genome. Plant J 82:393–412
41. Johnson M, Zaretskaya I, Raytselis Y, Merezhuk Y, McGinnis S, Madden TL (2008) NCBI BLAST: a better web interface. Nucleic Acids Res 36(suppl_2):W5–W9
42. Karimi K, Zamani A (2013) Mucor indicus: biology and industrial application perspectives: a review. Biotechnol Adv 31:466–481
43. Kaya D, Dilek FB, Goekcay CF (2007) Reuse of lagoon effluents in agriculture by post-treatment in a step feed dual treatment process. Desalination 215:29–36
44. Kesaano M, Sims RC (2014) Algal biofilm based technology for wastewater treatment. Algal Res 5:231–240
45. Kida K, Yamadaki M, Asno S, Nakata T, Sonoda Y (1989) The effect of aeration on stability of continuous ethanol fermentation by a flocculating yeast. J Ferment Bioeng 68(107–111):14
46. Kindle KL (1990) High-frequency nuclear transformation of Chlamydomonas reinhardtii. Proc Natl Acad Sci 87(3):1228–1232
47. Leach AR (2001) Molecular Modeling: principles and applications, 2nd edn. Pearson Education EMA, Sussex
48. Lee AK, Lewis DM, Ashman PJ (2013) Harvesting of marine microalgae by electroflocculation: the energetics, plant design, and economics. Appl Energy 108:45–53
49. Lee J, Cho DH, Ramanan R, Kim BH, Oh HM et al (2013) Microalgae-associated bacteria play a key role in the flocculation of Chlorella vulgaris. Bioresour Technol 131:195–201
50. Lee YK (2001) Microalgal mass culture systems and methods: their limitation and potential. J Appl Phycol 13:307–315
51. Li Q, Zhao XQ, Chang AK, Zhang QM, Bai FW (2012) Ethanol-induced yeast flocculation directed by the promoter of TPS1 encoding trehalose-6-phosphate synthase 1 for efficient ethanol production. Metab Eng 14:1–8

52. Li T, Lin GY, Podola B, Melkonian M (2015) Continuous removal of zinc from wastewater and mine dump leachate by a microalgal biofilm PSBR. J Hazard Mater 297:112–118
53. Machado MD, Santos MSF, Gouveia C, Soares HMVM, Soares EV (2008) Removal of heavy metals using a brewer's yeast strain of Saccharomyces cerevisiae: the flocculation as a separation process. Bioresour Technol 99:2107–2115
54. Manheim D, Nelson Y (2013) Settling and bioflocculation of two species of algae used in wastewater treatment and algae biomass production. Environ Prog Sustain Energy 32:946–954
55. Manuell AL, Beligni MV, Elder JH, Siefker DT, Tran M, Weber A, McDonald TL, Mayfield SP (2007) Robust expression of a bioactive mammalian protein in Chlamydomonas chloroplast. Plant Biotechnol J 5(3):402–412
56. Mason LB, Amrhein C, Goodson CC, Matsumoto MR, Anderson MA (2005) Reducing sediment and phosphorus in tributary waters with alum and polyacrylamide. J Environ Qual 34:1998–2004
57. Mata TM, Martins AA, Caetano NS (2010) Microalgae for biodiesel production and other applications: a review. Renew Sust Energ Rev 14(1):217–232
58. Mayfield SP, Franklin SE, Lerner RA (2003) Expression and assembly of a fully active antibody in algae. Proc Natl Acad Sci U S A 100:438–442
59. Meyers B, Zaltsman A, Lacroix B, Kozlovsky SV, Krichevsky A (2010) Nuclear and plastid genetic engineering of plants: comparison of opportunities and challenges. Biotechnol Adv 28(6):747–756
60. Miki BLA, Poon NH, James AP, Seligy VL (1982) Possible mechanism for flocculation interactions governed by gene FLO1 in Saccharomyces cerevisiae. J Bacteriol 150:878–889
61. Milledge J, Heaven S (2013) A review of the harvesting of micro-algae for biofuel production. Rev Environ Sci Biotechnol 12:165–178
62. Miranda AF, Ramkumar N, Andriotis C, Höltkemeier T, Yasmin A, Rochfort S, Wlodkowic D et al (2017) Applications of microalgal biofilms for wastewater treatment and bioenergy production. Biotechnol Biofuels 10(1):120
63. Mulders SEV, Christianen E, Saerens SMG, Daenen L, Verbelen PJ, Willaert R, Verstrepen KJ, Delvaux FR (2009) Phenotypic diversity of Flo protein familymediated adhesion in Saccharomyces cerevisiae. FEMS Yeast Res 9:178–190
64. Muradov N (2014) Liberating energy from carbon: introduction to decarbonization. Springer, New York/Heidelberg/Dordrecht/London
65. Muradov N, Taha M, Miranda AF, Kadali K, Gujar A et al (2014) Dual application of duckweed and azolla plants for wastewater treatment and renewable fuels and petrochemicals production. Biotechnol Biofuels 7:1–17
66. Nazim M, Taha M, Miranda AF, Wrede D, Kadali K, Gujar A, Stevenson T, Ball AS, Mouradov A (2015) Fungal-assisted algal flocculation: application in wastewater treatment and biofuel production. Biotechnol Biofuels 8:24
67. Nichols ES, Swift RV, Amaro RE (2012) Rational prediction with molecular dynamics for hit identification. Curr Top Med Chem 12(18):2002–2012
68. Nonklang S, Ano A, Abdel-Banat BM, Saito Y, Hoshida H, Akada R (2009) Construction of flocculent Kluyveromyces marxianus strains suitable for high-temperature ethanol fermentation. Biosci Biotechnol Biochem 73(5):1090–1095
69. Oswald JW, Gotaas HB (1957) Photosynthesis in sewage treatment. Trans Am Soc Civ Eng 122:73–105
70. Oh HM, Lee SJ, Park MH, Kim HS, Kim HC, Yoon JH, Kwon GS, Yoon BD (2001) mHarvesting of *Chlorella vulgaris* using a bioflocculant from *Paenibacillus* sp. AM49. Biotechnol Lett 23(15):1229–1234
71. Patel JK, Alex Speers R, Lake JC (2011) Colloidal examination of Worts associated with premature yeast flocculation. J Am Soc Brew Chem 69(2):81
72. Pattin KA, Moore JH (2009) Role for protein-protein interaction databases in human genetics. Expert Rev Proteomics 6:647–659

73. Pires J, Alvim-Ferraz M, Martins F, Simoes M (2013) Wastewater treatment to enhance the economic viability of microalgae culture. Environ Sci Pollut Res 20:5096–5105
74. Powell RJ, Hill RT (2013) Rapid aggregation of biofuel-producing algae by the bacterium Bacillus sp strain RP1137. Appl Environ Microbiol 79:6093–6101
75. Pragya N, Pandey KK, Sahoo PK (2013) A review on harvesting, oil extraction and biofuels production technologies from microalgae. Renew Sust Energ Rev 24:159–171
76. Price MS, Classen JJ, Payne GA (2001) Aspergillus Niger absorbs copper and zinc from swine wastewater. Bioresour Technol 77:41–49
77. Purton S, Szaub JB, Wannathong T, Young R, Economou CK (2013) Genetic engineering of algal chloroplasts: progress and prospects. Russian J Plant Physiol 60:491–499
78. Ramos-Martinez EM, Fimognari L, Sakuragi Y (2017) High-yield secretion of recombinant proteins from the microalga Chlamydomonas reinhardtii. Plant Biotechnol J 15:1214–1224
79. Rasala BA, Mayfield SP (2011) The microalga Chlamydomonas reinhardtii as a platform for the production of human protein therapeutics. Bioeng Bugs 2:50–54
80. Rasala BA, Mayfield SP (2015) Photosynthetic biomanufacturing in green algae; production of recombinant proteins for industrial, nutritional,and medical uses. Photosynth Res 123:227–239
81. Rasala BA, Lee PA, Shen Z, Briggs SP, Mendez M, Mayfield SP (2012) Robust expression and secretion of Xylanase1 in Chlamydomonas reinhardtii by fusion to a selection gene and processing with the FMDV 2A peptide. PLoS One 7(8):e43349
82. Roeselers G, van Loosdrecht MCM, Muyzer G (2008) Phototrophic biofilms and their potential applications. J Appl Phycol 20(3):227–235
83. Salim S, Bosma R, Vermue MH, Wijffels RH (2011) Harvesting of microalgae by bio-flocculation. J Appl Phycol 23(5):849–855 15
84. Schenk PM, Thomas-Hall SR, Stephens E, Marx UC, Mussgnug JH et al (2008) Second generation biofuels: high-efficiency microalgae for biodiesel production. Bioenergy Res 1:20–43
85. Shahila M, Jacob J, May M, Kotula L, Thiyagarajan P, Johnson ME, Fung LW-M (2003) Structural analysis of the αN-terminal region of erythroid and nonerythroid spectrins by small-angle X-ray scattering. Biochemistry 42(49):14702–14710
86. Shamriz S, Ofoghi H (2016) Outlook in the application of Chlamydomonas reinhardtii chloroplast as a platform for recombinant protein production. Biotechnol Genet Eng Rev 32(1–2):92–106
87. Sharma KK, Garg S, Li Y, Malekizadeh A, Schenk PM (2013) Critical analysis of current microalgae dewatering techniques. Biofuels 4:397–407
88. Shi J, Podola B, Melkonian M (2014) Application of a prototype-scale twin-layer photobioreactor for effective N and P removal from different process stages of municipal wastewater by immobilized microalgae. Bioresour Technol 154:260–266
89. Soares EV, Duarte AA (2002) Addition of nutrients induce a fast loss of flocculation in starved cells of Saccharomyces cerevisiae. Biotechnol Lett 24:1957–1960
90. Stratford M (1989) Yeast flocculation: calcium specificity. Yeast 5:487–496
91. Sun M, Qian KX, Su N, Chang HY, Liu JX et al (2003) Foot-and-mouth disease virus VP1 protein fused with cholera toxin B subunit expressed in Chlamydomonas reinhardtii chloroplast. Biotechnol Lett 25:1087–1092
92. Suhre K, Sanejouand YH (2004) ElNemo: a normal mode web server for protein movement analysis and the generation of templates for molecular replacement. Nucleic Acids Res 32:W610–W614
93. Stephanie R, Hermann M, Berger IJ, Carrer H, Bock R (2001) Stable genetic transformation of tomato plastids and expression of a foreign protein in fruit. Nat Biotechnol 19:870–875
94. Szklarczyk D, Franceschini A, Kuhn M, Simonovic M, Roth A, Minguez P, Doerks T, Stark M, Muller J, Bork P, Jensen LJ, von Mering C (2011) The STRING database in 2011: functional interaction networks of proteins, globally integrated and scored. Nucleic Acids Res 39:D561–D568

95. Tan KWM, Lee YK (2017) Expression of the heterologous Dunaliella tertiolecta fatty acyl-ACP thioesterase leads to increased lipid production in Chlamydomonas reinhardtii. J Biotechnol 247:60–67
96. Teixeira JA, Mota M, Goma G (1990) Continuous ethanol production by a flocculating strain of Kluyveromyces marxianus: bioreactor performance. Bioprocess Eng 5:123–127
97. Tofalo R, Perpetuini G, Di Gianvito P, Schirone M, Corsetti A, Suzzi G (2014) Genetic diversity of FLO1 and FLO5 genes in wine flocculent Saccharomyces cerevisiae strains. Int J Food Microbiol 191:45–52
98. Vandamme D, Foubert I, Muylaert K (2013) Flocculation as a low-cost method for harvesting microalgae for bulk biomass production. Trends Biotechnol 31(4):233–239
99. Verstrepen KJ, Derdelinckx G, Delvaux FR, Winderickx J, Thevelein JM, Bauer FF, Pretorius IS (2001) Late fermentation expression of FLO1 in Saccharomyces cerevisiae. J Am Soc Brew Chem 59:69–76
100. Verstrepen KJ, Klis FM (2006) Flocculation, adhesion and biofilm formation in yeasts. Mol Microbiol 60(1):5–15
101. Wan C, Asraful Alam M, Zhao X-Q, Zhang X-Y, Guo S-L, Ho S-H, Chang J-S, Bai F-W (2015) Current progress and future prospect of microalgal biomass harvest using various flocculation technologies. Bioresour Technol 184:251–257
102. Wang J, Chen C (2006) Biosorption of heavy metals by Saccharomyces cerevisiae: a review. Biotechnol Adv 24:427–451
103. Wang PI, Marcotte EM (2010) It's the machine that matters: predicting gene function and phenotype from protein networks. J Proteome 73:2277–2289
104. Wrede D, Taha M, Miranda AF, Kadali K, Stevenson T, Ball AS, Mouradov A (2014) Co-cultivation of fungal and microalgal cells as an efficient system for harvesting microalgal cells, lipid production and wastewater treatment. PLoS One 9(11):e113497
105. Wu YH, Hu HY, Yu Y, Zhang TY, Zhu SF et al (2014) Microalgal species for sustainable biomass/lipid production using wastewater as resource: a review. Renew Sust Energ Rev 33:675–688
106. Yang ZQ, Li YN, Chen F, Li D, Zhang ZF et al (2006) Expression of human soluble TRAIL in Chlamydomonas reinhardtii chloroplast. Chin Sci Bull 51:1703–1709
107. Yusuke Y, Takashi S (2014) Recent advances in the study of chloroplast gene expression and its evolution., front. Plant Sci 5:61
108. Zhang GM, Wang B, Zhang PY, Wang L, Wang H (2006) Removal of algae by sonication–coagulation. J Environ Sci Health A Tox Hazard Subst Environ Eng 41:1379–1390
109. Zhou WG, Cheng YL, Li Y, Wan YQ, Liu YH et al (2012) Novel fungal Pelletization-assisted technology for Algae Harvesting and Wastewater Treatment. Appl Biochem Biotechnol 167:214–228
110. Zhou WG, Min M, Hu B, Ma XC, Liu YH et al (2013) Filamentous fungi assisted bio-flocculation: a novel alternative technique for harvesting heterotrophic and autotrophic microalgal cells. Sep Purif Technol 107:158–165
111. Zhao XQ, Li Q, He L, Li F, Que W, Bai FW (2012) Exploration of a natural reservoir of flocculating genes from various Saccharomyces cerevisiae strains and improved ethanol fermentation using stable genetically engineered flocculating yeast strains. Process Biochem 47:1612–1619

Phycoremediation: Can It Address Major Issues in Conventional Systems? 19

V. Sivasubramanian

Abstract

Conventional treatment systems employ huge amounts of hazardous chemicals in the treatment of industrial and domestic waste waters. A variety of chemicals are added to the effluent to manage the major parameters, which causes more problems during disposal. In majority of textile and leather industries, water is recycled through R/O process. The R/O reject disposal is a major issue, and installation of multiple-effect evaporators (MEE) is recommended which is more expensive and energy-intensive. Can treatment technologies employing algae provide alternate viable solutions to these and other problems faced by conventional systems? This is the question which will be answered through this article.

Phycoremediation technology, using algae for waste treatment, has become very popular now. More and more industries are coming forward to adopt this technology [1]. Like any other bioremediation process, this technology also has lots of advantages which will benefit the industry and environment by reducing the operation costs and reducing/removing the toxicity and avoiding chemicals which are usually added in conventional treatment processes. The methodology involves three stages: (1) lab-scale trials, (2) pilot-scale trial at industry site, and (3) scale up. The design of the scaled-up system is custom designed based on infrastructure and facilities already available in the industries.

The success of phycoremediation depends mainly on acceptance. If this new technology addresses drawbacks of other conventional systems, this is accepted immediately. Phycospectrum Environmental Research Centre (PERC), Chennai, India, has implemented algal technology in several industries in India and other

V. Sivasubramanian (✉)
Phycospectrum Environmental Research Centre (PERC),
132, AK Block, 7th Main Road, Anna Nagar, Chennai 600040, India

L. B. Sukla et al. (eds.), *The Role of Microalgae in Wastewater Treatment*,
https://doi.org/10.1007/978-981-13-1586-2_19

countries. Vertical roof tank system with improved illumination/aeration panel system developed by PERC is being installed with huge success rate in various industries. PERC has recently presented this technology at a meeting with Welsh Water, Wales, UK, and is in the process of signing an agreement to implement a demonstration plant at Swansea, UK.

Although removal of nutrients from waste water by algae is well established, Phycospectrum Environmental Research Centre (PERC) is the first organization in India to take micro-algal technology to industries with considerable amount of success. The first large-scale phycoremediation (algae-based remediation) plant was established during 2006 in a chemical industry (SNAP) in India. The major problem addressed in this industry was pH correction which was done conventionally using huge amount of sodium hydroxide ending up in generation of sludge. Algal technology has helped this industry to correct pH without using any chemicals, and since the total dissolved solids (TDS) of the effluent is around 40,000 mg/L, entire effluent is evaporated without generating any sludge. PERC, after the success with SNAP industry, has installed a number of pilot and large plants in India.

19.1 PERC's Projects in Colombia

Phycospectrum joint with CORE BIOTECH, Colombia (PHYCORE), has successfully demonstrated through a 20 KL pilot plant installed at an oil drilling site of Pacific Energy, to handle petrochemical waste and efficiently reduce nutrient load, coliform bacteria and traces of aromatic hydrocarbons. PHYCORE has also installed a demo plant at a slaughter house in Barranquilla, Colombia, recently.

19.2 Can Multiple-Effect Evaporators (MEE) Be Replaced with Algal Technology?

A multiple-effect evaporator (MEE) is recommended by pollution control agencies to evaporate industrial effluents and R/O (reverse osmosis) rejects with very high total dissolved solids (TDS), and it results in huge amounts of hazardous solid waste. In India MEE is extensively used in R/O reject management of textile and leather industries. The major problems with MEE are the costs of installation and operation and finally management of solid waste generated due to evaporation. Because of these reasons, industries have been looking for alternate technology to handle R/O rejects. Phycospectrum Environmental Research Centre (PERC), India, has recently developed and successfully demonstrated micro-algae-based technology at Brintons Carpets Asia Pvt. Ltd., India, very successfully (Figs. 19.1, 19.2 and 19.3).

The R/O reject from Brintons has a TDS of 20,000 to 25,000 mg/L, and this was continuously loaded into a 10 KL demonstration plant in which selected microalgae were also grown. The effluent with algae was allowed to run on a slope for capturing sunlight and heat and evaporate faster. After a short initial salt build-up, the system stabilized with no further increase in TDS, and further addition never

Fig. 19.1 Slope tank for R/O reject treatment showing algal growth at Brintons Carpets

	Algae	MEE
CAPEX (Equipment + Civil)	Rs. 6 lacs/(m3/day) (approx.)	Rs. 2.5 lacs/(m3/day) (approx.)
OPEX (Evaporation)	Rs.0.33 lacs/m3/Year @Rs. 100/m3/day	Rs.3.3 lacs/m3/year @Rs. 1000/m3/day
Land	100 m2/m3	Not applicable
Maintenance	Requires minor maintenance .	Requires Regular maintenance.
Rain Water Harvesting @ 1 m annual rainfall	Simultaneous Rain Water Harvesting; To provide source for 0.25 m3/m3 of effluent evaporated	Not Applicable
Sludge Formed	Algal sludge formed can give significant benefits to clients.	25.5 tons of sludge formed per m3 per annum@ 35000 ppm TDS and 50% solid content.
Carbon	Consumes 2 – 2.5 kWh per m3. Will sequester 35 tons of carbon per m3 annually.	Will consume 30 kWh per m3.

Fig. 19.2 Comparison of MEE and Algae Technology (based on 100 M3/day plant)

increased TDS. Algal biomass produced was collected on regular basis from the slope roof which is being analysed for possible utilization.

The system at Brintons Carpets, India, is stabilized and started producing almost 0.75 g of dry algal biomass per litre per day, and water is getting evaporated resulting in zero sludge. This system is being scaled up now. This successful implementation has proved the efficiency of algae-based evaporation system in replacing MEE handling high TDS waste streams like R/O rejects of various industries and more specifically textile and leather industries.

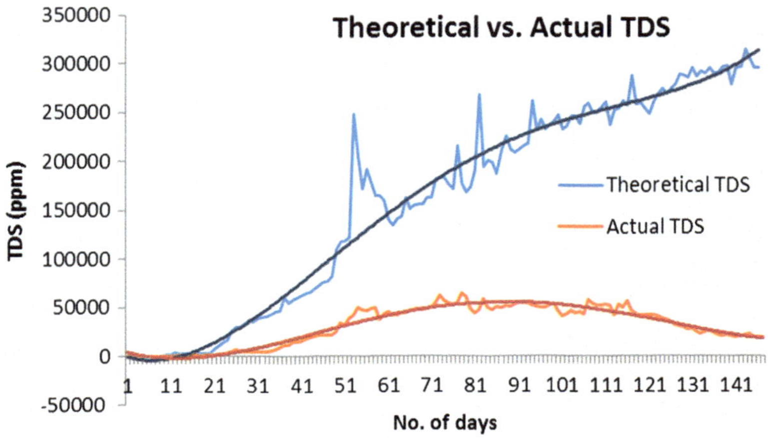

Fig. 19.3 Brintons Carpets – Actual TDS in tank varies between 20,000 and 60,000 ppm (Theoretical TDS based on actual TDS load is around 350,000 ppm)

19.3 Can Micro-Algal Technology Replace Bacterial Bioremediation?

Pasupati Acrylon Ltd. is one of the largest producers of acrylic fibre in India, with a range of options of products ranging from just raw white and bleached fibre to tow-dyed fabric. Recently, the company established a gel-dyeing unit to produce fabric of a much superior nature. Because the gel dye is a liquid, to prevent microbial activity, significant amounts of antimicrobial agents are present in the dye. Bacterial system failed because of the toxic effluent. The industry has an activated sludge process of 7800 M^3, but addition of just a 100 M^3 of raw gel dye effluent in this reactor would upset the reactor for days. A more advanced and robust treatment system was needed for tackling this toxicity (Figs. 19.4, 19.5, 19.6, and 19.7).

PERC developed micro-algal technology, and a scaled-up plant (550 KL with 3 m depth) was commissioned last month. In a continuous flow of 3 M^3/h or 76 M^3/day, a complete removal of toxicity from effluent was achieved with more than 90% COD reduction.

19.4 PMF (Pulsed Magnetic Field) Technology to Improve Algal Growth and Efficiency of Effluent Treatment

PERC has been working on PMF (pulsed magnetic field) technology to improve algal growth and enhance remediation efficiency at industrial effluent treatment plants. PERC in collaboration with Madras Institute of Magnetobiology (MIM),

Fig. 19.4 Slope – illumination – aeration panel of algal ETP – large scale at Pasupati Acrylon

Fig. 19.5 Showing algal growth in the reaction tank and on the slope surface – Pasupati Acrylon

Fig. 19.6 Comparison of treated effluent with raw effluent – Pasupati Acrylon

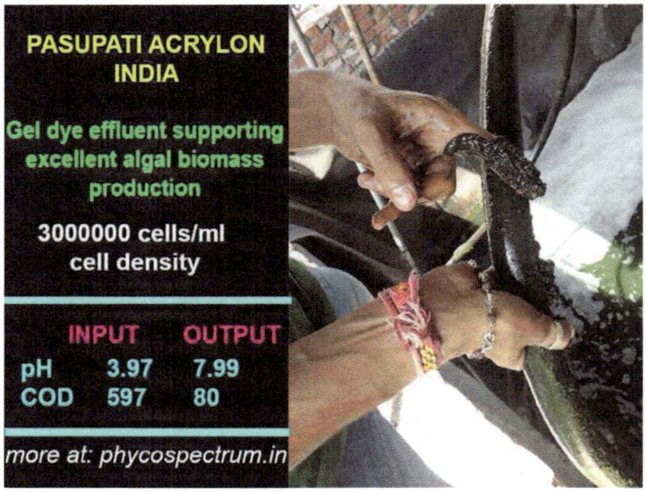

Fig. 19.7 Comparison of raw effluent with treated effluent in scaled-up plant at Pasupati Acrylon

India, has been studying this phenomenon for the last 6 years and developed an optimized technology for various algal processes.

Pulsed magnetic field enclosure has been installed in a 35 KL micro-algal raceway facility at Bharathidasan University, India, as part of Indo-UK joint research project on algal biofuels. This PMF enclosure is expected to enhance algal biomass productivity and increase oil content through an optimization protocol which is in progress right now. The idea of installing world's first ever PMF unit to a raceway pond is based on PERC's (Phycospectrum Environmental Research Centre) collaborative project with a technical support from Madras Institute of Magnetobiology, India, which was supported by MNRE (Ministry of New and Renewable Energy), Ministry of Science and Technology, Government of India, during 2011. The summary of the findings of this project is posted at MNRE website: -mnre.gov.in/file-manager/UserFiles/bio-fuel/R_D_biofuel-Developement-of-hybridised.... production.pdf. MNRE's support was given to PERC based on a research article published in *Journal of Algal Biomass Utilization* [2].

The findings of this project include optimization of sinusoidal magnetic field technology in developing a magnetic field-based technology to enhance biomass production and increase oil production in the micro-algae tested (Figs 19.8 and 19.9).

19.5 PMF Helps to Improve Micro-Algae-Based Effluent Treatment

Trials carried out by PERC with effluent from a textile industry revealed that application of PMF technology enhances remediation efficiency of micro-algae-based phycoremediation process (Table 19.1). Based on these trials, a field-scale PMF

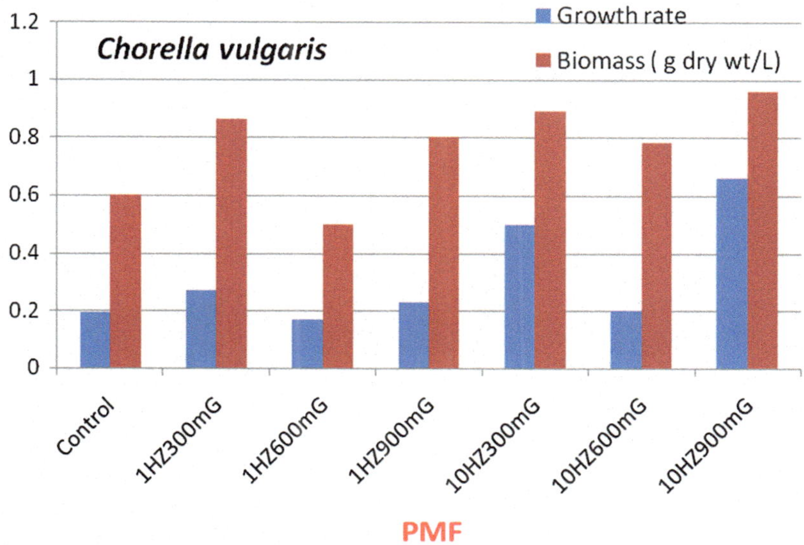

Fig. 19.8 Growth and biomass productivity of *Chlorella vulgaris* exposed to PMF

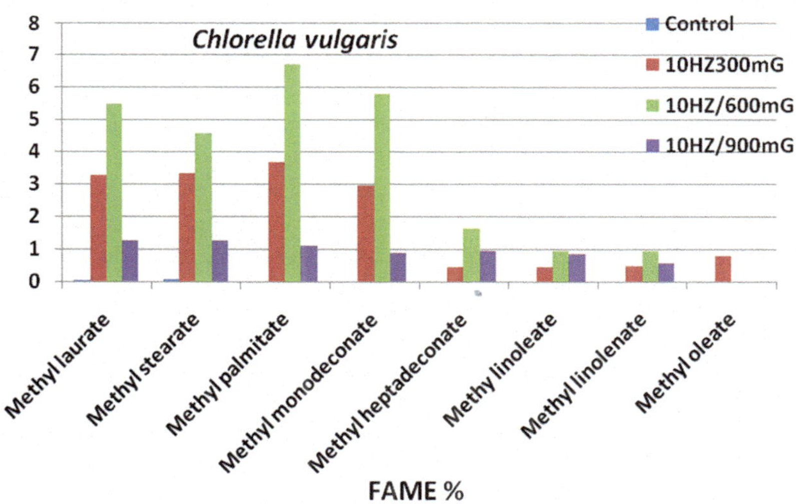

Fig. 19.9 Effect of PMF (10 HZ) on FAME % of *Chlorella vulgaris*

Table 19.1 Showing enhanced efficiency of phycoremediation of textile industry effluent when exposed to pulsed magnetic field (PMF)

Parameters	Raw textile effluent	Textile effluent treated with algae	Textile effluent treated with algae + PMF
Conductivity	12,130	12,290	10,790
pH	**2.82**	**3.10**	**7.05**
Nitrate	77	70.6	67.5
Phosphate	4.3	9.7	1.6
TDS	6897	7003	6178
TSS	7	11.8	9.6
COD	**202**	**220**	**55**
BOD	**50**	**62**	**12**
Chloride	3185	3328	2940
Sulphate	**519**	**537**	**243**

enclosure is being developed to be installed at effluent treatment facility at Pasupati Acrylon, India. This is expected to improve the efficiency of micro-algae-based system already in operation.

PMF technology when properly optimized can be applied to multivarious algae-based processes like oil production, nutraceutical production, industrial effluent treatment, etc. PERC has installed the first large-scale pulsed magnetic field enclosure in a 35 KL algae raceway pond. Similar PMF units are going to be installed in large-scale algae-based effluent treatment plants developed by PERC.

Phycoremediation technology has lots of advantages over conventional systems of treatment. Micro-algae are safe to handle and require minimal maintenance and operation cost. If algal technology works, there will be more than 90% reduction in operation cost when compared to conventional technologies. Added advantage is the production of valuable algal biomass which can become excellent feedstock for production of various products which will benefit the society.

References

1. Sivasubramanian V (2016) Phycoremediation and business prospects. In: Prasad MNV (ed) Bioremediation and bioeconomy. Elsevier Inc, pp 421–456
2. Sivasubramanian V, Subramanian VV, Leela Priya T, Murali R (2010) Application of pulsed magnetic field in improving the quality of algal biomass. J Algal Biomass Utln 1(4):1–9

Printed by Printforce, the Netherlands